The Model Theory of Mathematics

Contents

Chapter 1

Model theory

This article is about the mathematical discipline. For the informal notion in other parts of mathematics and science, see Mathematical model.

In mathematics, **model theory** is the study of classes of mathematical structures (e.g. groups, fields, graphs, universes of set theory) from the perspective of mathematical logic. The objects of study are models of theories in a formal language. We call a set of sentences in a formal language a **theory**; a **model** of a theory is a structure (e.g. an interpretation) that satisfies the sentences of that theory.

Model theory recognises and is intimately concerned with a duality: It examines semantical elements (meaning and truth) by means of syntactical elements (formulas and proofs) of a corresponding language. To quote the first page of Chang & Keisler (1990):[1]

universal algebra + logic = **model theory**.

Model theory developed rapidly during the 1990s, and a more modern definition is provided by Wilfrid Hodges (1997):

model theory = algebraic geometry − fields,

although model theorists are also interested in the study of fields. Other nearby areas of mathematics include combinatorics, number theory, arithmetic dynamics, analytic functions, and non-standard analysis.

In a similar way to proof theory, model theory is situated in an area of interdisciplinarity among mathematics, philosophy, and computer science. The most prominent professional organization in the field of model theory is the Association for Symbolic Logic.

1.1   Branches of model theory

This article focuses on finitary first order model theory of infinite structures. Finite model theory, which concentrates on finite structures, diverges significantly from the study of infinite structures in both the problems studied and the techniques used. Model theory in higher-order logics or infinitary logics is hampered by the fact that completeness does not in general hold for these logics. However, a great deal of study has also been done in such languages.

Informally, model theory can be divided into classical model theory, model theory applied to groups and fields, and geometric model theory. A missing subdivision is computable model theory, but this can arguably be viewed as an independent subfield of logic.

Examples of early theorems from classical model theory include Gödel's completeness theorem, the upward and downward Löwenheim–Skolem theorems, Vaught's two-cardinal theorem, Scott's isomorphism theorem, the omitting types theorem, and the Ryll-Nardzewski theorem. Examples of early results from model theory applied to fields are Tarski's elimination of quantifiers for real closed fields, Ax's theorem on pseudo-finite fields, and Robinson's development of non-standard analysis. An important step in the evolution of classical model theory occurred with the birth of stability theory (through Morley's theorem on uncountably categorical theories and Shelah's classification program), which developed a calculus of independence and rank based on syntactical conditions satisfied by theories.

During the last several decades applied model theory has repeatedly merged with the more pure stability theory. The result of this synthesis is called geometric model theory in this article (which is taken to include o-minimality, for example, as well as classical geometric stability theory). An example of a theorem from geometric model theory is Hrushovski's proof of the Mordell–Lang conjecture for function fields. The ambition of geometric model theory is to provide a *geography of mathematics* by embarking on a detailed study of definable sets in various mathematical structures, aided by the substantial tools developed in the study of pure model theory.

1.2   Universal algebra

Main article: Universal algebra

Fundamental concepts in universal algebra are signatures σ and σ-algebras. Since these concepts are formally defined in the article on structures, the present article can contend itself with an informal introduction which consists in examples of how these terms are used.

> The standard signature of rings is $\sigma_{ring} = \{\times, +, -, 0, 1\}$, where $\times$ and $+$ are binary, $-$ is unary, and 0 and 1 are nullary.
>
> The standard signature of semirings is $\sigma_{smr} = \{\times, +, 0, 1\}$, where the arities are as above.
>
> The standard signature of groups (with multiplicative notation) is $\sigma_{grp} = \{\times, ^{-1}, 1\}$, where $\times$ is binary, $^{-1}$ is unary and 1 is nullary.
>
> The standard signature of monoids is $\sigma_{mnd} = \{\times, 1\}$.
>
> A ring is a σ_{ring}-structure which satisfies the identities $u + (v + w) = (u + v) + w$, $u + v = v + u$, $u + 0 = u$, $u + (-u) = 0$, $u \times (v \times w) = (u \times v) \times w$, $u \times 1 = u$, $1 \times u = u$, $u \times (v + w) = (u \times v) + (u \times w)$ and $(v + w) \times u = (v \times u) + (w \times u)$.
>
> A group is a σ_{grp}-structure which satisfies the identities $u \times (v \times w) = (u \times v) \times w$, $u \times 1 = u$, $1 \times u = u$, $u \times u^{-1} = 1$ and $u^{-1} \times u = 1$.
>
> A monoid is a σ_{mnd}-structure which satisfies the identities $u \times (v \times w) = (u \times v) \times w$, $u \times 1 = u$ and $1 \times u = u$.
>
> A semigroup is a $\{\times\}$-structure which satisfies the identity $u \times (v \times w) = (u \times v) \times w$.
>
> A magma is just a $\{\times\}$-structure.

This is a very efficient way to define most classes of algebraic structures, because there is also the concept of σ-homomorphism, which correctly specializes to the usual notions of homomorphism for groups, semigroups, magmas and rings. For this to work, the signature must be chosen well.

Terms such as the σ_{ring}-term $t(u, v, w)$ given by $(u + (v \times w)) + (-1)$ are used to define identities $t = t'$, but also to construct free algebras. An equational class is a class of structures which, like the examples above and many others, is defined as the class of all σ-structures which satisfy a certain set of identities. Birkhoff's theorem states:

> A class of σ-structures is an equational class if and only if it is not empty and closed under subalgebras, homomorphic images, and direct products.

An important non-trivial tool in universal algebra are ultraproducts $\Pi_{i \in I} A_i / U$, where I is an infinite set indexing a system of σ-structures A_i, and U is an ultrafilter on I.

While model theory is generally considered a part of mathematical logic, universal algebra, which grew out of Alfred North Whitehead's (1898) work on abstract algebra, is part of algebra. This is reflected by their respective MSC classifications. Nevertheless model theory can be seen as an extension of universal algebra.

1.3 Finite model theory

Main article: Finite model theory

Finite model theory is the area of model theory which has the closest ties to universal algebra. Like some parts of universal algebra, and in contrast with the other areas of model theory, it is mainly concerned with finite algebras, or more generally, with finite σ-structures for signatures σ which may contain relation symbols as in the following example:

> The standard signature for graphs is $\sigma_{\mathrm{grph}}=\{E\}$, where E is a binary relation symbol.
>
> A graph is a σ_{grph}-structure satisfying the sentences $\forall u \forall v (uEv \rightarrow vEu)$ and $\forall u \neg (uEu)$.

A σ-homomorphism is a map that commutes with the operations and preserves the relations in σ. This definition gives rise to the usual notion of graph homomorphism, which has the interesting property that a bijective homomorphism need not be invertible. Structures are also a part of universal algebra; after all, some algebraic structures such as ordered groups have a binary relation <. What distinguishes finite model theory from universal algebra is its use of more general logical sentences (as in the example above) in place of identities. (In a model-theoretic context an identity $t=t'$ is written as a sentence $\forall u_1 u_2 \ldots u_n (t = t')$.)

The logics employed in finite model theory are often substantially more expressive than first-order logic, the standard logic for model theory of infinite structures.

1.4 First-order logic

Main article: First-order logic

Whereas universal algebra provides the semantics for a signature, logic provides the syntax. With terms, identities and quasi-identities, even universal algebra has some limited syntactic tools; first-order logic is the result of making quantification explicit and adding negation into the picture.

A first-order **formula** is built out of atomic formulas such as $R(f(x,y),z)$ or $y = x + 1$ by means of the Boolean connectives $\neg, \wedge, \vee, \rightarrow$ and prefixing of quantifiers $\forall v$ or $\exists v$. A sentence is a formula in which each occurrence of a variable is in the scope of a corresponding quantifier. Examples for formulas are φ (or φ(x) to mark the fact that at most x is an unbound variable in φ) and ψ defined as follows:

$$\varphi \;=\; \forall u \forall v (\exists w (x \times w = u \times v) \rightarrow (\exists w (x \times w = u) \vee \exists w (x \times w = v))) \wedge x \neq 0 \wedge x \neq 1,$$

$$\psi \;=\; \forall u \forall v ((u \times v = x) \rightarrow (u = x) \vee (v = x)) \wedge x \neq 0 \wedge x \neq 1.$$

(Note that the equality symbol has a double meaning here.) It is intuitively clear how to translate such formulas into mathematical meaning. In the σ_{smr}-structure $\mathcal{N}$ of the natural numbers, for example, an element n **satisfies** the formula φ if and only if n is a prime number. The formula ψ similarly defines irreducibility. Tarski gave a rigorous definition, sometimes called "Tarski's definition of truth", for the satisfaction relation $\models$, so that one easily proves:

$$\mathcal{N} \models \varphi(n) \iff n$$

$$\mathcal{N} \models \psi(n) \iff n$$

A set T of sentences is called a (first-order) theory. A theory is **satisfiable** if it has a **model** $\mathcal{M} \models T$, i.e. a structure (of the appropriate signature) which satisfies all the sentences in the set T. Consistency of a theory is usually defined in a syntactical way, but in first-order logic by the completeness theorem there is no need to distinguish between satisfiability and consistency. Therefore model theorists often use "consistent" as a synonym for "satisfiable".

A theory is called **categorical** if it determines a structure up to isomorphism, but it turns out that this definition is not useful, due to serious restrictions in the expressivity of first-order logic. The Löwenheim–Skolem theorem implies that for every theory T[2] which has an infinite model and for every infinite cardinal number κ, there is a model $\mathcal{M} \models T$ such that the number of elements of $\mathcal{M}$ is exactly κ. Therefore only finitary structures can be described by a categorical theory.

Lack of expressivity (when compared to higher logics such as second-order logic) has its advantages, though. For model theorists, the Löwenheim–Skolem theorem is an important practical tool rather than the source of Skolem's paradox. In a certain sense made precise by Lindström's theorem, first-order logic is the most expressive logic for which both the Löwenheim–Skolem theorem and the compactness theorem hold.

As a corollary (i.e., its contrapositive), the compactness theorem says that every unsatisfiable first-order theory has a finite unsatisfiable subset. This theorem is of central importance in infinite model theory, where the words "by compactness" are commonplace. One way to prove it is by means of ultraproducts. An alternative proof uses the completeness theorem, which is otherwise reduced to a marginal role in most of modern model theory.

1.5   Axiomatizability, elimination of quantifiers, and model-completeness

The first step, often trivial, for applying the methods of model theory to a class of mathematical objects such as groups, or trees in the sense of graph theory, is to choose a signature σ and represent the objects as σ-structures. The next step is to show that the class is an elementary class, i.e. axiomatizable in first-order logic (i.e. there is a theory T such that a σ-structure is in the class if and only if it satisfies T). E.g. this step fails for the trees, since connectedness cannot be expressed in first-order logic. Axiomatizability ensures that model theory can speak about the right objects. Quantifier elimination can be seen as a condition which ensures that model theory does not say too much about the objects.

A theory T has quantifier elimination if every first-order formula $\varphi(x_1,...,xn)$ over its signature is equivalent modulo T to a first-order formula $\psi(x_1,...,xn)$ without quantifiers, i.e. $\forall x_1 \ldots \forall x_n(\phi(x_1, \ldots, x_n) \leftrightarrow \psi(x_1, \ldots, x_n))$ holds in all models of T. For example the theory of algebraically closed fields in the signature $\sigma_{\text{ring}}=(\times,+,-,0,1)$ has quantifier elimination because every formula is equivalent to a Boolean combination of equations between polynomials.

A substructure of a σ-structure is a subset of its domain, closed under all functions in its signature σ, which is regarded as a σ-structure by restricting all functions and relations in σ to the subset. An embedding of a σ-structure $\mathcal{A}$ into another σ-structure $\mathcal{B}$ is a map f: A $\rightarrow$ B between the domains which can be written as an isomorphism of $\mathcal{A}$ with a substructure of $\mathcal{B}$. Every embedding is an injective homomorphism, but the converse holds only if the signature contains no relation symbols.

If a theory does not have quantifier elimination, one can add additional symbols to its signature so that it does. Early model theory spent much effort on proving axiomatizability and quantifier elimination results for specific theories, especially in algebra. But often instead of quantifier elimination a weaker property suffices:

A theory T is called model-complete if every substructure of a model of T which is itself a model of T is an elementary substructure. There is a useful criterion for testing whether a substructure is an elementary substructure, called the Tarski–Vaught test. It follows from this criterion that a theory T is model-complete if and only if every first-order formula $\varphi(x_1,...,xn)$ over its signature is equivalent modulo T to an existential first-order formula, i.e. a formula of the following form:

$$\exists v_1 \ldots \exists v_m \psi(x_1, \ldots, x_n, v_1, \ldots, v_m)$$

where ψ is quantifier free. A theory that is not model-complete may or may not have a **model completion**, which is a related model-complete theory that is not, in general, an extension of the original theory. A more general notion is that of **model companions**.

1.6 Categoricity

As observed in the section on first-order logic, first-order theories cannot be categorical, i.e. they cannot describe a unique model up to isomorphism, unless that model is finite. But two famous model-theoretic theorems deal with the weaker notion of κ-categoricity for a cardinal κ. A theory T is called **κ-categorical** if any two models of T that are of cardinality κ are isomorphic. It turns out that the question of κ-categoricity depends critically on whether κ is bigger than the cardinality of the language (i.e. $\aleph_0 + |\sigma|$, where $|\sigma|$ is the cardinality of the signature). For finite or countable signatures this means that there is a fundamental difference between $\aleph_0$-cardinality and κ-cardinality for uncountable κ.

A few characterizations of $\aleph_0$-categoricity include:

> For a complete first-order theory T in a finite or countable signature the following conditions are equivalent:
>
> 1. T is $\aleph_0$-categorical.
> 2. For every natural number n, the Stone space $Sn(T)$ is finite.
> 3. For every natural number n, the number of formulas $\varphi(x_1, ..., x_n)$ in n free variables, up to equivalence modulo T, is finite.

This result, due independently to Engeler, Ryll-Nardzewski and Svenonius, is sometimes referred to as the Ryll-Nardzewski theorem.

Further, $\aleph_0$-categorical theories and their countable models have strong ties with oligomorphic groups. They are often constructed as Fraïssé limits.

Michael Morley's highly non-trivial result that (for countable languages) there is only *one* notion of uncountable categoricity was the starting point for modern model theory, and in particular classification theory and stability theory:

> Morley's categoricity theorem
>
> If a first-order theory T in a finite or countable signature is κ-categorical for some uncountable cardinal κ, then T is κ-categorical for all uncountable cardinals κ.

Uncountably categorical (i.e. κ-categorical for all uncountable cardinals κ) theories are from many points of view the most well-behaved theories. A theory that is both $\aleph_0$-categorical and uncountably categorical is called **totally categorical**.

1.7 Model theory and set theory

Set theory (which is expressed in a countable language), if it is consistent, has a countable model; this is known as Skolem's paradox, since there are sentences in set theory which postulate the existence of uncountable sets and yet these sentences are true in our countable model. Particularly the proof of the independence of the continuum hypothesis requires considering sets in models which appear to be uncountable when viewed from *within* the model, but are countable to someone *outside* the model.

The model-theoretic viewpoint has been useful in set theory; for example in Kurt Gödel's work on the constructible universe, which, along with the method of forcing developed by Paul Cohen can be shown to prove the (again philosophically interesting) independence of the axiom of choice and the continuum hypothesis from the other axioms of set theory.

In the other direction, model theory itself can be formalized within ZFC set theory. The development of the fundamentals of model theory (such as the compactness theorem) rely on the axiom of choice, or more exactly the Boolean prime ideal theorem. Other results in model theory depend on set-theoretic axioms beyond the standard ZFC framework. For example, if the Continuum Hypothesis holds then every countable model has an ultrapower which is saturated (in its own cardinality). Similarly, if the Generalized Continuum Hypothesis holds then every model has a saturated elementary extension. Neither of these results are provable in ZFC alone. Finally, some questions arising from model theory (such as compactness for infinitary logics) have been shown to be equivalent to large cardinal axioms.

1.8   Other basic notions of model theory

1.8.1   Reducts and expansions

Main article: Reduct

A field or a vector space can be regarded as a (commutative) group by simply ignoring some of its structure. The corresponding notion in model theory is that of a **reduct** of a structure to a subset of the original signature. The opposite relation is called an *expansion* - e.g. the (additive) group of the rational numbers, regarded as a structure in the signature $\{+,0\}$ can be expanded to a field with the signature $\{\times,+,1,0\}$ or to an ordered group with the signature $\{+,0,<\}$.

Similarly, if σ' is a signature that extends another signature σ, then a complete σ'-theory can be restricted to σ by intersecting the set of its sentences with the set of σ-formulas. Conversely, a complete σ-theory can be regarded as a σ'-theory, and one can extend it (in more than one way) to a complete σ'-theory. The terms reduct and expansion are sometimes applied to this relation as well.

1.8.2   Interpretability

Main article: Interpretation (model theory)

Given a mathematical structure, there are very often associated structures which can be constructed as a quotient of part of the original structure via an equivalence relation. An important example is a quotient group of a group.

One might say that to understand the full structure one must understand these quotients. When the equivalence relation is definable, we can give the previous sentence a precise meaning. We say that these structures are **interpretable**.

A key fact is that one can translate sentences from the language of the interpreted structures to the language of the original structure. Thus one can show that if a structure M interprets another whose theory is undecidable, then M itself is undecidable.

1.8.3   Using the compactness and completeness theorems

Gödel's completeness theorem (not to be confused with his incompleteness theorems) says that a theory has a model if and only if it is consistent, i.e. no contradiction is proved by the theory. This is the heart of model theory as it lets us answer questions about theories by looking at models and vice versa. One should not confuse the completeness theorem with the notion of a complete theory. A complete theory is a theory that contains every sentence or its negation. Importantly, one can find a complete consistent theory extending any consistent theory. However, as shown by Gödel's incompleteness theorems only in relatively simple cases will it be possible to have a complete consistent theory that is also recursive, i.e. that can be described by a recursively enumerable set of axioms. In particular, the theory of natural numbers has no recursive complete and consistent theory. Non-recursive theories are of little practical use, since it is undecidable if a proposed axiom is indeed an axiom, making proof-checking a supertask.

The compactness theorem states that a set of sentences S is satisfiable if every finite subset of S is satisfiable. In the context of proof theory the analogous statement is trivial, since every proof can have only a finite number of antecedents used in the proof. In the context of model theory, however, this proof is somewhat more difficult. There are two well known proofs, one by Gödel (which goes via proofs) and one by Malcev (which is more direct and allows us to restrict the cardinality of the resulting model).

Model theory is usually concerned with first-order logic, and many important results (such as the completeness and compactness theorems) fail in second-order logic or other alternatives. In first-order logic all infinite cardinals look the same to a language which is countable. This is expressed in the Löwenheim–Skolem theorems, which state that any countable theory with an infinite model 𝔄 has models of all infinite cardinalities (at least that of the language) which agree with 𝔄 on all sentences, i.e. they are 'elementarily equivalent'.

1.8.4 Types

Main article: Type (model theory)

Fix an L-structure M, and a natural number n. The set of definable subsets of M^n over some parameters A is a Boolean algebra. By Stone's representation theorem for Boolean algebras there is a natural dual notion to this. One can consider this to be the topological space consisting of maximal consistent sets of formulae over A. We call this the space of (complete) n-types over A, and write $S_n(A)$.

Now consider an element $m \in M^n$. Then the set of all formulae ϕ with parameters in A in free variables $x_1, \ldots, x_n$ so that $M \models \phi(m)$ is consistent and maximal such. It is called the *type* of m over A.

One can show that for any n-type p, there exists some elementary extension N of M and some $a \in N^n$ so that p is the type of a over A.

Many important properties in model theory can be expressed with types. Further many proofs go via constructing models with elements that contain elements with certain types and then using these elements.

Illustrative Example: Suppose M is an algebraically closed field. The theory has quantifier elimination. This allows us to show that a type is determined exactly by the polynomial equations it contains. Thus the space of n-types over a subfield A is bijective with the set of prime ideals of the polynomial ring $A[x_1, \ldots, x_n]$. This is the same set as the spectrum of $A[x_1, \ldots, x_n]$. Note however that the topology considered on the type space is the constructible topology: a set of types is basic open iff it is of the form $\{p : f(x) = 0 \in p\}$ or of the form $\{p : f(x) \neq 0 \in p\}$. This is finer than the Zariski topology.

1.9 History

Model theory as a subject has existed since approximately the middle of the 20th century. However some earlier research, especially in mathematical logic, is often regarded as being of a model-theoretical nature in retrospect. The first significant result in what is now model theory was a special case of the downward Löwenheim–Skolem theorem, published by Leopold Löwenheim in 1915. The compactness theorem was implicit in work by Thoralf Skolem,[3] but it was first published in 1930, as a lemma in Kurt Gödel's proof of his completeness theorem. The Löwenheim–Skolem theorem and the compactness theorem received their respective general forms in 1936 and 1941 from Anatoly Maltsev.

1.10 See also

1.11 Notes

[1] Chang and Keisler, p. 1.

[2] In a countable signature. The theorem has a straightforward generalization to uncountable signatures.

[3] "All three commentators [i.e. Vaught, van Heijenoort and Dreben] agree that both the completeness and compactness theorems were implicit in Skolem 1923...." [Dawson, J. W. (1993). "The compactness of first-order logic:from gödel to lindström". *History and Philosophy of Logic* **14**: 15. doi:10.1080/01445349308837208.]

1.12 References

1.12.1 Canonical textbooks

- Chang, Chen Chung; Keisler, H. Jerome (1990) [1973]. *Model Theory*. Studies in Logic and the Foundations of Mathematics (3rd ed.). Elsevier. ISBN 978-0-444-88054-3.

- Hodges, Wilfrid (1997). *A shorter model theory*. Cambridge: Cambridge University Press. ISBN 978-0-521-58713-6.

- Marker, David (2002). *Model Theory: An Introduction*. Graduate Texts in Mathematics 217. Springer. ISBN 0-387-98760-6.

1.12.2 Other textbooks

- Bell, John L.; Slomson, Alan B. (2006) [1969]. *Models and Ultraproducts: An Introduction* (reprint of 1974 ed.). Dover Publications. ISBN 0-486-44979-3.

- Ebbinghaus, Heinz-Dieter; Flum, Jörg; Thomas, Wolfgang (1994). *Mathematical Logic*. Springer. ISBN 0-387-94258-0.

- Hinman, Peter G. (2005). *Fundamentals of Mathematical Logic*. A K Peters. ISBN 1-56881-262-0.

- Hodges, Wilfrid (1993). *Model theory*. Cambridge University Press. ISBN 0-521-30442-3.

- Manzano, Maria (1999). *Model theory*. Oxford University Press. ISBN 0-19-853851-0.

- Poizat, Bruno (2000). *A Course in Model Theory*. Springer. ISBN 0-387-98655-3.

- Rautenberg, Wolfgang (2010). *A Concise Introduction to Mathematical Logic* (3rd ed.). New York: Springer Science+Business Media. doi:10.1007/978-1-4419-1221-3. ISBN 978-1-4419-1220-6.

- Rothmaler, Philipp (2000). *Introduction to Model Theory* (new ed.). Taylor & Francis. ISBN 90-5699-313-5.

- Ziegler, Martin; Tent, Katrin (2012). *A Course in Model Theory*. Cambridge University Press. ISBN 9780521763240.

1.12.3 Free online texts

- Chatzidakis, Zoe (2001). *Introduction to Model Theory* (PDF). pp. 26 pages.

- Pillay, Anand (2002). *Lecture Notes – Model Theory* (PDF). pp. 61 pages.

- Hazewinkel, Michiel, ed. (2001), "Model theory", *Encyclopedia of Mathematics*, Springer, ISBN 978-1-55608-010-4

- Hodges, Wilfrid, *Model theory*. The Stanford Encyclopedia Of Philosophy, E. Zalta (ed.).

- Hodges, Wilfrid, *First-order Model theory*. The Stanford Encyclopedia Of Philosophy, E. Zalta (ed.).

- Simmons, Harold (2004), *An introduction to Good old fashioned model theory*. Notes of an introductory course for postgraduates (with exercises).

- J. Barwise and S. Feferman (editors), Model-Theoretic Logics, Perspectives in Mathematical Logic, Volume 8, New York: Springer-Verlag, 1985.

Chapter 2

Mathematical model

Not to be confused with the same term used in model theory, a branch of mathematical logic. An artifact that is used to illustrate a mathematical idea may also be called a mathematical model, the usage of which is the reverse of the sense explained in this article.

A **mathematical model** is a description of a system using mathematical concepts and language. The process of developing a mathematical model is termed **mathematical modeling**. Mathematical models are used in the natural sciences (such as physics, biology, earth science, meteorology) and engineering disciplines (such as computer science, artificial intelligence), as well as in the social sciences (such as economics, psychology, sociology, political science). Physicists, engineers, statisticians, operations research analysts, and economists use mathematical models most extensively. A model may help to explain a system and to study the effects of different components, and to make predictions about behaviour.

Mathematical models can take many forms, including but not limited to dynamical systems, statistical models, differential equations, or game theoretic models. These and other types of models can overlap, with a given model involving a variety of abstract structures. In general, mathematical models may include logical models. In many cases, the quality of a scientific field depends on how well the mathematical models developed on the theoretical side agree with results of repeatable experiments. Lack of agreement between theoretical mathematical models and experimental measurements often leads to important advances as better theories are developed.

2.1   Model classifications in mathematics

Mathematical models are usually composed of relationships and *variables*. Relationships can be described by *operators*, such as algebraic operators, functions, differential operators, etc. Variables are abstractions of system parameters of interest, that can be quantified. Several classification critera can be used for mathematical models according to their structure:

- **Linear vs. nonlinear:** If all the operators in a mathematical model exhibit linearity, the resulting mathematical model is defined as linear. A model is considered to be nonlinear otherwise. The definition of linearity and nonlinearity is dependent on context, and linear models may have nonlinear expressions in them. For example, in a statistical linear model, it is assumed that a relationship is linear in the parameters, but it may be nonlinear in the predictor variables. Similarly, a differential equation is said to be linear if it can be written with linear differential operators, but it can still have nonlinear expressions in it. In a mathematical programming model, if the objective functions and constraints are represented entirely by linear equations, then the model is regarded as a linear model. If one or more of the objective functions or constraints are represented with a nonlinear equation, then the model is known as a nonlinear model.
  Nonlinearity, even in fairly simple systems, is often associated with phenomena such as chaos and irreversibility. Although there are exceptions, nonlinear systems and models tend to be more difficult to study than linear ones.

A common approach to nonlinear problems is linearization, but this can be problematic if one is trying to study aspects such as irreversibility, which are strongly tied to nonlinearity.

- **Static vs. dynamic:** A *dynamic* model accounts for time-dependent changes in the state of the system, while a *static* (or steady-state) model calculates the system in equilibrium, and thus is time-invariant. Dynamic models typically are represented by differential equations.

- **Explicit vs. implicit:** If all of the input parameters of the overall model are known, and the output parameters can be calculated by a finite series of computations (known as linear programming, not to be confused with *linearity* as described above), the model is said to be *explicit*. But sometimes it is the *output* parameters which are known, and the corresponding inputs must be solved for by an iterative procedure, such as Newton's method (if the model is linear) or Broyden's method (if non-linear). For example, a jet engine's physical properties such as turbine and nozzle throat areas can be explicitly calculated given a design thermodynamic cycle (air and fuel flow rates, pressures, and temperatures) at a specific flight condition and power setting, but the engine's operating cycles at other flight conditions and power settings cannot be explicitly calculated from the constant physical properties.

- **Discrete vs. continuous:** A discrete model treats objects as discrete, such as the particles in a molecular model or the states in a statistical model; while a continuous model represents the objects in a continuous manner, such as the velocity field of fluid in pipe flows, temperatures and stresses in a solid, and electric field that applies continuously over the entire model due to a point charge.

- **Deterministic vs. probabilistic (stochastic):** A deterministic model is one in which every set of variable states is uniquely determined by parameters in the model and by sets of previous states of these variables; therefore, a deterministic model always performs the same way for a given set of initial conditions. Conversely, in a stochastic model—usually called a "statistical model"—randomness is present, and variable states are not described by unique values, but rather by probability distributions.

- **Deductive, inductive, or floating:** A deductive model is a logical structure based on a theory. An inductive model arises from empirical findings and generalization from them. The floating model rests on neither theory nor observation, but is merely the invocation of expected structure. Application of mathematics in social sciences outside of economics has been criticized for unfounded models.[1] Application of catastrophe theory in science has been characterized as a floating model.[2]

2.2   Significance in the natural sciences

Mathematical models are of great importance in the natural sciences, particularly in physics. Physical theories are almost invariably expressed using mathematical models.

Throughout history, more and more accurate mathematical models have been developed. Newton's laws accurately describe many everyday phenomena, but at certain limits relativity theory and quantum mechanics must be used, even these do not apply to all situations and need further refinement. It is possible to obtain the less accurate models in appropriate limits, for example relativistic mechanics reduces to Newtonian mechanics at speeds much less than the speed of light. Quantum mechanics reduces to classical physics when the quantum numbers are high. For example, the de Broglie wavelength of a tennis ball is insignificantly small, so classical physics is a good approximation to use in this case.

It is common to use idealized models in physics to simplify things. Massless ropes, point particles, ideal gases and the particle in a box are among the many simplified models used in physics. The laws of physics are represented with simple equations such as Newton's laws, Maxwell's equations and the Schrödinger equation. These laws are such as a basis for making mathematical models of real situations. Many real situations are very complex and thus modeled approximate on a computer, a model that is computationally feasible to compute is made from the basic laws or from approximate models made from the basic laws. For example, molecules can be modeled by molecular orbital models that are approximate solutions to the Schrödinger equation. In engineering, physics models are often made by mathematical methods such as finite element analysis.

Different mathematical models use different geometries that are not necessarily accurate descriptions of the geometry of the universe. Euclidean geometry is much used in classical physics, while special relativity and general relativity are examples of theories that use geometries which are not Euclidean.

2.3 Some applications

Since prehistorical times simple models such as maps and diagrams have been used.

Often when engineers analyze a system to be controlled or optimized, they use a mathematical model. In analysis, engineers can build a descriptive model of the system as a hypothesis of how the system could work, or try to estimate how an unforeseeable event could affect the system. Similarly, in control of a system, engineers can try out different control approaches in simulations.

A mathematical model usually describes a system by a set of variables and a set of equations that establish relationships between the variables. Variables may be of many types; real or integer numbers, boolean values or strings, for example. The variables represent some properties of the system, for example, measured system outputs often in the form of signals, timing data, counters, and event occurrence (yes/no). The actual model is the set of functions that describe the relations between the different variables.

2.4 Building blocks

In business and engineering, mathematical models may be used to maximize a certain output. The system under consideration will require certain inputs. The system relating inputs to outputs depends on other variables too: decision variables, state variables, exogenous variables, and random variables.

Decision variables are sometimes known as independent variables. Exogenous variables are sometimes known as parameters or constants. The variables are not independent of each other as the state variables are dependent on the decision, input, random, and exogenous variables. Furthermore, the output variables are dependent on the state of the system (represented by the state variables).

Objectives and constraints of the system and its users can be represented as functions of the output variables or state variables. The objective functions will depend on the perspective of the model's user. Depending on the context, an objective function is also known as an *index of performance*, as it is some measure of interest to the user. Although there is no limit to the number of objective functions and constraints a model can have, using or optimizing the model becomes more involved (computationally) as the number increases.

For example, in economics students often apply linear algebra when using input-output models. Complicated mathematical models that have many variables may be consolidated by use of vectors where one symbol represents several variables.

2.5 A priori information

Mathematical modeling problems are often classified into black box or white box models, according to how much a priori information on the system is available. A black-box model is a system of which there is no a priori information available. A white-box model (also called glass box or clear box) is a system where all necessary information is available. Practically all systems are somewhere between the black-box and white-box models, so this concept is useful only as an intuitive guide for deciding which approach to take.

Usually it is preferable to use as much a priori information as possible to make the model more accurate. Therefore, the white-box models are usually considered easier, because if you have used the information correctly, then the model will behave correctly. Often the a priori information comes in forms of knowing the type of functions relating different variables. For example, if we make a model of how a medicine works in a human system, we know that usually the amount of medicine in the blood is an exponentially decaying function. But we are still left with several unknown parameters; how rapidly does the medicine amount decay, and what is the initial amount of medicine in blood? This example is therefore not a completely white-box model. These parameters have to be estimated through some means before one can use the model.

In black-box models one tries to estimate both the functional form of relations between variables and the numerical parameters in those functions. Using a priori information we could end up, for example, with a set of functions that probably could describe the system adequately. If there is no a priori information we would try to use functions as general as possible to cover all different models. An often used approach for black-box models are neural networks which usually do not make assumptions about incoming data. Alternatively the NARMAX (Nonlinear AutoRegressive Moving Average model with eXogenous inputs) algorithms which were developed as part of nonlinear system identification [3] can be used to select the model terms, determine the model structure, and estimate the unknown parameters in the presence of correlated and nonlinear noise. The advantage of NARMAX models compared to neural networks is that NARMAX produces models that can be written down and related to the underlying process, whereas neural networks produce an approximation that is opaque.

2.5.1 Subjective information

Sometimes it is useful to incorporate subjective information into a mathematical model. This can be done based on intuition, experience, or expert opinion, or based on convenience of mathematical form. Bayesian statistics provides a theoretical framework for incorporating such subjectivity into a rigorous analysis: we specify a prior probability distribution (which can be subjective), and then update this distribution based on empirical data.

An example of when such approach would be necessary is a situation in which an experimenter bends a coin slightly and tosses it once, recording whether it comes up heads, and is then given the task of predicting the probability that the next flip comes up heads. After bending the coin, the true probability that the coin will come up heads is unknown; so the experimenter would need to make a decision (perhaps by looking at the shape of the coin) about what prior distribution to use. Incorporation of such subjective information might be important to get an accurate estimate of the probability.

2.6 Complexity

In general, model complexity involves a trade-off between simplicity and accuracy of the model. Occam's razor is a principle particularly relevant to modeling; the essential idea being that among models with roughly equal predictive power, the simplest one is the most desirable. While added complexity usually improves the realism of a model, it can make the model difficult to understand and analyze, and can also pose computational problems, including numerical instability. Thomas Kuhn argues that as science progresses, explanations tend to become more complex before a Paradigm shift offers radical simplification.

For example, when modeling the flight of an aircraft, we could embed each mechanical part of the aircraft into our model and would thus acquire an almost white-box model of the system. However, the computational cost of adding such a huge amount of detail would effectively inhibit the usage of such a model. Additionally, the uncertainty would increase due to an overly complex system, because each separate part induces some amount of variance into the model. It is therefore usually appropriate to make some approximations to reduce the model to a sensible size. Engineers often can accept some approximations in order to get a more robust and simple model. For example, Newton's classical mechanics is an approximated model of the real world. Still, Newton's model is quite sufficient for most ordinary-life situations, that is, as long as particle speeds are well below the speed of light, and we study macro-particles only.

2.7 Training

Any model which is not pure white-box contains some parameters that can be used to fit the model to the system it is intended to describe. If the modeling is done by a neural network, the optimization of parameters is called *training*. In more conventional modeling through explicitly given mathematical functions, parameters are determined by curve fitting.

2.8 Model evaluation

A crucial part of the modeling process is the evaluation of whether or not a given mathematical model describes a system accurately. This question can be difficult to answer as it involves several different types of evaluation.

2.8.1 Fit to empirical data

Usually the easiest part of model evaluation is checking whether a model fits experimental measurements or other empirical data. In models with parameters, a common approach to test this fit is to split the data into two disjoint subsets: training data and verification data. The training data are used to estimate the model parameters. An accurate model will closely match the verification data even though these data were not used to set the model's parameters. This practice is referred to as cross-validation in statistics.

Defining a metric to measure distances between observed and predicted data is a useful tool of assessing model fit. In statistics, decision theory, and some economic models, a loss function plays a similar role.

While it is rather straightforward to test the appropriateness of parameters, it can be more difficult to test the validity of the general mathematical form of a model. In general, more mathematical tools have been developed to test the fit of statistical models than models involving differential equations. Tools from non-parametric statistics can sometimes be used to evaluate how well the data fit a known distribution or to come up with a general model that makes only minimal assumptions about the model's mathematical form.

2.8.2 Scope of the model

Assessing the scope of a model, that is, determining what situations the model is applicable to, can be less straightforward. If the model was constructed based on a set of data, one must determine for which systems or situations the known data is a "typical" set of data.

The question of whether the model describes well the properties of the system between data points is called interpolation, and the same question for events or data points outside the observed data is called extrapolation.

As an example of the typical limitations of the scope of a model, in evaluating Newtonian classical mechanics, we can note that Newton made his measurements without advanced equipment, so he could not measure properties of particles travelling at speeds close to the speed of light. Likewise, he did not measure the movements of molecules and other small particles, but macro particles only. It is then not surprising that his model does not extrapolate well into these domains, even though his model is quite sufficient for ordinary life physics.

2.8.3 Philosophical considerations

Many types of modeling implicitly involve claims about causality. This is usually (but not always) true of models involving differential equations. As the purpose of modeling is to increase our understanding of the world, the validity of a model rests not only on its fit to empirical observations, but also on its ability to extrapolate to situations or data beyond those originally described in the model. One can think of this as the differentiation between qualitative and quantitative predictions. One can also argue that a model is worthless unless it provides some insight which goes beyond what is already known from direct investigation of the phenomenon being studied.

An example of such criticism is the argument that the mathematical models of Optimal foraging theory do not offer insight that goes beyond the common-sense conclusions of evolution and other basic principles of ecology.[4]

2.9 Examples

- One of the popular examples in computer science is the mathematical models of various machines, an example is the deterministic finite automaton which is defined as an abstract mathematical concept, but due to the deterministic nature of a DFA, it is implementable in hardware and software for solving various specific problems. For example, the following is a DFA M with a binary alphabet, which requires that the input contains an even number of 0s.

$M = (Q, \Sigma, \delta, q_0, F)$ where

- $Q = \{S_1, S_2\}$,

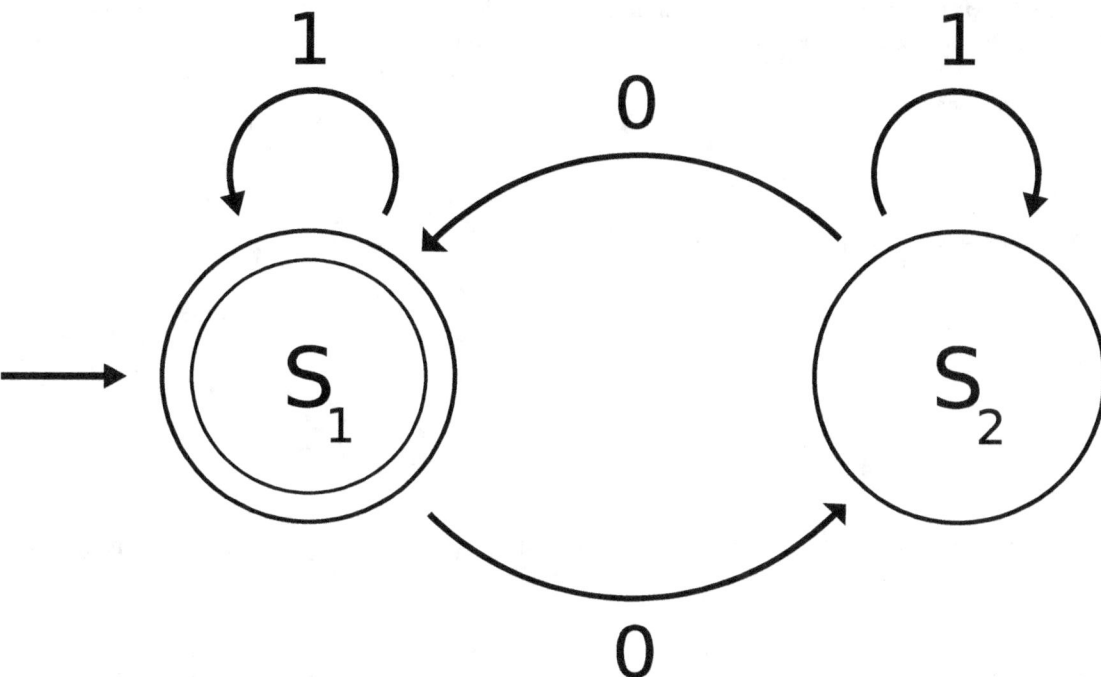

The state diagram for M

- $\Sigma = \{0, 1\}$,

- $q_0 = S_1$,

- $F = \{S_1\}$, and

- δ is defined by the following state transition table:

The state S_1 represents that there has been an even number of 0s in the input so far, while S_2 signifies an odd number. A 1 in the input does not change the state of the automaton. When the input ends, the state will show whether the input contained an even number of 0s or not. If the input did contain an even number of 0s, M will finish in state S_1, an accepting state, so the input string will be accepted.

The language recognized by M is the regular language given by the regular expression 1*(0 (1*) 0 (1*))*, where "*" is the Kleene star, e.g., 1* denotes any non-negative number (possibly zero) of symbols "1".

- Many everyday activities carried out without a thought are uses of mathematical models. A geographical map projection of a region of the earth onto a small, plane surface is a model[5] which can be used for many purposes such as planning travel.

- Another simple activity is predicting the position of a vehicle from its initial position, direction and speed of travel, using the equation that distance traveled is the product of time and speed. This is known as dead reckoning when used more formally. Mathematical modeling in this way does not necessarily require formal mathematics; animals have been shown to use dead reckoning.[6][7]

- *Population Growth.* A simple (though approximate) model of population growth is the Malthusian growth model. A slightly more realistic and largely used population growth model is the logistic function, and its extensions.

- *Model of a particle in a potential-field.* In this model we consider a particle as being a point of mass which describes a trajectory in space which is modeled by a function giving its coordinates in space as a function of time. The potential field is given by a function $V : \mathbb{R}^3 \to \mathbb{R}$ and the trajectory, that is a function $\mathbf{r} : \mathbb{R} \to \mathbb{R}^3$, is the solution of the differential equation:

$$-\frac{\mathrm{d}^2\mathbf{r}(t)}{\mathrm{d}t^2}m = \frac{\partial V[\mathbf{r}(t)]}{\partial x}\hat{\mathbf{x}} + \frac{\partial V[\mathbf{r}(t)]}{\partial y}\hat{\mathbf{y}} + \frac{\partial V[\mathbf{r}(t)]}{\partial z}\hat{\mathbf{z}},$$

that can be written also as:

$$m\frac{\mathrm{d}^2\mathbf{r}(t)}{\mathrm{d}t^2} = -\nabla V[\mathbf{r}(t)].$$

Note this model assumes the particle is a point mass, which is certainly known to be false in many cases in which we use this model; for example, as a model of planetary motion.

- *Model of rational behavior for a consumer.* In this model we assume a consumer faces a choice of n commodities labeled 1,2,...,n each with a market price $p_1, p_2,..., pn$. The consumer is assumed to have a *cardinal* utility function U (cardinal in the sense that it assigns numerical values to utilities), depending on the amounts of commodities x_1, x_2,..., xn consumed. The model further assumes that the consumer has a budget M which is used to purchase a vector $x_1, x_2,..., xn$ in such a way as to maximize $U(x_1, x_2,..., xn)$. The problem of rational behavior in this model then becomes an optimization problem, that is:

  $\max U(x_1, x_2, \ldots, x_n)$
  subject to:
  $\sum_{i=1}^{n} p_i x_i \leq M.$
  $x_i \geq 0 \quad \forall i \in \{1, 2, \ldots, n\}$

  This model has been used in general equilibrium theory, particularly to show existence and Pareto efficiency of economic equilibria. However, the fact that this particular formulation assigns *numerical values* to levels of satisfaction is the source of criticism (and even ridicule). However, it is not an essential ingredient of the theory and again this is an idealization.

- *Neighbour-sensing model* explains the mushroom formation from the initially chaotic fungal network.

- *Computer science*: models in Computer Networks, data models, surface model,...

- *Mechanics*: movement of rocket model,...

Modeling requires selecting and identifying relevant aspects of a situation in the real world.

2.10   See also

- Agent-based model

- Cliodynamics

- Computer simulation

- Conceptual model

- Decision engineering

- Grey box model

- Mathematical biology

- Mathematical diagram

- Mathematical models in physics

- Mathematical psychology

- Mathematical sociology

- Microscale and macroscale models

- Statistical Model

- System identification

- TK Solver - Rule Based Modeling

2.11   References

[1] Andreski, Stanislav (1972). *Social Sciences as Sorcery*. St. Martin's Press. ISBN 0-14-021816-5.

[2] Truesdell, Clifford (1984). *An Idiot's Fugitive Essays on Science*. Springer. pp. 121–7. ISBN 3-540-90703-3.

[3] Billings S.A. (2013), *Nonlinear System Identification: NARMAX Methods in the Time, Frequency, and Spatio-Temporal Domains*, Wiley.

[4] Pyke, G. H. (1984). "Optimal Foraging Theory: A Critical Review". *Annual Review of Ecology and Systematics* **15**: 523–575. doi:10.1146/annurev.es.15.110184.002515.

[5] landinfo.com, definition of map projection

[6] Gallistel (1990). *The Organization of Learning*. Cambridge: The MIT Press. ISBN 0-262-07113-4.

[7] Whishaw, I. Q.; Hines, D. J.; Wallace, D. G. (2001). "Dead reckoning (path integration) requires the hippocampal formation: Evidence from spontaneous exploration and spatial learning tasks in light (allothetic) and dark (idiothetic) tests". *Behavioural Brain Research* **127** (1–2): 49–69. doi:10.1016/S0166-4328(01)00359-X. PMID 11718884.

2.12   Further reading

2.12.1   Books

- Aris, Rutherford [1978] (1994). *Mathematical Modelling Techniques*, New York: Dover. ISBN 0-486-68131-9

- Bender, E.A. [1978] (2000). *An Introduction to Mathematical Modeling*, New York: Dover. ISBN 0-486-41180-X

- Gershenfeld, N. (1998) *The Nature of Mathematical Modeling*, Cambridge University Press ISBN 0-521-57095-6 .

- Lin, C.C. & Segel, L.A. (1988). *Mathematics Applied to Deterministic Problems in the Natural Sciences*, Philadelphia: SIAM. ISBN 0-89871-229-7

2.12.2 Specific applications

- Korotayev A., Malkov A., Khaltourina D. (2006). *Introduction to Social Macrodynamics: Compact Macromodels of the World System Growth*. Moscow: Editorial URSS ISBN 5-484-00414-4 .

- Peierls, R. (1980). "Model-making in physics". *Contemporary Physics* **21**: 3–17. doi:10.1080/00107518008210938.

- *An Introduction to Infectious Disease Modelling* by Emilia Vynnycky and Richard G White.

2.13 External links

General reference material

- Patrone, F. Introduction to modeling via differential equations, with critical remarks.

- Plus teacher and student package: Mathematical Modelling. Brings together all articles on mathematical modeling from *Plus Magazine*, the online mathematics magazine produced by the Millennium Mathematics Project at the University of Cambridge.

Philosophical background

- Frigg, R. and S. Hartmann, Models in Science, in: The Stanford Encyclopedia of Philosophy, (Spring 2006 Edition)

- Griffiths, E. C. (2010) What is a model?

Chapter 3

Structure (mathematical logic)

In universal algebra and in model theory, a **structure** consists of a set along with a collection of finitary operations, and relations that are defined on it.

Universal algebra studies structures that generalize the algebraic structures such as groups, rings, fields and vector spaces. The term **universal algebra** is used for structures with no relation symbols.[1]

Model theory has a different scope that encompasses more arbitrary theories, including foundational structures such as models of set theory. From the model-theoretic point of view, structures are the objects used to define the semantics of first-order logic. For a given theory in model theory, a structure is called a **model**, if it satisfies the defining axioms of that theory, although it is sometimes disambiguated as a *semantic model* when one discusses the notion in the more general setting of mathematical models. Logicians sometimes refer to structures as interpretations.[2]

In database theory, structures with no functions are studied as models for relational databases, in the form of relational models.

3.1 Definition

See also: Model theory § Universal algebra and Universal algebra § Basic idea

Formally, a **structure** can be defined as a triple $\mathcal{A} = (A, \sigma, I)$ consisting of a **domain** A, a signature σ, and an **interpretation function** I that indicates how the signature is to be interpreted on the domain. To indicate that a structure has a particular signature σ one can refer to it as a σ-structure.

3.1.1 Domain

The domain of a structure is an arbitrary set; it is also called the **underlying set** of the structure, its **carrier** (especially in universal algebra), or its **universe** (especially in model theory). In classical first-order logic, the definition of a structure prohibits the empty domain.[3]

Sometimes the notation $\mathrm{dom}(\mathcal{A})$ or $|\mathcal{A}|$ is used for the domain of $\mathcal{A}$, but often no notational distinction is made between a structure and its domain. (I.e. the same symbol $\mathcal{A}$ refers both to the structure and its domain.)[4]

3.1.2 Signature

Main article: Signature (logic)

The signature of a structure consists of a set of **function symbols** and **relation symbols** along with a function that ascribes to each symbol s a natural number $n = \text{ar}(s)$ which is called the **arity** of s because it is the arity of the interpretation of s.

Since the signatures that arise in algebra often contain only function symbols, a signature with no relation symbols is called an **algebraic signature**. A structure with such a signature is also called an **algebra**; this should not be confused with the notion of an algebra over a field.

3.1.3 Interpretation function

Main article: Interpretation (model theory)

The **interpretation function** I of $\mathcal{A}$ assigns functions and relations to the symbols of the signature. Each function symbol f of arity n is assigned an n-ary function $f^{\mathcal{A}} = I(f)$ on the domain. Each relation symbol R of arity n is assigned an n-ary relation $R^{\mathcal{A}} = I(R) \subseteq A^{\text{ar}(R)}$ on the domain. A nullary function symbol c is called a **constant symbol**, because its interpretation $I(c)$ can be identified with a constant element of the domain.

When a structure (and hence an interpretation function) is given by context, no notational distinction is made between a symbol s and its interpretation $I(s)$. For example if f is a binary function symbol of $\mathcal{A}$, one simply writes $f : \mathcal{A}^2 \to \mathcal{A}$ rather than $f^{\mathcal{A}} : |\mathcal{A}|^2 \to |\mathcal{A}|$.

3.1.4 Examples

The standard signature σf for fields consists of two binary function symbols $+$ and $\times$, a unary function symbol $-$, and the two constant symbols 0 and 1. Thus a structure (algebra) for this signature consists of a set of elements A together with two binary functions, a unary function, and two distinguished elements; but there is no requirement that it satisfy any of the field axioms. The rational numbers Q, the real numbers R and the complex numbers C, like any other field, can be regarded as σ-structures in an obvious way:

$$\mathcal{Q} = (Q, \sigma_f, I_{\mathcal{Q}})$$

$$\mathcal{R} = (R, \sigma_f, I_{\mathcal{R}})$$

$$\mathcal{C} = (C, \sigma_f, I_{\mathcal{C}})$$

where

> $I_{\mathcal{Q}}(+)\colon Q \times Q \to Q$ is addition of rational numbers,
> $I_{\mathcal{Q}}(\times)\colon Q \times Q \to Q$ is multiplication of rational numbers,
> $I_{\mathcal{Q}}(-)\colon Q \to Q$ is the function that takes each rational number x to -x, and
> $I_{\mathcal{Q}}(0) \in Q$ is the number 0 and
> $I_{\mathcal{Q}}(1) \in Q$ is the number 1;

and $I_{\mathcal{R}}$ and $I_{\mathcal{C}}$ are similarly defined.

But the ring Z of integers, which is not a field, is also a σf-structure in the same way. In fact, there is no requirement that *any* of the field axioms hold in a σf-structure.

A signature for ordered fields needs an additional binary relation such as $<$ or $\le$, and therefore structures for such a signature are not algebras, even though they are of course algebraic structures in the usual, loose sense of the word.

The ordinary signature for set theory includes a single binary relation $\in$. A structure for this signature consists of a set of elements and an interpretation of the $\in$ relation as a binary relation on these elements.

3.2   Induced substructures and closed subsets

$\mathcal{A}$ is called an (induced) substructure of $\mathcal{B}$ if

- $\mathcal{A}$ and $\mathcal{B}$ have the same signature $\sigma(\mathcal{A}) = \sigma(\mathcal{B})$;

- the domain of $\mathcal{A}$ is contained in the domain of $\mathcal{B}$: $|\mathcal{A}| \subseteq |\mathcal{B}|$; and

- the interpretations of all function and relation symbols agree on $|\mathcal{B}|$.

The usual notation for this relation is $\mathcal{A} \subseteq \mathcal{B}$.

A subset $B \subseteq |\mathcal{A}|$ of the domain of a structure $\mathcal{A}$ is called **closed** if it is closed under the functions of $\mathcal{A}$, i.e. if the following condition is satisfied: for every natural number n, every n-ary function symbol f (in the signature of $\mathcal{A}$) and all elements $b_1, b_2, \ldots, b_n \in B$, the result of applying f to the n-tuple $b_1 b_2 \ldots b_n$ is again an element of B: $f(b_1, b_2, \ldots, b_n) \in B$.

For every subset $B \subseteq |\mathcal{A}|$ there is a smallest closed subset of $|\mathcal{A}|$ that contains B. It is called the closed subset **generated** by B, or the **hull** of B, and denoted by $\langle B \rangle$ or $\langle B \rangle_{\mathcal{A}}$. The operator $\langle \rangle$ is a finitary closure operator on the set of subsets of $|\mathcal{A}|$.

If $\mathcal{A} = (A, \sigma, I)$ and $B \subseteq A$ is a closed subset, then (B, σ, I') is an induced substructure of $\mathcal{A}$, where I' assigns to every symbol of σ the restriction to B of its interpretation in $\mathcal{A}$. Conversely, the domain of an induced substructure is a closed subset.

The closed subsets (or induced substructures) of a structure form a lattice. The meet of two subsets is their intersection. The join of two subsets is the closed subset generated by their union. Universal algebra studies the lattice of substructures of a structure in detail.

3.2.1   Examples

Let $\sigma = \{+, \times, -, 0, 1\}$ be again the standard signature for fields. When regarded as σ-structures in the natural way, the rational numbers form a substructure of the real numbers, and the real numbers form a substructure of the complex numbers. The rational numbers are the smallest substructure of the real (or complex) numbers that also satisfies the field axioms.

The set of integers gives an even smaller substructure of the real numbers which is not a field. Indeed, the integers are the substructure of the real numbers generated by the empty set, using this signature. The notion in abstract algebra that corresponds to a substructure of a field, in this signature, is that of a subring, rather than that of a subfield.

The most obvious way to define a graph is a structure with a signature σ consisting of a single binary relation symbol E. The vertices of the graph form the domain of the structure, and for two vertices a and b, $(a, b) \in$ E means that a and b are connected by an edge. In this encoding, the notion of induced substructure is more restrictive than the notion of subgraph. For example, let G be a graph consisting of two vertices connected by an edge, and let H be the graph consisting of the same vertices but no edges. H is a subgraph of G, but not an induced substructure. The notion in graph theory that corresponds to induced substructures is that of induced subgraphs.

3.3   Homomorphisms and embeddings

See also: Universal algebra § Basic constructions

3.3.1   Homomorphisms

Given two structures $\mathcal{A}$ and $\mathcal{B}$ of the same signature σ, a **(σ-)homomorphism** from $\mathcal{A}$ to $\mathcal{B}$ is a map $h : |\mathcal{A}| \to |\mathcal{B}|$ that preserves the functions and relations. More precisely:

- For every n-ary function symbol f of σ and any elements $a_1, a_2, \ldots, a_n \in |\mathcal{A}|$, the following equation holds:

$$h(f(a_1, a_2, \ldots, a_n)) = f(h(a_1), h(a_2), \ldots, h(a_n))$$

- For every n-ary relation symbol R of σ and any elements $a_1, a_2, \ldots, a_n \in |\mathcal{A}|$, the following implication holds:

$$(a_1, a_2, \ldots, a_n) \in R \implies (h(a_1), h(a_2), \ldots, h(a_n)) \in R$$

The notation for a homomorphism h from $\mathcal{A}$ to $\mathcal{B}$ is $h : \mathcal{A} \to \mathcal{B}$.

For every signature σ there is a concrete category σ-**Hom** which has σ-structures as objects and σ-homomorphisms as morphisms.

A homomorphism $h : \mathcal{A} \to \mathcal{B}$ is sometimes called **strong** if for every n-ary relation symbol R and any elements $b_1, b_2, \ldots, b_n \in |\mathcal{B}|$ such that $(b_1, b_2, \ldots, b_n) \in R$, there are $a_1, a_2, \ldots, a_n \in |\mathcal{A}|$ such that $(a_1, a_2, \ldots, a_n) \in R$ and $b_1 = h(a_1), b_2 = h(a_2), \ldots, b_n = h(a_n)$. The strong homomorphisms give rise to a subcategory of σ-**Hom**.

3.3.2 Embeddings

A (σ-)homomorphism $h : \mathcal{A} \to \mathcal{B}$ is called a (σ-)**embedding** if it is one-to-one and

- for every n-ary relation symbol R of σ and any elements $a_1, a_2, \ldots, a_n$, the following equivalence holds:

$$(a_1, a_2, \ldots, a_n) \in R \iff (h(a_1), h(a_2), \ldots, h(a_n)) \in R$$

Thus an embedding is the same thing as a strong homomorphism which is one-to-one. The category σ-**Emb** of σ-structures and σ-embeddings is a concrete subcategory of σ-**Hom**.

Induced substructures correspond to subobjects in σ-**Emb**. If σ has only function symbols, σ-**Emb** is the subcategory of monomorphisms of σ-**Hom**. In this case induced substructures also correspond to subobjects in σ-**Hom**.

3.3.3 Example

As seen above, in the standard encoding of graphs as structures the induced substructures are precisely the induced subgraphs. However, a homomorphism between graphs is the same thing as a homomorphism between the two structures coding the graph. In the example of the previous section, even though the subgraph H of G is not induced, the identity map id: $H \to G$ is a homomorphism. This map is in fact a monomorphism in the category σ-**Hom**, and therefore H is a subobject of G which is not an induced substructure.

3.3.4 Homomorphism problem

The following problem is known as the *homomorphism problem*:

> Given two finite structures $\mathcal{A}$ and $\mathcal{B}$ of a finite relational signature, find a homomorphism $h : \mathcal{A} \to \mathcal{B}$ or show that no such homomorphism exists.

Every constraint satisfaction problem (CSP) has a translation into the homomorphism problem.[5] Therefore the complexity of CSP can be studied using the methods of finite model theory.

Another application is in database theory, where a relational model of a database is essentially the same thing as a relational structure. It turns out that a conjunctive query on a database can be described by another structure in the same signature as the database model. A homomorphism from the relational model to the structure representing the query is the same thing as a solution to the query. This shows that the conjunctive query problem is also equivalent to the homomorphism problem.

3.4 Structures and first-order logic

See also: Model theory § First-order logic and Model theory § Axiomatizability, elimination of quantifiers, and model-completeness

Structures are sometimes referred to as "first-order structures". This is misleading, as nothing in their definition ties them to any specific logic, and in fact they are suitable as semantic objects both for very restricted fragments of first-order logic such as that used in universal algebra, and for second-order logic. In connection with first-order logic and model theory, structures are often called **models**, even when the question "models of what?" has no obvious answer.

3.4.1 Satisfaction relation

Each first-order structure $\mathcal{M}$ has a **satisfaction relation** $\mathcal{M} \vDash \phi$ defined for all formulas ϕ in the language consisting of the language of $\mathcal{M}$ together with a constant symbol for each element of M, which is interpreted as that element. This relation is defined inductively using Tarski's T-schema.

A structure $\mathcal{M}$ is said to be a **model** of a theory T if the language of $\mathcal{M}$ is the same as the language of T and every sentence in T is satisfied by $\mathcal{M}$. Thus, for example, a "ring" is a structure for the language of rings that satisfies each of the ring axioms, and a model of ZFC set theory is a structure in the language of set theory that satisfies each of the ZFC axioms.

3.4.2 Definable relations

An n-ary relation R on the universe M of a structure $\mathcal{M}$ is said to be **definable** (or **explicitly definable**, or $\emptyset$ -**definable**) if there is a formula $\varphi(x_1,...,xn)$ such that

$$R = \{(a_1, \ldots, a_n) \in M^n : \mathcal{M} \vDash \phi(a_1, \ldots, a_n)\}.$$

In other words, R is definable if and only if there is a formula φ such that

$$(a_1, \ldots, a_n) \in R \Leftrightarrow \mathcal{M} \vDash \phi(a_1, \ldots, a_n)$$

is correct.

An important special case is the definability of specific elements. An element m of M is definable in $\mathcal{M}$ if and only if there is a formula $\varphi(x)$ such that

$$\mathcal{M} \vDash \forall x(x = m \leftrightarrow \phi(x)).$$

Definability with parameters

A relation R is said to be **definable with parameters** (or $|\mathcal{M}|$ -**definable**) if there is a formula φ with parameters from $\mathcal{M}$ such that R is definable using φ. Every element of a structure is definable using the element itself as a parameter.

It should be noted that some authors use *definable* to mean *definable without parameters*, while other authors mean *definable with parameters*. Broadly speaking, the convention that *definable* means *definable without parameters* is more common amongst set theorists, while the opposite convention is more common amongst model theorists.

Implicit definability

Recall from above that an *n*-ary relation R on the universe M of a structure $\mathcal{M}$ is explicitly definable if there is a formula $\varphi(x_1,...,xn)$ such that

$$R = \{(a_1,\ldots,a_n) \in M^n : \mathcal{M} \vDash \phi(a_1,\ldots,a_n)\}$$

Here the formula φ used to define a relation R must be over the signature of $\mathcal{M}$ and so φ may not mention R itself, since R is not in the signature of $\mathcal{M}$. If there is a formula φ in the extended language containing the language of $\mathcal{M}$ and a new symbol R, and the relation R is the only relation on $\mathcal{M}$ such that $\mathcal{M} \vDash \phi$, then R is said to be **implicitly definable** over $\mathcal{M}$.

By Beth's theorem, every implicitly definable relation is explicitly definable.

3.5 Many-sorted structures

Structures as defined above are sometimes called **one-sorted structures** to distinguish them from the more general **many-sorted structures**. A many-sorted structure can have an arbitrary number of domains. The **sorts** are part of the signature, and they play the role of names for the different domains. Many-sorted signatures also prescribe on which sorts the functions and relations of a many-sorted structure are defined. Therefore the arities of function symbols or relation symbols must be more complicated objects such as tuples of sorts rather than natural numbers.

Vector spaces, for example, can be regarded as two-sorted structures in the following way. The two-sorted signature of vector spaces consists of two sorts V (for vectors) and S (for scalars) and the following function symbols:

If V is a vector space over a field F, the corresponding two-sorted structure $\mathcal{V}$ consists of the vector domain $|\mathcal{V}|_V = V$, the scalar domain $|\mathcal{V}|_S = F$, and the obvious functions, such as the vector zero $0_V^{\mathcal{V}} = 0 \in |\mathcal{V}|_V$, the scalar zero $0_S^{\mathcal{V}} = 0 \in |\mathcal{V}|_S$, or scalar multiplication $\times^{\mathcal{V}} : |\mathcal{V}|_S \times |\mathcal{V}|_V \to |\mathcal{V}|_V$.

Many-sorted structures are often used as a convenient tool even when they could be avoided with a little effort. But they are rarely defined in a rigorous way, because it is straightforward and tedious (hence unrewarding) to carry out the generalization explicitly.

In most mathematical endeavours, not much attention is paid to the sorts. A many-sorted logic however naturally leads to a type theory. As Bart Jacobs puts it: "A logic is always a logic over a type theory." This emphasis in turn leads to categorical logic because a logic over a type theory categorically corresponds to one ("total") category, capturing the logic, being fibred over another ("base") category, capturing the type theory.[6]

3.6 Other generalizations

3.6.1 Partial algebras

Both universal algebra and model theory study classes of (structures or) algebras that are defined by a signature and a set of axioms. In the case of model theory these axioms have the form of first-order sentences. The formalism of universal algebra is much more restrictive; essentially it only allows first-order sentences that have the form of universally quantified equations between terms, e.g. $x\,y\,(x + y = y + x)$. One consequence is that the choice of a signature is more significant in universal algebra than it is in model theory. For example the class of groups, in the signature consisting of the binary function symbol $\times$ and the constant symbol 1, is an elementary class, but it is not a variety. Universal algebra solves this problem by adding a unary function symbol $^{-1}$.

In the case of fields this strategy works only for addition. For multiplication it fails because 0 does not have a multiplicative inverse. An ad hoc attempt to deal with this would be to define $0^{-1} = 0$. (This attempt fails, essentially because with this definition $0 \times 0^{-1} = 1$ is not true.) Therefore one is naturally led to allow partial functions, i.e., functions that are defined only on a subset of their domain. However, there are several obvious ways to generalize notions such as substructure, homomorphism and identity.

3.6.2 Structures for typed languages

In type theory, there are many sorts of variables, each of which has a **type**. Types are inductively defined; given two types δ and σ there is also a type $\sigma \rightarrow \delta$ that represents functions from objects of type σ to objects of type δ. A structure for a typed language (in the ordinary first-order semantics) must include a separate set of objects of each type, and for a function type the structure must have complete information about the function represented by each object of that type.

3.6.3 Higher-order languages

Main article: Second-order logic

There is more than one possible semantics for higher-order logic, as discussed in the article on second-order logic. When using full higher-order semantics, a structure need only have a universe for objects of type 0, and the T-schema is extended so that a quantifier over a higher-order type is satisfied by the model if and only if it is disquotationally true. When using first-order semantics, an additional sort is added for each higher-order type, as in the case of a many sorted first order language.

3.6.4 Structures that are proper classes

In the study of set theory and category theory, it is sometimes useful to consider structures in which the domain of discourse is a proper class instead of a set. These structures are sometimes called **class models** to distinguish them from the "set models" discussed above. When the domain is a proper class, each function and relation symbol may also be represented by a proper class.

In Bertrand Russell's Principia Mathematica, structures were also allowed to have a proper class as their domain.

3.7 See also

- Mathematical structure

- Algebraic structure

3.8 Notes

[1] Some authors refer to structures as "algebras" when generalizing universal algebra to allow relations as well as functions.

[2] Hodges, Wilfrid (2009). "Functional Modelling and Mathematical Models". In Meijers, Anthonie. *Philosophy of technology and engineering sciences*. Handbook of the Philosophy of Science **9**. Elsevier. ISBN 978-0-444-51667-1.

[3] This is similar to the definition of a prime number in elementary number theory, which has been carefully chosen so that the irreducible number 1 is not considered prime. The convention that the domain of a structure may not be empty is particularly important in logic, because several common inference rules, notably, universal instantiation, are not sound when empty structures are permitted. A logical system that allows the empty domain is known as an inclusive logic.

[4] As a consequence of these conventions, the notation $|\mathcal{A}|$ may also be used to refer to the cardinality of the domain of $\mathcal{A}$. In practice this never leads to confusion.

[5] Jeavons, Peter; Cohen, David; Pearson, Justin (1998), "Constraints and universal algebra", *Annals of Mathematics and Artificial Intelligence* **24**: 51–67, doi:10.1023/A:1018941030227.

[6] Jacobs, Bart (1999), *Categorical Logic and Type Theory*, Elsevier, pp. 1–4

3.9 References

- Burris, Stanley N.; Sankappanavar, H. P. (1981), *A Course in Universal Algebra*, Berlin, New York: Springer-Verlag

- Chang, Chen Chung; Keisler, H. Jerome (1989) [1973], *Model Theory*, Elsevier, ISBN 978-0-7204-0692-4

- Diestel, Reinhard (2005) [1997], *Graph Theory*, Graduate Texts in Mathematics **173** (3rd ed.), Berlin, New York: Springer-Verlag, ISBN 978-3-540-26183-4

- Ebbinghaus, Heinz-Dieter; Flum, Jörg; Thomas, Wolfgang (1994), *Mathematical Logic* (2nd ed.), New York: Springer, ISBN 978-0-387-94258-2

- Hinman, P. (2005), *Fundamentals of Mathematical Logic*, A K Peters, ISBN 978-1-56881-262-5

- Hodges, Wilfrid (1993), *Model theory*, Cambridge: Cambridge University Press, ISBN 978-0-521-30442-9

- Hodges, Wilfrid (1997), *A shorter model theory*, Cambridge: Cambridge University Press, ISBN 978-0-521-58713-6

- Marker, David (2002), *Model Theory: An Introduction*, Berlin, New York: Springer-Verlag, ISBN 978-0-387-98760-6

- Poizat, Bruno (2000), *A Course in Model Theory: An Introduction to Contemporary Mathematical Logic*, Berlin, New York: Springer-Verlag, ISBN 978-0-387-98655-5

- Rautenberg, Wolfgang (2010), *A Concise Introduction to Mathematical Logic* (3rd ed.), New York: Springer Science+Business Media, doi:10.1007/978-1-4419-1221-3, ISBN 978-1-4419-1220-6

- Rothmaler, Philipp (2000), *Introduction to Model Theory*, London: CRC Press, ISBN 978-90-5699-313-9

3.10 External links

- Semantics section in Classical Logic (an entry of Stanford Encyclopedia of Philosophy)

Chapter 4

Group (mathematics)

This article is about basic notions of groups in mathematics. For a more advanced treatment, see Group theory.

In mathematics, a **group** is an algebraic structure consisting of a set of elements together with an operation that combines any two elements to form a third element. The operation satisfies four conditions called the group axioms, namely closure, associativity, identity and invertibility. One of the most familiar examples of a group is the set of integers together with the addition operation; the addition of any two integers forms another integer. The abstract formalization of the group axioms, detached as it is from the concrete nature of any particular group and its operation, allows entities with highly diverse mathematical origins in abstract algebra and beyond to be handled in a flexible way, while retaining their essential structural aspects. The ubiquity of groups in numerous areas within and outside mathematics makes them a central organizing principle of contemporary mathematics.[1][2]

Groups share a fundamental kinship with the notion of symmetry. For example, a symmetry group encodes symmetry features of a geometrical object: the group consists of the set of transformations that leave the object unchanged and the operation of combining two such transformations by performing one after the other. Lie groups are the symmetry groups used in the Standard Model of particle physics; Point groups are used to help understand symmetry phenomena in molecular chemistry; and Poincaré groups can express the physical symmetry underlying special relativity.

The concept of a group arose from the study of polynomial equations, starting with Évariste Galois in the 1830s. After contributions from other fields such as number theory and geometry, the group notion was generalized and firmly established around 1870. Modern group theory—an active mathematical discipline—studies groups in their own right.[a][-] To explore groups, mathematicians have devised various notions to break groups into smaller, better-understandable pieces, such as subgroups, quotient groups and simple groups. In addition to their abstract properties, group theorists also study the different ways in which a group can be expressed concretely (its group representations), both from a theoretical and a computational point of view. A theory has been developed for finite groups, which culminated with the classification of finite simple groups announced in 1983.[aa][-] Since the mid-1980s, geometric group theory, which studies finitely generated groups as geometric objects, has become a particularly active area in group theory.

4.1 Definition and illustration

4.1.1 First example: the integers

One of the most familiar groups is the set of integers **Z** which consists of the numbers

..., −4, −3, −2, −1, 0, 1, 2, 3, 4, ...,[3] together with addition.

The following properties of integer addition serve as a model for the abstract group axioms given in the definition below.

1. For any two integers a and b, the sum $a + b$ is also an integer. Thus, adding two integers never yields some other type of number, such as a fraction. This property is known as *closure* under addition.

The manipulations of this Rubik's Cube form the Rubik's Cube group.

2. For all integers a, b and c, $(a + b) + c = a + (b + c)$. Expressed in words, adding a to b first, and then adding the result to c gives the same final result as adding a to the sum of b and c, a property known as *associativity*.

3. If a is any integer, then $0 + a = a + 0 = a$. Zero is called the *identity element* of addition because adding it to any integer returns the same integer.

4. For every integer a, there is an integer b such that $a + b = b + a = 0$. The integer b is called the *inverse element* of the integer a and is denoted $-a$.

The integers, together with the operation $+$, form a mathematical object belonging to a broad class sharing similar structural aspects. To appropriately understand these structures as a collective, the following abstract definition is developed.

4.1.2   Definition

[T]he axioms for a group are short and natural... Yet somehow hidden behind these axioms is the monster simple group, a huge and extraordinary mathematical object, which appears to rely on numerous bizarre coincidences to exist. The axioms for groups give no obvious hint that anything like this exists.

Richard Borcherds in *Mathematicians: An Outer View of the Inner World* [4]

A group is a set, *G*, together with an operation • (called the *group law* of *G*) that combines any two elements *a* and *b* to form another element, denoted *a* • *b* or *ab*. To qualify as a group, the set and operation, (*G*, •), must satisfy four requirements known as the *group axioms*:[5]

Closure For all *a*, *b* in *G*, the result of the operation, *a* • *b*, is also in *G*.[b][·]

Associativity For all *a*, *b* and *c* in *G*, (*a* • *b*) • *c* = *a* • (*b* • *c*).

Identity element There exists an element *e* in *G*, such that for every element *a* in *G*, the equation *e* • *a* = *a* • *e* = *a* holds. Such an element is unique (see below), and thus one speaks of *the* identity element.

Inverse element For each *a* in *G*, there exists an element *b* in *G* such that *a* • *b* = *b* • *a* = *e*, where *e* is the identity element.

The result of an operation may depend on the order of the operands. In other words, the result of combining element *a* with element *b* need not yield the same result as combining element *b* with element *a*; the equation

$$a \cdot b = b \cdot a$$

may not always be true. This equation always holds in the group of integers under addition, because *a* + *b* = *b* + *a* for any two integers (commutativity of addition). Groups for which the commutativity equation *a* • *b* = *b* • *a* always holds are called *abelian groups* (in honor of Niels Abel). The symmetry group described in the following section is an example of a group that is not abelian.

The identity element of a group *G* is often written as 1 or $1G$,[6] a notation inherited from the multiplicative identity. The identity element may also be written as 0, especially if the group operation is denoted by +, in which case the group is called an additive group. The identity element can also be written as *id*.

The set *G* is called the *underlying set* of the group (*G*, •). Often the group's underlying set *G* is used as a short name for the group (*G*, •). Along the same lines, shorthand expressions such as "a subset of the group *G*" or "an element of group *G*" are used when what is actually meant is "a subset of the underlying set *G* of the group (*G*, •)" or "an element of the underlying set *G* of the group (*G*, •)". Usually, it is clear from the context whether a symbol like *G* refers to a group or to an underlying set.

4.1.3   Second example: a symmetry group

Two figures in the plane are congruent if one can be changed into the other using a combination of rotations, reflections, and translations. Any figure is congruent to itself. However, some figures are congruent to themselves in more than one way, and these extra congruences are called symmetries. A square has eight symmetries. These are:

- the identity operation leaving everything unchanged, denoted id;

- rotations of the square around its center by 90° clockwise, 180° clockwise, and 270° clockwise, denoted by r_1, r_2 and r_3, respectively;

- reflections about the vertical and horizontal middle line (f_h and f_v), or through the two diagonals (f_d and f_c).

These symmetries are represented by functions. Each of these functions sends a point in the square to the corresponding point under the symmetry. For example, r_1 sends a point to its rotation 90° clockwise around the square's center, and f_h sends a point to its reflection across the square's vertical middle line. Composing two of these symmetry functions gives another symmetry function. These symmetries determine a group called the dihedral group of degree 4 and denoted D_4. The underlying set of the group is the above set of symmetry functions, and the group operation is function composition.[7] Two symmetries are combined by composing them as functions, that is, applying the first one to the square, and the second one to the result of the first application. The result of performing first a and then b is written symbolically *from right to left* as

$b \cdot a$ ("apply the symmetry b after performing the symmetry a").

The right-to-left notation is the same notation that is used for composition of functions.

The group table on the right lists the results of all such compositions possible. For example, rotating by 270° clockwise (r_3) and then reflecting horizontally (f_h) is the same as performing a reflection along the diagonal (f_d). Using the above symbols, highlighted in blue in the group table:

$f_h \cdot r_3 = f_d$.

Given this set of symmetries and the described operation, the group axioms can be understood as follows:

1. The closure axiom demands that the composition $b \cdot a$ of any two symmetries a and b is also a symmetry. Another example for the group operation is

   $r_3 \cdot f_h = f_c$,

   i.e. rotating 270° clockwise after reflecting horizontally equals reflecting along the counter-diagonal (f_c). Indeed every other combination of two symmetries still gives a symmetry, as can be checked using the group table.

2. The associativity constraint deals with composing more than two symmetries: Starting with three elements a, b and c of D_4, there are two possible ways of using these three symmetries in this order to determine a symmetry of the square. One of these ways is to first compose a and b into a single symmetry, then to compose that symmetry with c. The other way is to first compose b and c, then to compose the resulting symmetry with a. The associativity condition

   $(a \cdot b) \cdot c = a \cdot (b \cdot c)$

   means that these two ways are the same, i.e., a product of many group elements can be simplified in any grouping. For example, $(f_d \cdot f_v) \cdot r_2 = f_d \cdot (f_v \cdot r_2)$ can be checked using the group table at the right

   While associativity is true for the symmetries of the square and addition of numbers, it is not true for all operations. For instance, subtraction of numbers is not associative: $(7 - 3) - 2 = 2$ is not the same as $7 - (3 - 2) = 6$.

3. The identity element is the symmetry id leaving everything unchanged: for any symmetry a, performing id after a (or a after id) equals a, in symbolic form,

   $\text{id} \cdot a = a$,
   $a \cdot \text{id} = a$.

4. An inverse element undoes the transformation of some other element. Every symmetry can be undone: each of the following transformations—identity id, the reflections f_h, f_v, f_d, f_c and the 180° rotation r_2—is its own inverse, because performing it twice brings the square back to its original orientation. The rotations r_3 and r_1 are each other's inverses, because rotating 90° and then rotation 270° (or vice versa) yields a rotation over 360° which leaves the square unchanged. In symbols,

   $f_h \cdot f_h = \text{id}$,
   $r_3 \cdot r_1 = r_1 \cdot r_3 = \text{id}$.

In contrast to the group of integers above, where the order of the operation is irrelevant, it does matter in D_4: $f_h \cdot r_1 = f_c$ but $r_1 \cdot f_h = f_d$. In other words, D_4 is not abelian, which makes the group structure more difficult than the integers introduced first.

4.2   History

Main article: History of group theory

The modern concept of an abstract group developed out of several fields of mathematics.[8][9][10] The original motivation for group theory was the quest for solutions of polynomial equations of degree higher than 4. The 19th-century French mathematician Évariste Galois, extending prior work of Paolo Ruffini and Joseph-Louis Lagrange, gave a criterion for the solvability of a particular polynomial equation in terms of the symmetry group of its roots (solutions). The elements of such a Galois group correspond to certain permutations of the roots. At first, Galois' ideas were rejected by his contemporaries, and published only posthumously.[11][12] More general permutation groups were investigated in particular by Augustin Louis Cauchy. Arthur Cayley's *On the theory of groups, as depending on the symbolic equation $\theta^n = 1$* (1854) gives the first abstract definition of a finite group.[13]

Geometry was a second field in which groups were used systematically, especially symmetry groups as part of Felix Klein's 1872 Erlangen program.[14] After novel geometries such as hyperbolic and projective geometry had emerged, Klein used group theory to organize them in a more coherent way. Further advancing these ideas, Sophus Lie founded the study of Lie groups in 1884.[15]

The third field contributing to group theory was number theory. Certain abelian group structures had been used implicitly in Carl Friedrich Gauss' number-theoretical work *Disquisitiones Arithmeticae* (1798), and more explicitly by Leopold Kronecker.[16] In 1847, Ernst Kummer made early attempts to prove Fermat's Last Theorem by developing groups describing factorization into prime numbers.[17]

The convergence of these various sources into a uniform theory of groups started with Camille Jordan's *Traité des substitutions et des équations algébriques* (1870).[18] Walther von Dyck (1882) introduced the idea of specifying a group by means of generators and relations, and was also the first to give an axiomatic definition of an "abstract group", in the terminology of the time.[19] As of the 20th century, groups gained wide recognition by the pioneering work of Ferdinand Georg Frobenius and William Burnside, who worked on representation theory of finite groups, Richard Brauer's modular representation theory and Issai Schur's papers.[20] The theory of Lie groups, and more generally locally compact groups was studied by Hermann Weyl, Élie Cartan and many others.[21] Its algebraic counterpart, the theory of algebraic groups, was first shaped by Claude Chevalley (from the late 1930s) and later by the work of Armand Borel and Jacques Tits.[22]

The University of Chicago's 1960–61 Group Theory Year brought together group theorists such as Daniel Gorenstein, John G. Thompson and Walter Feit, laying the foundation of a collaboration that, with input from numerous other mathematicians, classified all finite simple groups in 1982. This project exceeded previous mathematical endeavours by its sheer size, in both length of proof and number of researchers. Research is ongoing to simplify the proof of this classification.[23] These days, group theory is still a highly active mathematical branch, impacting many other fields.[a][>]

4.3   Elementary consequences of the group axioms

Basic facts about all groups that can be obtained directly from the group axioms are commonly subsumed under *elementary group theory*.[24] For example, repeated applications of the associativity axiom show that the unambiguity of

$$a \bullet b \bullet c = (a \bullet b) \bullet c = a \bullet (b \bullet c)$$

generalizes to more than three factors. Because this implies that parentheses can be inserted anywhere within such a series of terms, parentheses are usually omitted.[25]

The axioms may be weakened to assert only the existence of a left identity and left inverses. Both can be shown to be actually two-sided, so the resulting definition is equivalent to the one given above.[26]

4.3.1 Uniqueness of identity element and inverses

Two important consequences of the group axioms are the uniqueness of the identity element and the uniqueness of inverse elements. There can be only one identity element in a group, and each element in a group has exactly one inverse element. Thus, it is customary to speak of *the* identity, and *the* inverse of an element.[27]

To prove the uniqueness of an inverse element of a, suppose that a has two inverses, denoted b and c, in a group $(G, \bullet)$. Then

The two extremal terms b and c are equal, since they are connected by a chain of equalities. In other words, there is only one inverse element of a. Similarly, to prove that the identity element of a group is unique, assume G is a group with two identity elements e and f. Then $e = e \bullet f = f$, hence e and f are equal.

4.3.2 Division

In groups, it is possible to perform division: given elements a and b of the group G, there is exactly one solution x in G to the equation $x \bullet a = b$.[27] In fact, right multiplication of the equation by a^{-1} gives the solution $x = x \bullet a \bullet a^{-1} = b \bullet a^{-1}$. Similarly there is exactly one solution y in G to the equation $a \bullet y = b$, namely $y = a^{-1} \bullet b$. In general, x and y need not agree.

A consequence of this is that multiplying by a group element g is a bijection. Specifically, if g is an element of the group G, there is a bijection from G to itself called *left translation* by g sending $h \in G$ to $g \bullet h$. Similarly, *right translation* by g is a bijection from G to itself sending h to $h \bullet g$. If G is abelian, left and right translation by a group element are the same.

4.4 Basic concepts

Further information: Glossary of group theory

To understand groups beyond the level of mere symbolic manipulations as above, more structural concepts have to be employed.[c][b] There is a conceptual principle underlying all of the following notions: to take advantage of the structure offered by groups (which sets, being "structureless", do not have), constructions related to groups have to be *compatible* with the group operation. This compatibility manifests itself in the following notions in various ways. For example, groups can be related to each other via functions called group homomorphisms. By the mentioned principle, they are required to respect the group structures in a precise sense. The structure of groups can also be understood by breaking them into pieces called subgroups and quotient groups. The principle of "preserving structures"—a recurring topic in mathematics throughout—is an instance of working in a category, in this case the category of groups.[28]

4.4.1 Group homomorphisms

Main article: Group homomorphism

Group homomorphisms[g][b] are functions that preserve group structure. A function $a: G \to H$ between two groups $(G, \bullet)$ and $(H, *)$ is called a *homomorphism* if the equation

$$a(g \bullet k) = a(g) * a(k)$$

holds for all elements g, k in G. In other words, the result is the same when performing the group operation after or before applying the map a. This requirement ensures that $a(1G) = 1H$, and also $a(g)^{-1} = a(g^{-1})$ for all g in G. Thus a group homomorphism respects all the structure of G provided by the group axioms.[29]

Two groups G and H are called *isomorphic* if there exist group homomorphisms a: $G \rightarrow H$ and b: $H \rightarrow G$, such that applying the two functions one after another in each of the two possible orders gives the identity functions of G and H. That is, $a(b(h)) = h$ and $b(a(g)) = g$ for any g in G and h in H. From an abstract point of view, isomorphic groups carry the same information. For example, proving that $g \bullet g = 1G$ for some element g of G is equivalent to proving that $a(g) * a(g) = 1H$, because applying a to the first equality yields the second, and applying b to the second gives back the first.

4.4.2   Subgroups

Main article: Subgroup

Informally, a *subgroup* is a group H contained within a bigger one, G.[30] Concretely, the identity element of G is contained in H, and whenever h_1 and h_2 are in H, then so are $h_1 \bullet h_2$ and h_1^{-1}, so the elements of H, equipped with the group operation on G restricted to H, indeed form a group.

In the example above, the identity and the rotations constitute a subgroup $R = \{\text{id}, r_1, r_2, r_3\}$, highlighted in red in the group table above: any two rotations composed are still a rotation, and a rotation can be undone by (i.e. is inverse to) the complementary rotations 270° for 90°, 180° for 180°, and 90° for 270° (note that rotation in the opposite direction is not defined). The subgroup test is a necessary and sufficient condition for a subset H of a group G to be a subgroup: it is sufficient to check that $g^{-1}h \in H$ for all elements g, $h \in H$. Knowing the subgroups is important in understanding the group as a whole.[d[>]]

Given any subset S of a group G, the subgroup generated by S consists of products of elements of S and their inverses. It is the smallest subgroup of G containing S.[31] In the introductory example above, the subgroup generated by r_2 and f_v consists of these two elements, the identity element id and $f_h = f_v \bullet r_2$. Again, this is a subgroup, because combining any two of these four elements or their inverses (which are, in this particular case, these same elements) yields an element of this subgroup.

4.4.3   Cosets

Main article: Coset

In many situations it is desirable to consider two group elements the same if they differ by an element of a given subgroup. For example, in D_4 above, once a reflection is performed, the square never gets back to the r_2 configuration by just applying the rotation operations (and no further reflections), i.e. the rotation operations are irrelevant to the question whether a reflection has been performed. Cosets are used to formalize this insight: a subgroup H defines left and right cosets, which can be thought of as translations of H by arbitrary group elements g. In symbolic terms, the *left* and *right* cosets of H containing g are

$$gH = \{g \bullet h : h \in H\} \text{ and } Hg = \{h \bullet g : h \in H\}, \text{ respectively.}^{[32]}$$

The cosets of any subgroup H form a partition of G; that is, the union of all left cosets is equal to G and two left cosets are either equal or have an empty intersection.[33] The first case $g_1H = g_2H$ happens precisely when $g_1^{-1} \bullet g_2 \in H$, i.e. if the two elements differ by an element of H. Similar considerations apply to the right cosets of H. The left and right cosets of H may or may not be equal. If they are, i.e. for all g in G, $gH = Hg$, then H is said to be a *normal subgroup*.

In D_4, the introductory symmetry group, the left cosets gR of the subgroup R consisting of the rotations are either equal to R, if g is an element of R itself, or otherwise equal to $U = f_cR = \{f_c, f_v, f_d, f_h\}$ (highlighted in green). The subgroup R is also normal, because $f_cR = U = Rf_c$ and similarly for any element other than f_c. (In fact, in the case of D_4, observe that all such cosets are equal, such that $f_hR = f_vR = f_dR = f_cR$.)

4.4.4   Quotient groups

Main article: Quotient group

In some situations the set of cosets of a subgroup can be endowed with a group law, giving a *quotient group* or *factor group*. For this to be possible, the subgroup has to be normal. Given any normal subgroup N, the quotient group is defined by

$$G \,/\, N = \{gN, g \in G\}, \text{ "} G \text{ modulo } N \text{".}^{[34]}$$

This set inherits a group operation (sometimes called coset multiplication, or coset addition) from the original group G: $(gN) \bullet (hN) = (gh)N$ for all g and h in G. This definition is motivated by the idea (itself an instance of general structural considerations outlined above) that the map $G \to G \,/\, N$ that associates to any element g its coset gN be a group homomorphism, or by general abstract considerations called universal properties. The coset $eN = N$ serves as the identity in this group, and the inverse of gN in the quotient group is $(gN)^{-1} = (g^{-1})N$.[e][>]

The elements of the quotient group $D_4 \,/\, R$ are R itself, which represents the identity, and $U = f_v R$. The group operation on the quotient is shown at the right. For example, $U \bullet U = f_v R \bullet f_v R = (f_v \bullet f_v)R = R$. Both the subgroup $R = \{id, r_1, r_2, r_3\}$, as well as the corresponding quotient are abelian, whereas D_4 is not abelian. Building bigger groups by smaller ones, such as D_4 from its subgroup R and the quotient $D_4 \,/\, R$ is abstracted by a notion called semidirect product.

Quotient groups and subgroups together form a way of describing every group by its *presentation*: any group is the quotient of the free group over the *generators* of the group, quotiented by the subgroup of *relations*. The dihedral group D_4, for example, can be generated by two elements r and f (for example, $r = r_1$, the right rotation and $f = f_v$ the vertical (or any other) reflection), which means that every symmetry of the square is a finite composition of these two symmetries or their inverses. Together with the relations

$$r^4 = f^2 = (r \bullet f)^2 = 1,^{[35]}$$

the group is completely described. A presentation of a group can also be used to construct the Cayley graph, a device used to graphically capture discrete groups.

Sub- and quotient groups are related in the following way: a subset H of G can be seen as an injective map $H \to G$, i.e. any element of the target has at most one element that maps to it. The counterpart to injective maps are surjective maps (every element of the target is mapped onto), such as the canonical map $G \to G \,/\, N$.[y][>] Interpreting subgroup and quotients in light of these homomorphisms emphasizes the structural concept inherent to these definitions alluded to in the introduction. In general, homomorphisms are neither injective nor surjective. Kernel and image of group homomorphisms and the first isomorphism theorem address this phenomenon.

4.5 Examples and applications

Main articles: Examples of groups and Applications of group theory

A periodic wallpaper pattern gives rise to a wallpaper group.

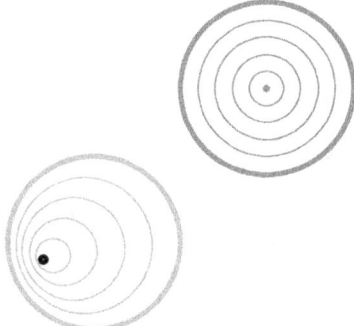

The fundamental group of a plane minus a point (bold) consists of loops around the missing point. This group is isomorphic to the integers.

Examples and applications of groups abound. A starting point is the group $\mathbf{Z}$ of integers with addition as group operation, introduced above. If instead of addition multiplication is considered, one obtains multiplicative groups. These groups are predecessors of important constructions in abstract algebra.

Groups are also applied in many other mathematical areas. Mathematical objects are often examined by associating groups to them and studying the properties of the corresponding groups. For example, Henri Poincaré founded what is now called algebraic topology by introducing the fundamental group.[36] By means of this connection, topological properties such as proximity and continuity translate into properties of groups.[b] For example, elements of the fundamental group are represented by loops. The second image at the right shows some loops in a plane minus a point. The blue loop is considered null-homotopic (and thus irrelevant), because it can be continuously shrunk to a point. The presence of the hole prevents the orange loop from being shrunk to a point. The fundamental group of the plane with a point deleted turns out to be infinite cyclic, generated by the orange loop (or any other loop winding once around the hole). This way, the fundamental group detects the hole.

In more recent applications, the influence has also been reversed to motivate geometric constructions by a group-theoretical background.[j] In a similar vein, geometric group theory employs geometric concepts, for example in the study of hyperbolic groups.[37] Further branches crucially applying groups include algebraic geometry and number theory.[38]

In addition to the above theoretical applications, many practical applications of groups exist. Cryptography relies on the combination of the abstract group theory approach together with algorithmical knowledge obtained in computational group theory, in particular when implemented for finite groups.[39] Applications of group theory are not restricted to mathematics; sciences such as physics, chemistry and computer science benefit from the concept.

4.5.1 Numbers

Many number systems, such as the integers and the rationals enjoy a naturally given group structure. In some cases, such as with the rationals, both addition and multiplication operations give rise to group structures. Such number systems are predecessors to more general algebraic structures known as rings and fields. Further abstract algebraic concepts such as modules, vector spaces and algebras also form groups.

Integers

The group of integers $\mathbf{Z}$ under addition, denoted $(\mathbf{Z}, +)$, has been described above. The integers, with the operation of multiplication instead of addition, $(\mathbf{Z}, \cdot)$ do *not* form a group. The closure, associativity and identity axioms are satisfied, but inverses do not exist: for example, $a = 2$ is an integer, but the only solution to the equation $a \cdot b = 1$ in this case is $b = 1/2$, which is a rational number, but not an integer. Hence not every element of $\mathbf{Z}$ has a (multiplicative) inverse.[k]

Rationals

The desire for the existence of multiplicative inverses suggests considering fractions

$$\frac{a}{b}.$$

Fractions of integers (with b nonzero) are known as rational numbers.[l] The set of all such fractions is commonly denoted $\mathbf{Q}$. There is still a minor obstacle for $(\mathbf{Q}, \cdot)$, the rationals with multiplication, being a group: because the rational number 0 does not have a multiplicative inverse (i.e., there is no x such that $x \cdot 0 = 1$), $(\mathbf{Q}, \cdot)$ is still not a group.

However, the set of all *nonzero* rational numbers $\mathbf{Q} \setminus \{0\} = \{q \in \mathbf{Q} \mid q \neq 0\}$ does form an abelian group under multiplication, denoted $(\mathbf{Q} \setminus \{0\}, \cdot)$.[m] Associativity and identity element axioms follow from the properties of integers. The closure requirement still holds true after removing zero, because the product of two nonzero rationals is never zero. Finally, the inverse of a/b is b/a, therefore the axiom of the inverse element is satisfied.

The rational numbers (including 0) also form a group under addition. Intertwining addition and multiplication operations yields more complicated structures called rings and—if division is possible, such as in $\mathbf{Q}$—fields, which occupy a central position in abstract algebra. Group theoretic arguments therefore underlie parts of the theory of those entities.[n]

4.5.2 Modular arithmetic

In modular arithmetic, two integers are added and then the sum is divided by a positive integer called the *modulus*. The result of modular addition is the remainder of that division. For any modulus, n, the set of integers from 0 to $n - 1$ forms a group under modular addition: the inverse of any element a is $n - a$, and 0 is the identity element. This is familiar from the addition of hours on the face of a clock: if the hour hand is on 9 and is advanced 4 hours, it ends up on 1, as shown at the right. This is expressed by saying that $9 + 4$ equals 1 "modulo 12" or, in symbols,

$$9 + 4 \equiv 1 \text{ modulo } 12.$$

The group of integers modulo n is written $\mathbf{Z}n$ or $\mathbf{Z}/n\mathbf{Z}$.

For any prime number p, there is also the multiplicative group of integers modulo p.[40] Its elements are the integers 1 to $p - 1$. The group operation is multiplication modulo p. That is, the usual product is divided by p and the remainder of this division is the result of modular multiplication. For example, if $p = 5$, there are four group elements 1, 2, 3, 4. In this group, $4 \cdot 4 = 1$, because the usual product 16 is equivalent to 1, which divided by 5 yields a remainder of 1. for 5 divides $16 - 1 = 15$, denoted

$$16 \equiv 1 \pmod 5.$$

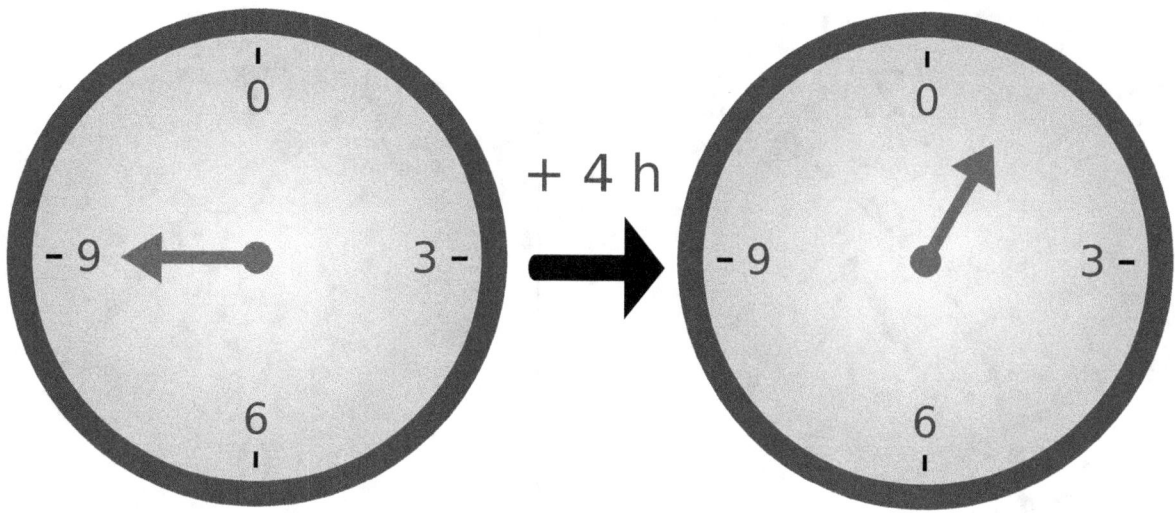

The hours on a clock form a group that uses addition modulo 12. Here 9 + 4 = 1

The primality of p ensures that the product of two integers neither of which is divisible by p is not divisible by p either, hence the indicated set of classes is closed under multiplication.[o][ᵖ] The identity element is 1, as usual for a multiplicative group, and the associativity follows from the corresponding property of integers. Finally, the inverse element axiom requires that given an integer a not divisible by p, there exists an integer b such that

$$a \cdot b \equiv 1 \pmod p, \text{ i.e. } p \text{ divides the difference } a \cdot b - 1.$$

The inverse b can be found by using Bézout's identity and the fact that the greatest common divisor $\gcd(a, p)$ equals 1.[41] In the case $p = 5$ above, the inverse of 4 is 4, and the inverse of 3 is 2, as $3 \cdot 2 = 6 \equiv 1 \pmod 5$. Hence all group axioms are fulfilled. Actually, this example is similar to $(\mathbf{Q} \setminus \{0\}, \cdot)$ above: it consists of exactly those elements in $\mathbf{Z}/p\mathbf{Z}$ that have a multiplicative inverse.[42] These groups are denoted $\mathbf{F}p^{\times}$. They are crucial to public-key cryptography.[ᵖ]

4.5.3 Cyclic groups

Main article: Cyclic group
A *cyclic group* is a group all of whose elements are powers of a particular element a.[43] In multiplicative notation, the elements of the group are:

$$..., a^{-3}, a^{-2}, a^{-1}, a^0 = e, a, a^2, a^3, ...,$$

where a^2 means $a \bullet a$, and a^{-3} stands for $a^{-1} \bullet a^{-1} \bullet a^{-1} = (a \bullet a \bullet a)^{-1}$ etc.[h][ᵖ] Such an element a is called a generator or a primitive element of the group. In additive notation, the requirement for an element to be primitive is that each element of the group can be written as

$$..., -a-a, -a, 0, a, a+a, ...$$

In the groups $\mathbf{Z}/n\mathbf{Z}$ introduced above, the element 1 is primitive, so these groups are cyclic. Indeed, each element is expressible as a sum all of whose terms are 1. Any cyclic group with n elements is isomorphic to this group. A second example for cyclic groups is the group of n-th complex roots of unity, given by complex numbers z satisfying $z^n = 1$. These numbers can be visualized as the vertices on a regular n-gon, as shown in blue at the right for $n = 6$. The group operation is multiplication of complex numbers. In the picture, multiplying with z corresponds to a counter-clockwise rotation by $60°$.[44] Using some field theory, the group $\mathbf{F}p^{\times}$ can be shown to be cyclic: for example, if $p = 5$, 3 is a generator since $3^1 = 3$, $3^2 = 9 \equiv 4$, $3^3 \equiv 2$, and $3^4 \equiv 1$.

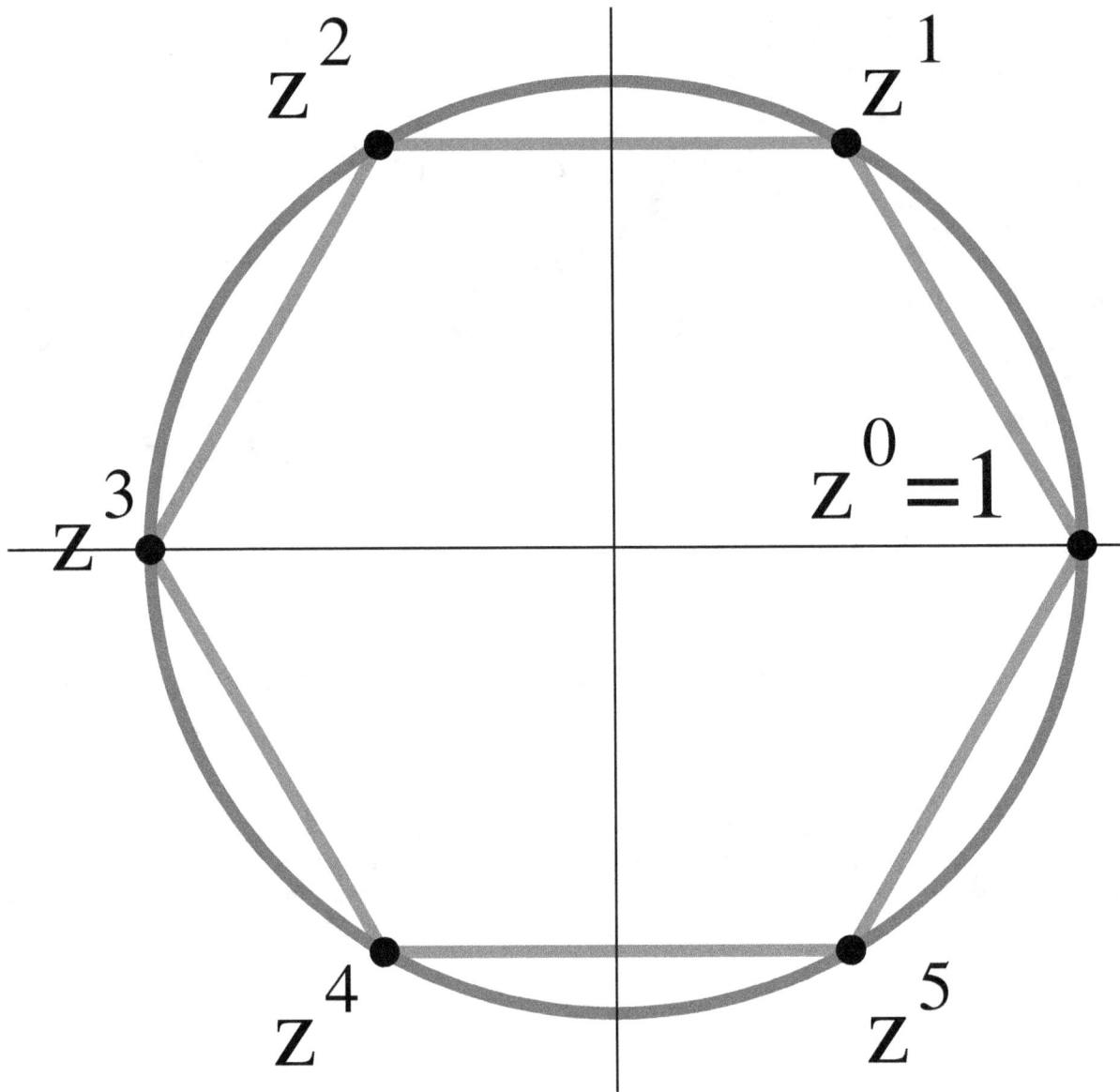

The 6th complex roots of unity form a cyclic group. z is a primitive element, but z^2 is not, because the odd powers of z are not a power of z^2.

Some cyclic groups have an infinite number of elements. In these groups, for every non-zero element *a*, all the powers of *a* are distinct; despite the name "cyclic group", the powers of the elements do not cycle. An infinite cyclic group is isomorphic to $(\mathbf{Z}, +)$, the group of integers under addition introduced above.[45] As these two prototypes are both abelian, so is any cyclic group.

The study of finitely generated abelian groups is quite mature, including the fundamental theorem of finitely generated abelian groups; and reflecting this state of affairs, many group-related notions, such as center and commutator, describe the extent to which a given group is not abelian.[46]

4.5.4 Symmetry groups

Main article: Symmetry group
See also: Molecular symmetry, Space group and Symmetry in physics

Symmetry groups are groups consisting of symmetries of given mathematical objects—be they of geometric nature, such as the introductory symmetry group of the square, or of algebraic nature, such as polynomial equations and their solutions.[47] Conceptually, group theory can be thought of as the study of symmetry.[b] Symmetries in mathematics greatly simplify the study of geometrical or analytical objects. A group is said to act on another mathematical object X if every group element performs some operation on X compatibly to the group law. In the rightmost example below, an element of order 7 of the (2,3,7) triangle group acts on the tiling by permuting the highlighted warped triangles (and the other ones, too). By a group action, the group pattern is connected to the structure of the object being acted on.

Rotations and reflections form the symmetry group of a great icosahedron.

In chemical fields, such as crystallography, space groups and point groups describe molecular symmetries and crystal symmetries. These symmetries underlie the chemical and physical behavior of these systems, and group theory enables simplification of quantum mechanical analysis of these properties.[48] For example, group theory is used to show that optical transitions between certain quantum levels cannot occur simply because of the symmetry of the states involved.

Not only are groups useful to assess the implications of symmetries in molecules, but surprisingly they also predict that molecules sometimes can change symmetry. The Jahn-Teller effect is a distortion of a molecule of high symmetry when it adopts a particular ground state of lower symmetry from a set of possible ground states that are related to each other by the symmetry operations of the molecule.[49][50]

Likewise, group theory helps predict the changes in physical properties that occur when a material undergoes a phase transition, for example, from a cubic to a tetrahedral crystalline form. An example is ferroelectric materials, where the change from a paraelectric to a ferroelectric state occurs at the Curie temperature and is related to a change from the high-symmetry paraelectric state to the lower symmetry ferroelectric state, accompanied by a so-called soft phonon mode, a vibrational lattice mode that goes to zero frequency at the transition.[51]

Such spontaneous symmetry breaking has found further application in elementary particle physics, where its occurrence is related to the appearance of Goldstone bosons.

Finite symmetry groups such as the Mathieu groups are used in coding theory, which is in turn applied in error correction of transmitted data, and in CD players.[52] Another application is differential Galois theory, which characterizes functions having antiderivatives of a prescribed form, giving group-theoretic criteria for when solutions of certain differential equations are well-behaved.u[>] Geometric properties that remain stable under group actions are investigated in (geometric) invariant theory.[53]

4.5.5   General linear group and representation theory

Main articles: General linear group and Representation theory
Matrix groups consist of matrices together with matrix multiplication. The *general linear group* GL(n, **R**) consists of all

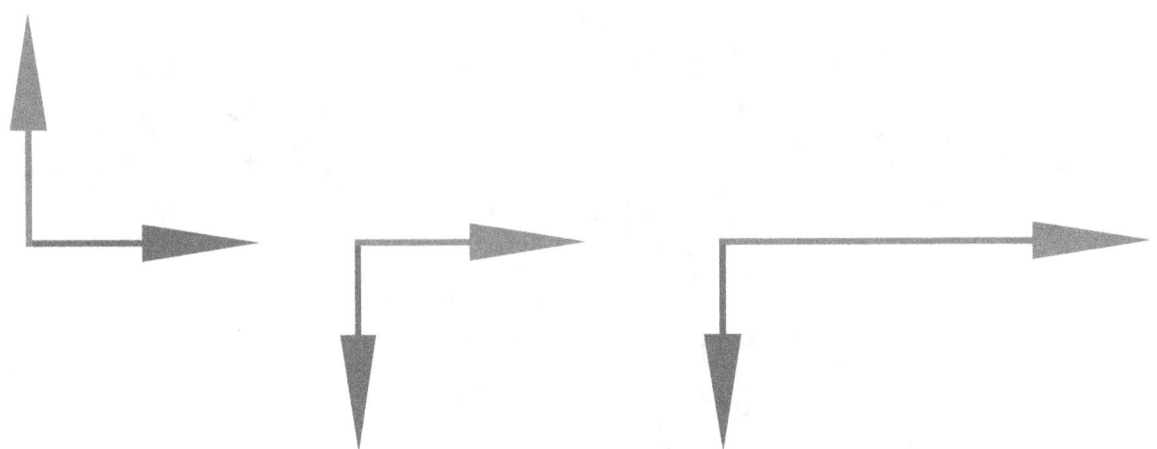

Two vectors (the left illustration) multiplied by matrices (the middle and right illustrations). The middle illustration represents a clockwise rotation by 90°, while the right-most one stretches the x-coordinate by factor 2.

invertible n-by-n matrices with real entries.[54] Its subgroups are referred to as *matrix groups* or *linear groups*. The dihedral group example mentioned above can be viewed as a (very small) matrix group. Another important matrix group is the special orthogonal group SO(n). It describes all possible rotations in n dimensions. Via Euler angles, rotation matrices are used in computer graphics.[55]

Representation theory is both an application of the group concept and important for a deeper understanding of groups.[56][57] It studies the group by its group actions on other spaces. A broad class of group representations are linear representations, i.e. the group is acting on a vector space, such as the three-dimensional Euclidean space **R**^{3}. A representation of G on an n-dimensional real vector space is simply a group homomorphism

$$\varrho\colon G \to \mathrm{GL}(n, \mathbf{R})$$

from the group to the general linear group. This way, the group operation, which may be abstractly given, translates to the multiplication of matrices making it accessible to explicit computations.w[>]

Given a group action, this gives further means to study the object being acted on.[x[·]] On the other hand, it also yields information about the group. Group representations are an organizing principle in the theory of finite groups, Lie groups, algebraic groups and topological groups, especially (locally) compact groups.[56][58]

4.5.6   Galois groups

Main article: Galois group

Galois groups were developed to help solve polynomial equations by capturing their symmetry features.[59][60] For example, the solutions of the quadratic equation $ax^2 + bx + c = 0$ are given by

$$x = \frac{-b \pm \sqrt{b^2 - 4ac}}{2a}.$$

Exchanging "+" and "−" in the expression, i.e. permuting the two solutions of the equation can be viewed as a (very simple) group operation. Similar formulae are known for cubic and quartic equations, but do *not* exist in general for degree 5 and higher.[61] Abstract properties of Galois groups associated with polynomials (in particular their solvability) give a criterion for polynomials that have all their solutions expressible by radicals, i.e. solutions expressible using solely addition, multiplication, and roots similar to the formula above.[62]

The problem can be dealt with by shifting to field theory and considering the splitting field of a polynomial. Modern Galois theory generalizes the above type of Galois groups to field extensions and establishes—via the fundamental theorem of Galois theory—a precise relationship between fields and groups, underlining once again the ubiquity of groups in mathematics.

4.6   Finite groups

Main article: Finite group

A group is called *finite* if it has a finite number of elements. The number of elements is called the order of the group.[63] An important class is the *symmetric groups* S_N, the groups of permutations of N letters. For example, the symmetric group on 3 letters S_3 is the group consisting of all possible orderings of the three letters ABC, i.e. contains the elements ABC, ACB, ..., up to CBA, in total 6 (or 3 factorial) elements. This class is fundamental insofar as any finite group can be expressed as a subgroup of a symmetric group S_N for a suitable integer N (Cayley's theorem). Parallel to the group of symmetries of the square above, S_3 can also be interpreted as the group of symmetries of an equilateral triangle.

The order of an element a in a group G is the least positive integer n such that $a^n = e$, where a^n represents

$$\underbrace{a \cdots a}_{n\text{factors}},$$

i.e. application of the operation • to n copies of a. (If • represents multiplication, then a^n corresponds to the nth power of a.) In infinite groups, such an n may not exist, in which case the order of a is said to be infinity. The order of an element equals the order of the cyclic subgroup generated by this element.

More sophisticated counting techniques, for example counting cosets, yield more precise statements about finite groups: Lagrange's Theorem states that for a finite group G the order of any finite subgroup H divides the order of G. The Sylow theorems give a partial converse.

The dihedral group (discussed above) is a finite group of order 8. The order of r_1 is 4, as is the order of the subgroup R it generates (see above). The order of the reflection elements f_v etc. is 2. Both orders divide 8, as predicted by Lagrange's theorem. The groups $\mathbf{F}_p^\times$ above have order $p - 1$.

4.6.1 Classification of finite simple groups

Main article: Classification of finite simple groups

Mathematicians often strive for a complete classification (or list) of a mathematical notion. In the context of finite groups, this aim leads to difficult mathematics. According to Lagrange's theorem, finite groups of order p, a prime number, are necessarily cyclic (abelian) groups $\mathbf{Z}p$. Groups of order p^2 can also be shown to be abelian, a statement which does not generalize to order p^3, as the non-abelian group D_4 of order $8 = 2^3$ above shows.[64] Computer algebra systems can be used to list small groups, but there is no classification of all finite groups.[q][>] An intermediate step is the classification of finite simple groups.[r][>] A nontrivial group is called *simple* if its only normal subgroups are the trivial group and the group itself.[s][>] The Jordan–Hölder theorem exhibits finite simple groups as the building blocks for all finite groups.[65] Listing all finite simple groups was a major achievement in contemporary group theory. 1998 Fields Medal winner Richard Borcherds succeeded to prove the monstrous moonshine conjectures, a surprising and deep relation of the largest finite simple sporadic group—the "monster group"—with certain modular functions, a piece of classical complex analysis, and string theory, a theory supposed to unify the description of many physical phenomena.[66]

4.7 Groups with additional structure

Many groups are simultaneously groups and examples of other mathematical structures. In the language of category theory, they are group objects in a category, meaning that they are objects (that is, examples of another mathematical structure) which come with transformations (called morphisms) that mimic the group axioms. For example, every group (as defined above) is also a set, so a group is a group object in the category of sets.

4.7.1 Topological groups

Main article: Topological group

Some topological spaces may be endowed with a group law. In order for the group law and the topology to interweave well, the group operations must be continuous functions, that is, $g \bullet h$, and g^{-1} must not vary wildly if g and h vary only little. Such groups are called *topological groups,* and they are the group objects in the category of topological spaces.[67] The most basic examples are the reals $\mathbf{R}$ under addition, $(\mathbf{R} \setminus \{0\}, \cdot)$, and similarly with any other topological field such as the complex numbers or p-adic numbers. All of these groups are locally compact, so they have Haar measures and can be studied via harmonic analysis. The former offer an abstract formalism of invariant integrals. Invariance means, in the case of real numbers for example:

$$\int f(x)\,dx = \int f(x+c)\,dx$$

for any constant c. Matrix groups over these fields fall under this regime, as do adele rings and adelic algebraic groups, which are basic to number theory.[68] Galois groups of infinite field extensions such as the absolute Galois group can also be equipped with a topology, the so-called Krull topology, which in turn is central to generalize the above sketched connection of fields and groups to infinite field extensions.[69] An advanced generalization of this idea, adapted to the needs of algebraic geometry, is the étale fundamental group.[70]

4.7.2 Lie groups

Main article: Lie group

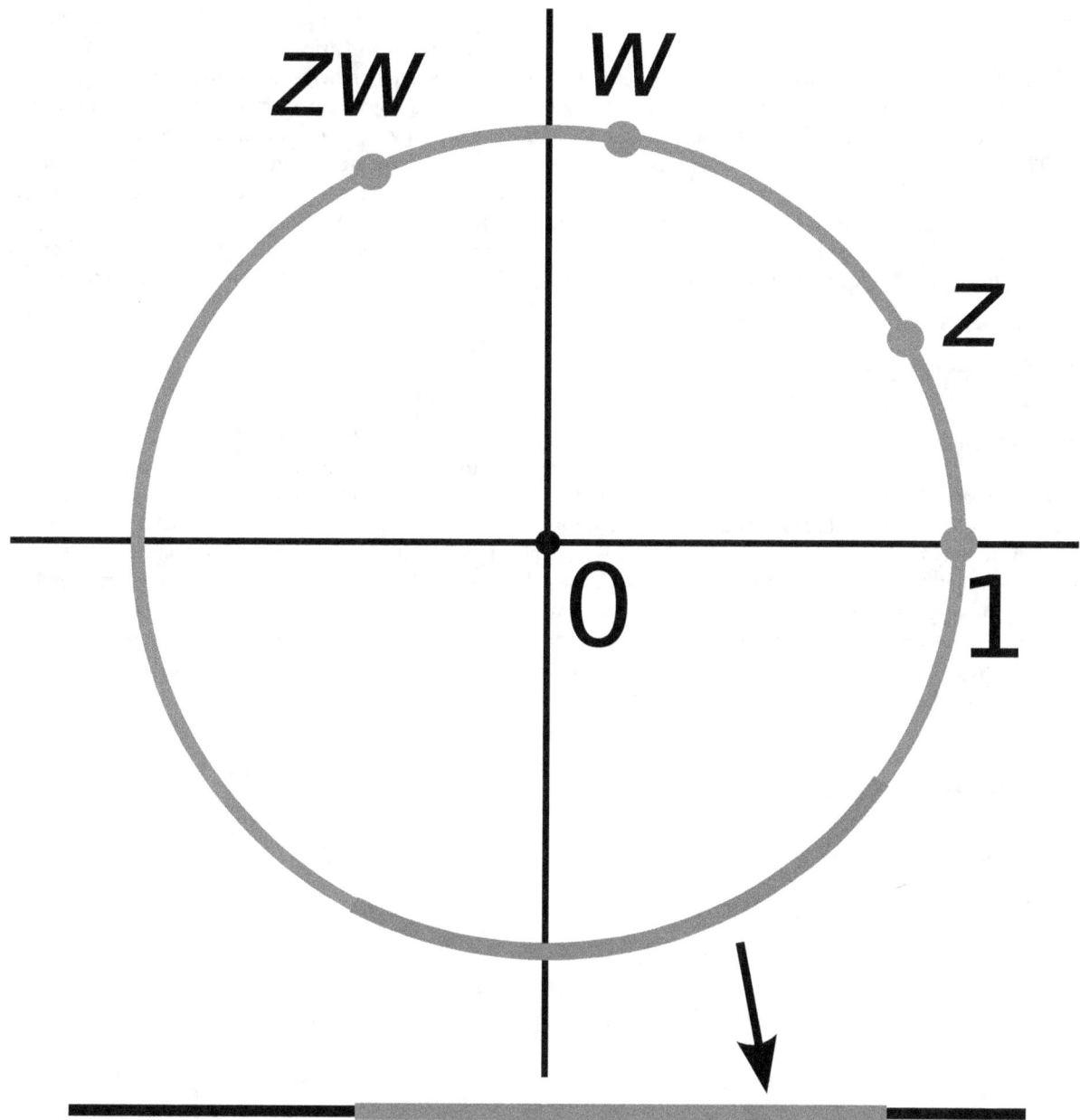

The unit circle in the complex plane under complex multiplication is a Lie group and, therefore, a topological group. It is topological since complex multiplication and division are continuous. It is a manifold and thus a Lie group, because every small piece, such as the red arc in the figure, looks like a part of the real line (shown at the bottom).

Lie groups (in honor of Sophus Lie) are groups which also have a manifold structure, i.e. they are spaces looking locally like some Euclidean space of the appropriate dimension.[71] Again, the additional structure, here the manifold structure, has to be compatible, i.e. the maps corresponding to multiplication and the inverse have to be smooth.

A standard example is the general linear group introduced above: it is an open subset of the space of all n-by-n matrices, because it is given by the inequality

$$\det (A) \neq 0,$$

where *A* denotes an *n*-by-*n* matrix.[72]

Lie groups are of fundamental importance in modern physics: Noether's theorem links continuous symmetries to conserved quantities.[73] Rotation, as well as translations in space and time are basic symmetries of the laws of mechanics. They can, for instance, be used to construct simple models—imposing, say, axial symmetry on a situation will typically lead to significant simplification in the equations one needs to solve to provide a physical description.[v] Another example are the Lorentz transformations, which relate measurements of time and velocity of two observers in motion relative to each other. They can be deduced in a purely group-theoretical way, by expressing the transformations as a rotational symmetry of Minkowski space. The latter serves—in the absence of significant gravitation—as a model of space time in special relativity.[74] The full symmetry group of Minkowski space, i.e. including translations, is known as the Poincaré group. By the above, it plays a pivotal role in special relativity and, by implication, for quantum field theories.[75] Symmetries that vary with location are central to the modern description of physical interactions with the help of gauge theory.[76]

4.8 Generalizations

In abstract algebra, more general structures are defined by relaxing some of the axioms defining a group.[28][77][78] For example, if the requirement that every element has an inverse is eliminated, the resulting algebraic structure is called a monoid. The natural numbers **N** (including 0) under addition form a monoid, as do the nonzero integers under multiplication ($\mathbf{Z} \setminus \{0\}, \cdot$), see above. There is a general method to formally add inverses to elements to any (abelian) monoid, much the same way as ($\mathbf{Q} \setminus \{0\}, \cdot$) is derived from ($\mathbf{Z} \setminus \{0\}, \cdot$), known as the Grothendieck group. Groupoids are similar to groups except that the composition $a \bullet b$ need not be defined for all a and b. They arise in the study of more complicated forms of symmetry, often in topological and analytical structures, such as the fundamental groupoid or stacks. Finally, it is possible to generalize any of these concepts by replacing the binary operation with an arbitrary *n*-ary one (i.e. an operation taking *n* arguments). With the proper generalization of the group axioms this gives rise to an *n*-ary group.[79] The table gives a list of several structures generalizing groups.

4.9 See also

- Abelian group

- Cyclic group

- Euclidean group

- Finitely presented group

- Free group

- Fundamental group

- Grothendieck group

- Group algebra

- Group ring

- Heap (mathematics)

- Nilpotent group

- Non-abelian group

- Quantum group

- Reductive group

- Solvable group

- Symmetry in physics

- Computational group theory

4.10   Notes

^ **a:** Mathematical Reviews lists 3,224 research papers on group theory and its generalizations written in 2005.

^ **aa:** The classification was announced in 1983, but gaps were found in the proof. See classification of finite simple groups for further information.

^ **b:** The closure axiom is already implied by the condition that • be a binary operation. Some authors therefore omit this axiom. However, group constructions often start with an operation defined on a superset, so a closure step is common in proofs that a system is a group. Lang 2002

^ **c:** See, for example, the books of Lang (2002, 2005) and Herstein (1996, 1975).

^ **d:** However, a group is not determined by its lattice of subgroups. See Suzuki 1951.

^ **e:** The fact that the group operation extends this canonically is an instance of a universal property.

^ **f:** For example, if G is finite, then the size of any subgroup and any quotient group divides the size of G, according to Lagrange's theorem.

^ **g:** The word homomorphism derives from Greek ὁμός—the same and μορφή—structure.

^ **h:** The additive notation for elements of a cyclic group would be $t • a$, t in $\mathbf{Z}$.

^ **i:** See the Seifert–van Kampen theorem for an example.

^ **j:** An example is group cohomology of a group which equals the singular homology of its classifying space.

^ **k:** Elements which do have multiplicative inverses are called units, see Lang 2002, §II.1, p. 84.

^ **l:** The transition from the integers to the rationals by adding fractions is generalized by the quotient field.

^ **m:** The same is true for any field F instead of $\mathbf{Q}$. See Lang 2005, §III.1, p. 86.

^ **n:** For example, a finite subgroup of the multiplicative group of a field is necessarily cyclic. See Lang 2002, Theorem IV.1.9. The notions of torsion of a module and simple algebras are other instances of this principle.

^ **o:** The stated property is a possible definition of prime numbers. See prime element.

^ **p:** For example, the Diffie-Hellman protocol uses the discrete logarithm.

^ **q:** The groups of order at most 2000 are known. Up to isomorphism, there are about 49 billion. See Besche, Eick & O'Brien 2001.

^ **r:** The gap between the classification of simple groups and the one of all groups lies in the extension problem, a problem too hard to be solved in general. See Aschbacher 2004, p. 737.

^ **s:** Equivalently, a nontrivial group is simple if its only quotient groups are the trivial group and the group itself. See Michler 2006, Carter 1989.

^ **t:** More rigorously, every group is the symmetry group of some graph; see Frucht's theorem, Frucht 1939.

^ **u:** More precisely, the monodromy action on the vector space of solutions of the differential equations is considered. See Kuga 1993, pp. 105–113.

^ **v:** See Schwarzschild metric for an example where symmetry greatly reduces the complexity of physical systems.

^ **w:** This was crucial to the classification of finite simple groups, for example. See Aschbacher 2004.

^ **x:** See, for example, Schur's Lemma for the impact of a group action on simple modules. A more involved example is the action of an absolute Galois group on étale cohomology.

^ **y:** Injective and surjective maps correspond to mono- and epimorphisms, respectively. They are interchanged when passing to the dual category.

4.11   Citations

[1] Herstein 1975, §2, p. 26

[2] Hall 1967, §1.1, p. 1: "The idea of a group is one which pervades the whole of mathematics both pure and applied."

[3] Lang 2005, App. 2, p. 360

[4] Cook, Mariana R. (2009), *Mathematicians: An Outer View of the Inner World*, Princeton, N.J.: Princeton University Press, p. 24, ISBN 9780691139517

[5] Herstein 1975, §2.1, p. 27

[6] Weisstein, Eric W., "Identity Element", *MathWorld*.

[7] Herstein 1975, §2.6, p. 54

[8] Wussing 2007

[9] Kleiner 1986

[10] Smith 1906

[11] Galois 1908

[12] Kleiner 1986, p. 202

[13] Cayley 1889

[14] Wussing 2007, §III.2

[15] Lie 1973

[16] Kleiner 1986, p. 204

[17] Wussing 2007, §I.3.4

[18] Jordan 1870

[19] von Dyck 1882

[20] Curtis 2003

[21] Mackey 1976

[22] Borel 2001

[23] Aschbacher 2004

[24] Ledermann 1953, §1.2, pp. 4–5

[25] Ledermann 1973, §I.1, p. 3

[26] Lang 2002, §I.2, p. 7

[27] Lang 2005, §II.1, p. 17

[28] Mac Lane 1998

[29] Lang 2005, §II.3, p. 34

[30] Lang 2005, §II.1, p. 19

[31] Ledermann 1973, §II.12, p. 39

[32] Lang 2005, §II.4, p. 41

[33] Lang 2002, §I.2, p. 12

[34] Lang 2005, §II.4, p. 45

[35] Lang 2002, §I.2, p. 9

[36] Hatcher 2002, Chapter I, p. 30

[37] Coornaert, Delzant & Papadopoulos 1990

[38] for example, class groups and Picard groups; see Neukirch 1999, in particular §§I.12 and I.13

[39] Seress 1997

[40] Lang 2005, Chapter VII

[41] Rosen 2000, p. 54 (Theorem 2.1)

[42] Lang 2005, §VIII.1, p. 292

[43] Lang 2005, §II.1, p. 22

[44] Lang 2005, §II.2, p. 26

[45] Lang 2005, §II.1, p. 22 (example 11)

[46] Lang 2002, §I.5, p. 26, 29

[47] Weyl 1952

[48] Conway, Delgado Friedrichs & Huson et al. 2001. See also Bishop 1993

[49] Bersuker, Isaac (2006), *The Jahn-Teller Effect*, Cambridge University Press, p. 2, ISBN 0-521-82212-2

[50] Jahn & Teller 1937

[51] Dove, Martin T (2003), *Structure and Dynamics: an atomic view of materials*, Oxford University Press, p. 265, ISBN 0-19-850678-3

[52] Welsh 1989

[53] Mumford, Fogarty & Kirwan 1994

[54] Lay 2003

[55] Kuipers 1999

[56] Fulton & Harris 1991

[57] Serre 1977

[58] Rudin 1990

[59] Robinson 1996, p. viii

[60] Artin 1998

[61] Lang 2002, Chapter VI (see in particular p. 273 for concrete examples)

[62] Lang 2002, p. 292 (Theorem VI.7.2)

[63] Kurzweil & Stellmacher 2004

[64] Artin 1991, Theorem 6.1.14. See also Lang 2002, p. 77 for similar results.

[65] Lang 2002, §I. 3, p. 22

[66] Ronan 2007

[67] Husain 1966

[68] Neukirch 1999

[69] Shatz 1972

[70] Milne 1980

[71] Warner 1983

[72] Borel 1991

[73] Goldstein 1980

[74] Weinberg 1972

[75] Naber 2003

[76] Becchi 1997

[77] Denecke & Wismath 2002

[78] Romanowska & Smith 2002

[79] Dudek 2001

4.12 References

4.12.1 General references

- Artin, Michael (1991), *Algebra*, Prentice Hall, ISBN 978-0-89871-510-1, Chapter 2 contains an undergraduate-level exposition of the notions covered in this article.

- Devlin, Keith (2000), *The Language of Mathematics: Making the Invisible Visible*, Owl Books, ISBN 978-0-8050-7254-9, Chapter 5 provides a layman-accessible explanation of groups.

- Fulton, William; Harris, Joe (1991), *Representation theory. A first course*, Graduate Texts in Mathematics, Readings in Mathematics **129**, New York: Springer-Verlag, ISBN 978-0-387-97495-8, MR 1153249, ISBN 978-0-387-97527-6.

- Hall, G. G. (1967), *Applied group theory*, American Elsevier Publishing Co., Inc., New York, MR 0219593, an elementary introduction.

- Herstein, Israel Nathan (1996), *Abstract algebra* (3rd ed.), Upper Saddle River, NJ: Prentice Hall Inc., ISBN 978-0-13-374562-7, MR 1375019.

- Herstein, Israel Nathan (1975), *Topics in algebra* (2nd ed.), Lexington, Mass.: Xerox College Publishing, MR 0356988.

- Lang, Serge (2002), *Algebra*, Graduate Texts in Mathematics **211** (Revised third ed.), New York: Springer-Verlag, ISBN 978-0-387-95385-4, MR 1878556

- Lang, Serge (2005), *Undergraduate Algebra* (3rd ed.), Berlin, New York: Springer-Verlag, ISBN 978-0-387-22025-3.

- Ledermann, Walter (1953), *Introduction to the theory of finite groups*, Oliver and Boyd, Edinburgh and London, MR 0054593.

- Ledermann, Walter (1973), *Introduction to group theory*, New York: Barnes and Noble, OCLC 795613.

- Robinson, Derek John Scott (1996), *A course in the theory of groups*, Berlin, New York: Springer-Verlag, ISBN 978-0-387-94461-6.

4.12.2 Special references

- Artin, Emil (1998), *Galois Theory*, New York: Dover Publications, ISBN 978-0-486-62342-9.

- Aschbacher, Michael (2004), "The Status of the Classification of the Finite Simple Groups" (PDF), *Notices of the American Mathematical Society* **51** (7): 736–740.

- Becchi, C. (1997), *Introduction to Gauge Theories*, p. 5211, arXiv:hep-ph/9705211, Bibcode:1997hep.ph....5211B.

- Besche, Hans Ulrich; Eick, Bettina; O'Brien, E. A. (2001), "The groups of order at most 2000", *Electronic Research Announcements of the American Mathematical Society* **7**: 1–4, doi:10.1090/S1079-6762-01-00087-7, MR 1826989.

- Bishop, David H. L. (1993), *Group theory and chemistry*, New York: Dover Publications, ISBN 978-0-486-67355-4.

- Borel, Armand (1991), *Linear algebraic groups*, Graduate Texts in Mathematics **126** (2nd ed.), Berlin, New York: Springer-Verlag, ISBN 978-0-387-97370-8, MR 1102012.

- Carter, Roger W. (1989), *Simple groups of Lie type*, New York: John Wiley & Sons, ISBN 978-0-471-50683-6.

- Conway, John Horton; Delgado Friedrichs, Olaf; Huson, Daniel H.; Thurston, William P. (2001), "On three-dimensional space groups", *Beiträge zur Algebra und Geometrie* **42** (2): 475–507, arXiv:math.MG/9911185, MR 1865535.

- Coornaert, M.; Delzant, T.; Papadopoulos, A. (1990), *Géométrie et théorie des groupes [Geometry and Group Theory]*, Lecture Notes in Mathematics (in French) **1441**, Berlin, New York: Springer-Verlag, ISBN 978-3-540-52977-4, MR 1075994.

- Denecke, Klaus; Wismath, Shelly L. (2002), *Universal algebra and applications in theoretical computer science*, London: CRC Press, ISBN 978-1-58488-254-1.

- Dudek, W.A. (2001), "On some old problems in n-ary groups", *Quasigroups and Related Systems* **8**: 15–36.

- Frucht, R. (1939), "Herstellung von Graphen mit vorgegebener abstrakter Gruppe [Construction of Graphs with Prescribed Group]", *Compositio Mathematica* (in German) **6**: 239–50.

- Goldstein, Herbert (1980), *Classical Mechanics* (2nd ed.), Reading, MA: Addison-Wesley Publishing, pp. 588–596, ISBN 0-201-02918-9.

- Hatcher, Allen (2002), *Algebraic topology*, Cambridge University Press, ISBN 978-0-521-79540-1.

- Husain, Taqdir (1966), *Introduction to Topological Groups*, Philadelphia: W.B. Saunders Company, ISBN 978-0-89874-193-3

- Jahn, H.; Teller, E. (1937), "Stability of Polyatomic Molecules in Degenerate Electronic States. I. Orbital Degeneracy", *Proceedings of the Royal Society A* **161** (905): 220–235, Bibcode:1937RSPSA.161..220J, doi:10.1098/rspa.1937.0142.

- Kuipers, Jack B. (1999), *Quaternions and rotation sequences—A primer with applications to orbits, aerospace, and virtual reality*, Princeton University Press, ISBN 978-0-691-05872-6, MR 1670862.

- Kuga, Michio (1993), *Galois' dream: group theory and differential equations*, Boston, MA: Birkhäuser Boston, ISBN 978-0-8176-3688-3, MR 1199112.

- Kurzweil, Hans; Stellmacher, Bernd (2004), *The theory of finite groups*, Universitext, Berlin, New York: Springer-Verlag, ISBN 978-0-387-40510-0, MR 2014408.

- Lay, David (2003), *Linear Algebra and Its Applications*, Addison-Wesley, ISBN 978-0-201-70970-4.

- Mac Lane, Saunders (1998), *Categories for the Working Mathematician* (2nd ed.), Berlin, New York: Springer-Verlag, ISBN 978-0-387-98403-2.

- Michler, Gerhard (2006), *Theory of finite simple groups*, Cambridge University Press, ISBN 978-0-521-86625-5.

- Milne, James S. (1980), *Étale cohomology*, Princeton University Press, ISBN 978-0-691-08238-7

- Mumford, David; Fogarty, J.; Kirwan, F. (1994), *Geometric invariant theory* **34** (3rd ed.), Berlin, New York: Springer-Verlag, ISBN 978-3-540-56963-3, MR 1304906.

- Naber, Gregory L. (2003), *The geometry of Minkowski spacetime*, New York: Dover Publications, ISBN 978-0-486-43235-9, MR 2044239.

- Neukirch, Jürgen (1999), *Algebraic Number Theory*, Grundlehren der mathematischen Wissenschaften **322**, Berlin: Springer-Verlag, ISBN 978-3-540-65399-8, Zbl 0956.11021, MR 1697859.

- Romanowska, A.B.; Smith, J.D.H. (2002), *Modes*, World Scientific, ISBN 978-981-02-4942-7.

- Ronan, Mark (2007), *Symmetry and the Monster: The Story of One of the Greatest Quests of Mathematics*, Oxford University Press, ISBN 978-0-19-280723-6.

- Rosen, Kenneth H. (2000), *Elementary number theory and its applications* (4th ed.), Addison-Wesley, ISBN 978-0-201-87073-2, MR 1739433.

- Rudin, Walter (1990), *Fourier Analysis on Groups*, Wiley Classics, Wiley-Blackwell, ISBN 0-471-52364-X.

- Seress, Ákos (1997), "An introduction to computational group theory", *Notices of the American Mathematical Society* **44** (6): 671–679, MR 1452069.

- Serre, Jean-Pierre (1977), *Linear representations of finite groups*, Berlin, New York: Springer-Verlag, ISBN 978-0-387-90190-9, MR 0450380.

- Shatz, Stephen S. (1972), *Profinite groups, arithmetic, and geometry*, Princeton University Press, ISBN 978-0-691-08017-8, MR 0347778

- Suzuki, Michio (1951), "On the lattice of subgroups of finite groups", *Transactions of the American Mathematical Society* **70** (2): 345–371, doi:10.2307/1990375, JSTOR 1990375.

- Warner, Frank (1983), *Foundations of Differentiable Manifolds and Lie Groups*, Berlin, New York: Springer-Verlag, ISBN 978-0-387-90894-6.

- Weinberg, Steven (1972), *Gravitation and Cosmology*, New York: John Wiley & Sons, ISBN 0-471-92567-5.

- Welsh, Dominic (1989), *Codes and cryptography*, Oxford: Clarendon Press, ISBN 978-0-19-853287-3.

- Weyl, Hermann (1952), *Symmetry*, Princeton University Press, ISBN 978-0-691-02374-8.

4.12.3 Historical references

See also: Historically important publications in group theory

- Borel, Armand (2001), *Essays in the History of Lie Groups and Algebraic Groups*, Providence, R.I.: American Mathematical Society, ISBN 978-0-8218-0288-5

- Cayley, Arthur (1889), *The collected mathematical papers of Arthur Cayley*, II (1851–1860), Cambridge University Press.

- O'Connor, J.J; Robertson, E.F. (1996), *The development of group theory*.

- Curtis, Charles W. (2003), *Pioneers of Representation Theory: Frobenius, Burnside, Schur, and Brauer*, History of Mathematics, Providence, R.I.: American Mathematical Society, ISBN 978-0-8218-2677-5.

- von Dyck, Walther (1882), "Gruppentheoretische Studien (Group-theoretical Studies)", *Mathematische Annalen* (in German) **20** (1): 1–44, doi:10.1007/BF01443322.

- Galois, Évariste (1908), Tannery, Jules, ed., *Manuscrits de Évariste Galois [Évariste Galois' Manuscripts]* (in French), Paris: Gauthier-Villars (Galois work was first published by Joseph Liouville in 1843).

- Jordan, Camille (1870), *Traité des substitutions et des équations algébriques [Study of Substitutions and Algebraic Equations]* (in French), Paris: Gauthier-Villars.

- Kleiner, Israel (1986), "The evolution of group theory: a brief survey", *Mathematics Magazine* **59** (4): 195–215, doi:10.2307/2690312, MR 863090.

- Lie, Sophus (1973), *Gesammelte Abhandlungen. Band 1 [Collected papers. Volume 1]* (in German), New York: Johnson Reprint Corp., MR 0392459.

- Mackey, George Whitelaw (1976), *The theory of unitary group representations*, University of Chicago Press, MR 0396826

- Smith, David Eugene (1906), *History of Modern Mathematics*, Mathematical Monographs, No. 1.

- Wussing, Hans (2007), *The Genesis of the Abstract Group Concept: A Contribution to the History of the Origin of Abstract Group Theory*, New York: Dover Publications, ISBN 978-0-486-45868-7.

Chapter 5

Field (mathematics)

This article is about fields in algebra. For fields in geometry, see Vector field. For other uses, see Field (disambiguation).

In abstract algebra, a **field** is a nonzero commutative division ring, or equivalently a ring whose nonzero elements form an abelian group under multiplication. As such it is an algebraic structure with notions of addition, subtraction, multiplication, and division satisfying the appropriate abelian group equations and distributive law. The most commonly used fields are the field of real numbers, the field of complex numbers, and the field of rational numbers, but there are also finite fields, fields of functions, algebraic number fields, p-adic fields, and so forth.

Any field may be used as the scalars for a vector space, which is the standard general context for linear algebra. The theory of field extensions (including Galois theory) involves the roots of polynomials with coefficients in a field; among other results, this theory leads to impossibility proofs for the classical problems of angle trisection and squaring the circle with a compass and straightedge, as well as a proof of the Abel–Ruffini theorem on the algebraic insolubility of quintic equations. In modern mathematics, the theory of fields (or **field theory**) plays an essential role in number theory and algebraic geometry.

As an algebraic structure, every field is a ring, but not every ring is a field. The most important difference is that fields allow for division (though not division by zero), while a ring need not possess multiplicative inverses; for example the integers form a ring, but $2x = 1$ has no solution in integers. Also, the multiplication operation in a field is required to be commutative. A ring in which division is possible but commutativity is not assumed (such as the quaternions) is called a *division ring* or *skew field*. (Historically, division rings were sometimes referred to as fields, while fields were called *commutative fields*.)

As a ring, a field may be classified as a specific type of integral domain, and can be characterized by the following (not exhaustive) chain of class inclusions:

> **Commutative rings ⊃ integral domains ⊃ integrally closed domains ⊃ unique factorization domains ⊃ principal ideal domains ⊃ Euclidean domains ⊃ fields ⊃ finite fields**

5.1   Definition and illustration

Intuitively, a field is a set F that is a commutative group with respect to two compatible operations, addition and multiplication (the latter excluding zero), with "compatible" being formalized by *distributivity*, and the caveat that the additive and the multiplicative identities are distinct ($0 \neq 1$).

The most common way to formalize this is by defining a *field* as a set together with two operations, usually called *addition* and *multiplication*, and denoted by + and ·, respectively, such that the following axioms hold (note that *subtraction* and *division* are defined in terms of the inverse operations of addition and multiplication):[note 1]

***Closure* of *F* under addition and multiplication** For all a, b in F, both $a + b$ and $a \cdot b$ are in F (or more formally, +

and $\cdot$ are binary operations on F).

Associativity **of addition and multiplication** For all a, b, and c in F, the following equalities hold: $a + (b + c) = (a + b) + c$ and $a \cdot (b \cdot c) = (a \cdot b) \cdot c$.

Commutativity **of addition and multiplication** For all a and b in F, the following equalities hold: $a + b = b + a$ and $a \cdot b = b \cdot a$.

Existence of additive and multiplicative *identity elements* There exists an element of F, called the *additive identity* element and denoted by 0, such that for all a in F, $a + 0 = a$. Likewise, there is an element, called the *multiplicative identity* element and denoted by 1, such that for all a in F, $a \cdot 1 = a$. To exclude the trivial ring, the additive identity and the multiplicative identity are required to be distinct.

Existence of *additive inverses* **and** *multiplicative inverses* For every a in F, there exists an element $-a$ in F, such that $a + (-a) = 0$. Similarly, for any a in F other than 0, there exists an element a^{-1} in F, such that $a \cdot a^{-1} = 1$. (The elements $a + (-b)$ and $a \cdot b^{-1}$ are also denoted $a - b$ and a/b, respectively.) In other words, *subtraction* and *division* operations exist.

Distributivity **of multiplication over addition** For all a, b and c in F, the following equality holds: $a \cdot (b + c) = (a \cdot b) + (a \cdot c)$.

A field is therefore an algebraic structure $\langle F, +, \cdot, -, {}^{-1}, 0, 1 \rangle$; of type $\langle 2, 2, 1, 1, 0, 0 \rangle$, consisting of two abelian groups:

- F under $+$, $-$, and 0;
- $F \setminus \{0\}$ under $\cdot$, $^{-1}$, and 1, with $0 \neq 1$,

with $\cdot$ distributing over $+$.[1]

5.1.1   First example: rational numbers

A simple example of a field is the field of rational numbers, consisting of numbers which can be written as fractions a/b, where a and b are integers, and $b \neq 0$. The additive inverse of such a fraction is simply $-a/b$, and the multiplicative inverse (provided that $a \neq 0$) is b/a. To see the latter, note that

$$\frac{b}{a} \cdot \frac{a}{b} = \frac{ba}{ab} = 1.$$

The abstractly required field axioms reduce to standard properties of rational numbers, such as the law of distributivity

$$\frac{a}{b} \cdot \left(\frac{c}{d} + \frac{e}{f} \right)$$

$$= \frac{a}{b} \cdot \left(\frac{c}{d} \cdot \frac{f}{f} + \frac{e}{f} \cdot \frac{d}{d} \right)$$

$$= \frac{a}{b} \cdot \left(\frac{cf}{df} + \frac{ed}{fd} \right) = \frac{a}{b} \cdot \frac{cf + ed}{df}$$

$$= \frac{a(cf + ed)}{bdf} = \frac{acf}{bdf} + \frac{aed}{bdf} = \frac{ac}{bd} + \frac{ae}{bf}$$

$$= \frac{a}{b} \cdot \frac{c}{d} + \frac{a}{b} \cdot \frac{e}{f},$$

or the law of commutativity and law of associativity.

5.1.2   Second example: a field with four elements

In addition to familiar number systems such as the rationals, there are other, less immediate examples of fields. The following example is a field consisting of four elements called O, I, A and B. The notation is chosen such that O plays the role of the additive identity element (denoted 0 in the axioms), and I is the multiplicative identity (denoted 1 above). One can check that all field axioms are satisfied. For example:

$$A \cdot (B + A) = A \cdot I = A, \text{ which equals } A \cdot B + A \cdot A = I + B = A, \text{ as required by the distributivity.}$$

The above field is called a finite field with four elements, and can be denoted $\mathbf{F}_4$. Field theory is concerned with understanding the reasons for the existence of this field, defined in a fairly ad-hoc manner, and describing its inner structure. For example, from a glance at the multiplication table, it can be seen that any non-zero element (i.e., I, A, and B) is a power of A: $A = A^1$, $B = A^2 = A \cdot A$, and finally $I = A^3 = A \cdot A \cdot A$. This is not a coincidence, but rather one of the starting points of a deeper understanding of (finite) fields.

5.1.3   Alternative axiomatizations

As with other algebraic structures, there exist alternative axiomatizations. Because of the relations between the operations, one can alternatively axiomatize a field by explicitly assuming that there are four binary operations (add, subtract, multiply, divide) with axioms relating these, or (by functional decomposition) in terms of two binary operations (add and multiply) and two unary operations (additive inverse and multiplicative inverse), or other variants.

The usual axiomatization in terms of the two operations of addition and multiplication is brief and allows the other operations to be defined in terms of these basic ones, but in other contexts, such as topology and category theory, it is important to include all operations as explicitly given, rather than implicitly defined (compare topological group). This is because without further assumptions, the implicitly defined inverses may not be continuous (in topology), or may not be able to be defined (in category theory). Defining an inverse requires that one is working with a set, not a more general object.

For a very economical axiomatization of the field of real numbers, whose primitives are merely a set $\mathbf{R}$ with $1 \in \mathbf{R}$, addition, and a binary relation, "<". See Tarski's axiomatization of the reals.

5.2   Related algebraic structures

The axioms imposed above resemble the ones familiar from other algebraic structures. For example, the existence of the binary operation "·", together with its commutativity, associativity, (multiplicative) identity element and inverses are precisely the axioms for an abelian group. In other words, for any field, the subset of nonzero elements $F \setminus \{0\}$, also often denoted $F^\times$, is an abelian group $(F^\times, \cdot)$ usually called multiplicative group of the field. Likewise $(F, +)$ is an abelian group. The structure of a field is hence the same as specifying such two group structures (on the same set), obeying the distributivity.

Important other algebraic structures such as rings arise when requiring only part of the above axioms. For example, if the requirement of commutativity of the multiplication operation · is dropped, one gets structures usually called division rings or *skew fields*.

5.2.1   Remarks

By elementary group theory, applied to the abelian groups $(F^\times, \cdot)$, and $(F, +)$, the additive inverse $-a$ and the multiplicative inverse a^{-1} are uniquely determined by a.

Similar direct consequences from the field axioms include

$$-(a \cdot b) = (-a) \cdot b = a \cdot (-b), \text{ in particular } -a = (-1) \cdot a$$

as well as

$$a \cdot 0 = 0.$$

Both can be shown by replacing b or c with 0 in the distributive property.

5.3   History

The concept of *field* was used implicitly by Niels Henrik Abel and Évariste Galois in their work on the solvability of polynomial equations with rational coefficients of degree five or higher.

In 1857, Karl von Staudt published his Algebra of Throws which provided a geometric model satisfying the axioms of a field.[2] This construction has been frequently recalled as a contribution to the foundations of mathematics.

In 1871, Richard Dedekind introduced, for a set of real or complex numbers which is closed under the four arithmetic operations, the German word *Körper*, which means "body" or "corpus" (to suggest an organically closed entity),[3] hence the common use of the letter K to denote a field. He also defined rings (then called *order* or *order-modul*), but the term *"a ring"* (*Zahlring*) was invented by Hilbert.[4] In 1893, Eliakim Hastings Moore called the concept "field" in English.[5][6]

In 1881, Leopold Kronecker defined what he called a "domain of rationality", which is indeed a field of polynomials in modern terms. In 1893, Heinrich M. Weber gave the first clear definition of an abstract field.[7] In 1910, Ernst Steinitz published the very influential paper *Algebraische Theorie der Körper* (English: Algebraic Theory of Fields).[8] In this paper he axiomatically studies the properties of fields and defines many important field theoretic concepts like prime field, perfect field and the transcendence degree of a field extension.

Emil Artin developed the relationship between groups and fields in great detail from 1928 through 1942.

5.4   Examples

5.4.1   Rationals and algebraic numbers

The field of rational numbers **Q** has been introduced above. A related class of fields very important in number theory are algebraic number fields. We will first give an example, namely the field **Q**(ζ) consisting of numbers of the form

$$a + b\zeta$$

with $a, b \in \mathbf{Q}$, where ζ is a primitive third root of unity, i.e., a complex number satisfying $\zeta^3 = 1$, $\zeta \neq 1$. This field extension can be used to prove a special case of Fermat's last theorem, which asserts the non-existence of rational nonzero solutions to the equation

$$x^3 + y^3 = z^3.$$

In the language of field extensions detailed below, **Q**(ζ) is a field extension of degree 2. Algebraic number fields are by definition finite field extensions of **Q**, that is, fields containing **Q** having finite dimension as a **Q**-vector space.

5.4.2   Reals, complex numbers, and *p*-adic numbers

Take the real numbers **R**, under the usual operations of addition and multiplication. When the real numbers are given the usual ordering, they form a *complete ordered field*; it is this structure which provides the foundation for most formal treatments of calculus.

The complex numbers **C** consist of expressions

$$a + bi$$

where i is the imaginary unit, i.e., a (non-real) number satisfying $i^2 = -1$. Addition and multiplication of real numbers are defined in such a way that all field axioms hold for **C**. For example, the distributive law enforces

$$(a + bi)\cdot(c + di) = ac + bci + adi + bdi^2,$$ which equals $ac-bd + (bc + ad)i.$

The real numbers can be constructed by completing the rational numbers, i.e., filling the "gaps": for example $\sqrt{2}$ is such a gap. By a formally very similar procedure, another important class of fields, the field of p-adic numbers $\mathbf{Q}p$ is built. It is used in number theory and p-adic analysis.

Hyperreal numbers and superreal numbers extend the real numbers with the addition of infinitesimal and infinite numbers.

5.4.3   Constructible numbers

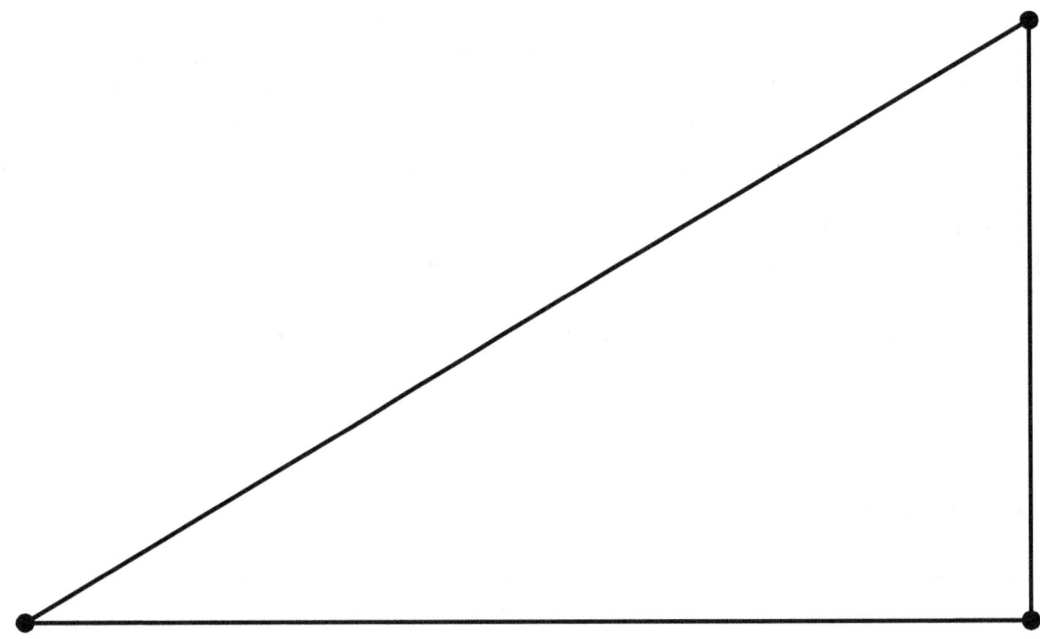

Given 0, 1, r_1 and r_2, the construction yields $r_1 \cdot r_2$

In antiquity, several geometric problems concerned the (in)feasibility of constructing certain numbers with compass and straightedge. For example, it was unknown to the Greeks that it is in general impossible to trisect a given angle. Using the field notion and field theory allows these problems to be settled. To do so, the field of constructible numbers is considered. It contains, on the plane, the points 0 and 1, and all complex numbers that can be constructed from these two by a finite number of construction steps using only compass and straightedge. This set, endowed with the usual addition and multiplication of complex numbers does form a field. For example, multiplying two (real) numbers r_1 and r_2 that have already been constructed can be done using construction at the right, based on the intercept theorem. This way, the obtained field F contains all rational numbers, but is bigger than $\mathbf{Q}$, because for any $f \in F$, the square root of f is also a constructible number.

A closely related concept is that of a Euclidean field, namely an ordered field whose positive elements are closed under square root. The real constructible numbers form the least Euclidean field, and the Euclidean fields are precisely the ordered extensions thereof.

5.4.4 Finite fields

Main article: Finite field

Finite fields (also called *Galois fields*) are fields with finitely many elements. The above introductory example F_4 is a field with four elements. F_2 consists of two elements, 0 and 1. This is the smallest field, because by definition a field has at least two distinct elements $1 \neq 0$. Interpreting the addition and multiplication in this latter field as XOR and AND operations, this field finds applications in computer science, especially in cryptography and coding theory.

In a finite field there is necessarily an integer n such that $1 + 1 + \cdots + 1$ (n repeated terms) equals 0. It can be shown that the smallest such n must be a prime number, called the *characteristic* of the field. If a (necessarily infinite) field has the property that $1 + 1 + \cdots + 1$ is never zero, for any number of summands, such as in $\mathbf{Q}$, for example, the characteristic is said to be zero.

A basic class of finite fields are the fields $\mathbf{F}p$ with p elements (p a prime number):

$$\mathbf{F}p = \mathbf{Z}/p\mathbf{Z} = \{0, 1, ..., p - 1\},$$

where the operations are defined by performing the operation in the set of integers $\mathbf{Z}$, dividing by p and taking the remainder; see modular arithmetic. A field K of characteristic p necessarily contains $\mathbf{F}p$,[9] and therefore may be viewed as a vector space over $\mathbf{F}p$, of finite dimension if K is finite. Thus a finite field K has prime power order, i.e., K has $q = p^n$ elements (where $n > 0$ is the number of elements in a basis of K over $\mathbf{F}p$). By developing more field theory, in particular the notion of the splitting field of a polynomial f over a field K, which is the smallest field containing K and all roots of f, one can show that two finite fields with the same number of elements are isomorphic, i.e., there is a one-to-one mapping of one field onto the other that preserves multiplication and addition. Thus we may speak of *the* finite field with q elements, usually denoted by $\mathbf{F}q$ or GF(q).

5.4.5 Archimedean fields

Main article: Archimedean field

An Archimedean field is an ordered field such that for each element there exists a finite expression $1 + 1 + \cdots + 1$ whose value is greater than that element, that is, there are no infinite elements. Equivalently, the field contains no infinitesimals; or, the field is isomorphic to a subfield of the reals. A necessary condition for an ordered field to be complete is that it be Archimedean, since in any non-Archimedean field there is neither a greatest infinitesimal nor a least positive rational, whence the sequence 1/2, 1/3, 1/4, ..., every element of which is greater than every infinitesimal, has no limit. (And since every proper subfield of the reals also contains such gaps, up to isomorphism the reals form the unique complete ordered field.)

5.4.6 Field of functions

Given a geometric object X, one can consider functions on such objects. Adding and multiplying them pointwise, i.e., $(f \cdot g)(x) = f(x) \cdot g(x)$ this leads to a field. However, for having multiplicative inverses, one has to consider partial functions, which, almost everywhere, are defined and have a non-zero value.

If X is an algebraic variety over a field F, then the rational functions $X \rightarrow F$ form a field, the function field of X. This field consists of the functions that are defined and are the quotient of two polynomial functions outside some subvariety. Likewise, if X is a Riemann surface, then the meromorphic functions $S \rightarrow \mathbf{C}$ form a field. Under certain circumstances, namely when S is compact, S can be reconstructed from this field.

5.4.7 Local and global fields

Another important distinction in the realm of fields, especially with regard to number theory, are local fields and global fields. Local fields are completions of global fields at a given place. For example, $\mathbf{Q}$ is a global field, and the attached local fields are $\mathbf{Q}p$ and $\mathbf{R}$ (Ostrowski's theorem). Algebraic number fields and function fields over $\mathbf{F}q$ are further global fields. Studying arithmetic questions in global fields may sometimes be done by looking at the corresponding questions locally—this technique is called local-global principle.

5.5 Some first theorems

- Every finite subgroup of the multiplicative group $F^\times$ is cyclic. This applies in particular to $\mathbf{F}q^\times$, it is cyclic of order $q - 1$. In the introductory example, a generator of $\mathbf{F}_4^\times$ is the element A.

- A integral domain is a field if and only if it has no ideals except $\{0\}$ and itself. Equivalently, an integral domain is a field if and only if its Krull dimension is 0.

- Isomorphism extension theorem

5.6 Constructing fields

5.6.1 Closure operations

Assuming the axiom of choice, for every field F, there exists a field F, called the algebraic closure of F, which contains F, is algebraic over F, which means that any element x of F satisfies a polynomial equation

$$f_n x^n + f_{n-1} x^{n-1} + \cdots + f_1 x + f_0 = 0, \text{ with coefficients } f_n, ..., f_0 \in F,$$

and is algebraically closed, i.e., any such polynomial does have at least one solution in F. The algebraic closure is unique up to isomorphism inducing the identity on F. However, in many circumstances in mathematics, it is not appropriate to treat F as being uniquely determined by F, since the isomorphism above is not itself unique. In these cases, one refers to such a F as *an* algebraic closure of F. A similar concept is the separable closure, containing all roots of separable polynomials, instead of all polynomials.

For example, if $F = \mathbf{Q}$, the algebraic closure $\mathbf{Q}$ is also called *field of algebraic numbers*. The field of algebraic numbers is an example of an algebraically closed field of characteristic zero; as such it satisfies the same first-order sentences as the field of complex numbers $\mathbf{C}$.

In general, all algebraic closures of a field are isomorphic. However, there is in general no preferable isomorphism between two closures. Likewise for separable closures.

5.6.2 Subfields and field extensions

A *subfield* is, informally, a small field contained in a bigger one. Formally, a subfield E of a field F is a subset containing 0 and 1, closed under the operations $+$, $-$, $\cdot$ and multiplicative inverses and with its own operations defined by restriction. For example, the real numbers contain several interesting subfields: the real algebraic numbers, the computable numbers and the rational numbers are examples.

The notion of field extension lies at the heart of field theory, and is crucial to many other algebraic domains. A field extension $F \,/\, E$ is simply a field F and a subfield $E \subset F$. Constructing such a field extension $F \,/\, E$ can be done by "adding new elements" or *adjoining elements* to the field E. For example, given a field E, the set $F = E(X)$ of rational functions, i.e., equivalence classes of expressions of the kind

$$\frac{p(X)}{q(X)},$$

where $p(X)$ and $q(X)$ are polynomials with coefficients in E, and q is not the zero polynomial, forms a field. This is the simplest example of a transcendental extension of E. It also is an example of a domain (the ring of polynomials E in this case) being embedded into its field of fractions $E(X)$.

The ring of formal power series $E[[X]]$ is also a domain, and again the (equivalence classes of) fractions of the form $p(X)/q(X)$ where p and q are elements of $E[[X]]$ form the field of fractions for $E[[X]]$. This field is actually the ring of Laurent series over the field E, denoted $E((X))$.

In the above two cases, the added symbol X and its powers did not interact with elements of E. It is possible however that the adjoined symbol may interact with E. This idea will be illustrated by adjoining an element to the field of real numbers **R**. As explained above, **C** is an extension of **R**. **C** can be obtained from **R** by adjoining the imaginary symbol i which satisfies $i^2 = -1$. The result is that **R**[i]=**C**. This is different from adjoining the symbol X to **R**, because in that case, the powers of X are all distinct objects, but here, $i^2=-1$ is actually an element of **R**.

Another way to view this last example is to note that i is a zero of the polynomial $p(X) = X^2 + 1$. The quotient ring $R[X]/(X^2+1)$ can be mapped onto **C** using the map $\overline{a+bX} \rightarrow a+ib$. Since the ideal (X^2+1) is generated by a polynomial irreducible over **R**, the ideal is maximal, hence the quotient ring is a field. This nonzero ring map from the quotient to **C** is necessarily an isomorphism of rings.

The above construction generalises to any irreducible polynomial in the polynomial ring $E[X]$, i.e., a polynomial $p(X)$ that cannot be written as a product of non-constant polynomials. The quotient ring $F = E[X] / (p(X))$, is again a field.

Alternatively, constructing such field extensions can also be done, if a bigger container is already given. Suppose given a field E, and a field G containing E as a subfield, for example G could be the algebraic closure of E. Let x be an element of G not in E. Then there is a smallest subfield of G containing E and x, denoted $F = E(x)$ and called *field extension F / E generated by x in G*.[10] Such extensions are also called *simple extensions*. Many extensions are of this type; see the primitive element theorem. For instance, **Q**(i) is the subfield of **C** consisting of all numbers of the form $a + bi$ where both a and b are rational numbers.

One distinguishes between extensions having various qualities. For example, an extension K of a field k is called *algebraic*, if every element of K is a root of some polynomial with coefficients in k. Otherwise, the extension is called *transcendental*. The aim of Galois theory is the study of *algebraic extensions* of a field.

5.6.3   Rings vs fields

Adding multiplicative inverses to an integral domain R yields the field of fractions of R. For example, the field of fractions of the integers **Z** is just **Q**. Also, the field $F(X)$ is the quotient field of the ring of polynomials $F[X]$.

Another method to obtain a field from a commutative ring R is taking the quotient R / m, where m is any maximal ideal of R. The above construction of $F = E[X] / (p(X))$, is an example, because the irreducibility of the polynomial $p(X)$ is equivalent to the maximality of the ideal generated by this polynomial. Another example are the finite fields $Fp = \mathbf{Z} / p\mathbf{Z}$.

5.6.4   Ultraproducts

If I is an index set, U is an ultrafilter on I, and Fi is a field for every i in I, the ultraproduct of the Fi with respect to U is a field.

For example, a non-principal ultraproduct of finite fields is a pseudo finite field; i.e., a PAC field having exactly one extension of any degree.

5.7   Galois theory

Main article: Galois theory

Galois theory aims to study the algebraic extensions of a field by studying the symmetry in the arithmetic operations of addition and multiplication. The fundamental theorem of Galois theory shows that there is a strong relation between the structure of the symmetry group and the set of algebraic extensions.

In the case where F / E is a finite (Galois) extension, Galois theory studies the algebraic extensions of E that are subfields of F. Such fields are called intermediate extensions. Specifically, the Galois group of F over E, denoted $\text{Gal}(F/E)$, is the group of field automorphisms of F that are trivial on E (i.e., the bijections $\sigma : F \to F$ that preserve addition and multiplication and that send elements of E to themselves), and the fundamental theorem of Galois theory states that there is a one-to-one correspondence between subgroups of $\text{Gal}(F/E)$ and the set of intermediate extensions of the extension F/E. The theorem, in fact, gives an explicit correspondence and further properties.

To study all (separable) algebraic extensions of E at once, one must consider the absolute Galois group of E, defined as the Galois group of the separable closure, E^{sep}, of E over E i.e., $\text{Gal}(E^{\text{sep}}/E)$. It is possible that the degree of this extension is infinite (as in the case of $E = \mathbf{Q}$). It is thus necessary to have a notion of Galois group for an infinite algebraic extension. The Galois group in this case is obtained as a "limit" (specifically an inverse limit) of the Galois groups of the finite Galois extensions of E. In this way, it acquires a topology.[note 2] The fundamental theorem of Galois theory can be generalized to the case of infinite Galois extensions by taking into consideration the topology of the Galois group, and in the case of E^{sep}/E it states that there this a one-to-one correspondence between *closed* subgroups of $\text{Gal}(E^{\text{sep}}/E)$ and the set of all separable algebraic extensions of E (technically, one only obtains those separable algebraic extensions of E that occur as subfields of the *chosen* separable closure E^{sep}, but since all separable closures of E are isomorphic, choosing a different separable closure would give the same Galois group and thus an "equivalent" set of algebraic extensions).

5.8   Generalizations

There are also proper classes with field structure, which are sometimes called **Fields**, with a capital F:

- The surreal numbers form a Field containing the reals, and would be a field except for the fact that they are a proper class, not a set.

- The nimbers form a Field. The set of nimbers with birthday smaller than 2^{2^n}, the nimbers with birthday smaller than any infinite cardinal are all examples of fields.

In a different direction, differential fields are fields equipped with a derivation. For example, the field $\mathbf{R}(X)$, together with the standard derivative of polynomials forms a differential field. These fields are central to differential Galois theory. Exponential fields, meanwhile, are fields equipped with an exponential function that provides a homomorphism between the additive and multiplicative groups within the field. The usual exponential function makes the real and complex numbers exponential fields, denoted $\mathbf{R}_{\text{exp}}$ and $\mathbf{C}_{\text{exp}}$ respectively.

Generalizing in a more categorical direction yields the field with one element and related objects.

5.8.1   Exponentiation

One does not in general study generalizations of fields with *three* binary operations. The familiar addition/subtraction, multiplication/division, exponentiation/root-extraction/logarithm operations from the natural numbers to the reals, each built up in terms of iteration of the last, mean that generalizing exponentiation as a binary operation is tempting, but has generally not proven fruitful; instead, an exponential field assumes a unary exponential function from the additive group to the multiplicative group, not a partially defined binary function. Note that the exponential operation of a^b is neither associative nor commutative, nor has a unique inverse (± 2 are both square roots of 4, for instance), unlike addition and multiplication, and further is not defined for many pairs—for example, $(-1)^{1/2} = \sqrt{-1}$ does not define a single number.

These all show that even for rational numbers exponentiation is not nearly as well-behaved as addition and multiplication, which is why one does not in general axiomatize exponentiation.

5.9 Applications

The concept of a field is of use, for example, in defining vectors and matrices, two structures in linear algebra whose components can be elements of an arbitrary field.

Finite fields are used in number theory, Galois theory, cryptography, coding theory and combinatorics; and again the notion of algebraic extension is an important tool.

5.10 See also

- Category of fields

- Glossary of field theory for more definitions in field theory.

- Heyting field

- Lefschetz principle

- Puiseux series

- Ring

- Vector space

- Vector spaces without fields

5.11 Notes

[1] That is, the axiom for addition only assumes a binary operation $+: F \times F \to F$, $a, b \mapsto a + b$. The axiom of inverse allows one to define a unary operation $-: F \to F$ $a \mapsto -a$ that sends an element to its negative (its additive inverse); this is not taken as given, but is implicitly defined in terms of addition as " $-a$ is the unique b such that $a + b = 0$ ", "implicitly" because it is defined in terms of solving an equation—and one then defines the binary operation of subtraction, also denoted by "−", as $-: F \times F \to F$, $a, b \mapsto a - b := a + (-b)$ in terms of addition and additive inverse. In the same way, one defines the binary operation of division $\div$ in terms of the assumed binary operation of multiplication and the implicitly defined operation of "reciprocal" (multiplicative inverse).

[2] As an inverse limit of finite discrete groups, it is equipped with the profinite topology, making it a profinite topological group

5.12 References

[1] Wallace, D A R (1998) *Groups, Rings, and Fields*, SUMS. Springer-Verlag: 151, Th. 2.

[2] Karl Georg Christian v. Staudt, *Beiträge zur Geometrie der Lage* (Contributions to the Geometry of Position), volume 2 (Nürnberg, (Germany): Bauer and Raspe, 1857). See: "Summen von Würfen" (sums of throws), pp. 166-171 ; "Produckte aus Würfen" (products of throws), pp. 171-176 ; "Potenzen von Würfen" (powers of throws), pp. 176-182.

[3] Peter Gustav Lejeune Dirichlet with R. Dedekind, *Vorlesungen über Zahlentheorie von P. G. Lejeune Dirichlet* (Lectures on Number Theory by P.G. Lejeune Dirichlet), 2nd ed., volume 1 (Braunschweig, Germany: Friedrich Vieweg und Sohn, 1871), p. 424. From page 424: *"Unter einem Körper wollen wir jedes System von unendlich vielen reellen oder complexen Zahlen verstehen, welches in sich so abgeschlossen und vollständig ist, dass die Addition, Subtraction, Multiplication und Division von je zwei dieser Zahlen immer wieder eine Zahl desselben Systems hervorbringt."* (By a "field" we will understand any system of

infinitely many real or complex numbers, which is so closed and complete that the addition, subtraction, multiplication, and division of any two of these numbers always again produces a number of the same system.)

[4] J J O'Connor and E F Robertson, *The development of Ring Theory*, September 2004.

[5] Moore, E. Hastings (1893), "A doubly-infinite system of simple groups", *Bulletin of the New York Mathematical Society* **3** (3): 73–78, doi:10.1090/S0002-9904-1893-00178-X, JFM 25.0198.01. From page 75: "Such a system of *s* marks [i.e., a finite field with *s* elements] we call a *field of order s*."

[6] *Earliest Known Uses of Some of the Words of Mathematics (F)*

[7] Fricke, Robert; Weber, Heinrich Martin (1924), *Lehrbuch der Algebra*, Vieweg, JFM 50.0042.03

[8] Steinitz, Ernst (1910), "Algebraische Theorie der Körper", *Journal für die reine und angewandte Mathematik* **137**: 167–309, doi:10.1515/crll.1910.137.167, ISSN 0075-4102, JFM 41.0445.03

[9] Jacobson (2009), p. 213

[10] Jacobson (2009), p. 213

5.13   Sources

- Artin, Michael (1991), *Algebra*, Prentice Hall, ISBN 978-0-13-004763-2, especially Chapter 13

- Allenby, R.B.J.T. (1991), *Rings, Fields and Groups*, Butterworth-Heinemann, ISBN 978-0-340-54440-2

- Blyth, T.S.; Robertson, E. F. (1985), *Groups, rings and fields: Algebra through practice*, Cambridge University Press. See especially Book 3 (ISBN 0-521-27288-2) and Book 6 (ISBN 0-521-27291-2).

- Jacobson, Nathan (2009), *Basic algebra* **1** (2nd ed.), Dover, ISBN 978-0-486-47189-1

- James Ax (1968), *The elementary theory of finite fields*, Ann. of Math. (2), **88**, 239–271

5.14   External links

- Hazewinkel, Michiel, ed. (2001), "Field", *Encyclopedia of Mathematics*, Springer, ISBN 978-1-55608-010-4

- Field Theory Q&A

- Fields at ProvenMath definition and basic properties.

- Field at PlanetMath.org.

Chapter 6

Graph (mathematics)

This article is about sets of vertices connected by edges. For graphs of mathematical functions, see Graph of a function. For other uses, see Graph (disambiguation).

In mathematics, and more specifically in graph theory, a **graph** is a representation of a set of objects where some pairs

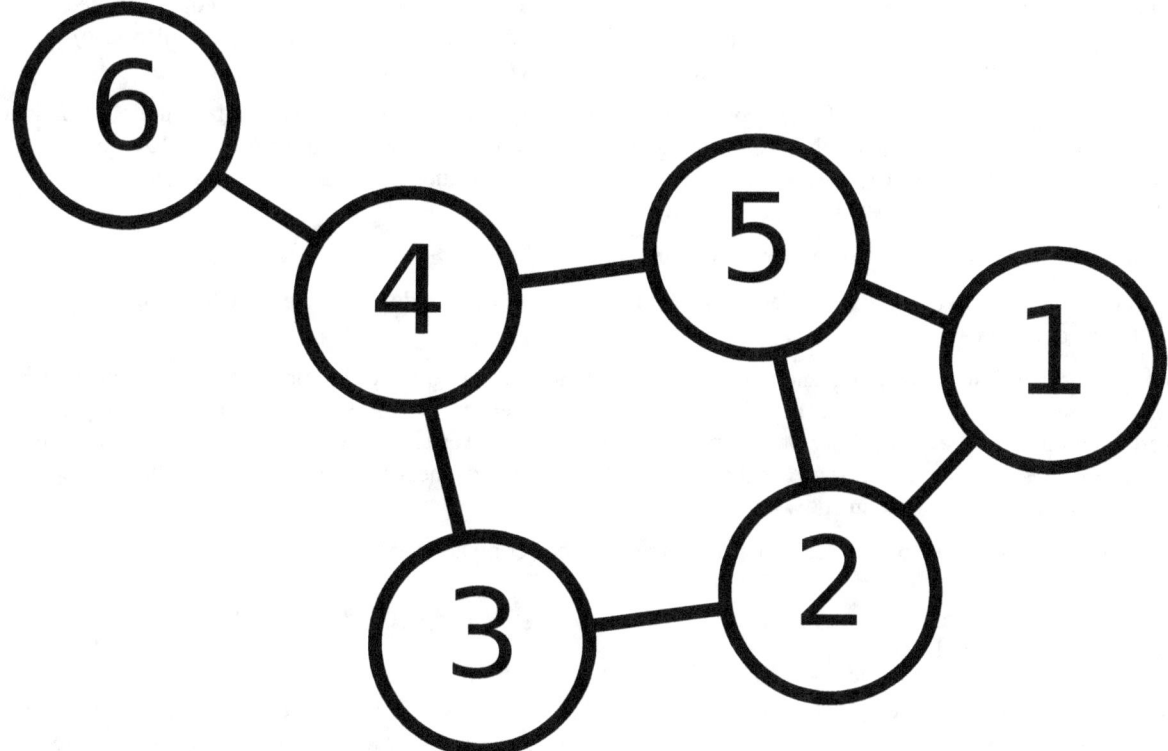

A drawing of a labeled graph on 6 vertices and 7 edges.

of objects are connected by links. The interconnected objects are represented by mathematical abstractions called *vertices* (also called *nodes* or *points*), and the links that connect some pairs of vertices are called *edges* (also called *arcs* or *lines*).[1] Typically, a graph is depicted in diagrammatic form as a set of dots for the vertices, joined by lines or curves for the edges. Graphs are one of the objects of study in discrete mathematics.

The edges may be directed or undirected. For example, if the vertices represent people at a party, and there is an edge between two people if they shake hands, then this is an undirected graph, because if person A shook hands with person

B, then person B also shook hands with person A. In contrast, if there is an edge from person A to person B when person A knows of person B, then this graph is directed, because knowledge of someone is not necessarily a symmetric relation (that is, one person knowing another person does not necessarily imply the reverse; for example, many fans may know of a celebrity, but the celebrity is unlikely to know of all their fans). The former type of graph is called an *undirected graph* and the edges are called *undirected edges* while the latter type of graph is called a *directed graph* and the edges are called *directed edges*.

Graphs are the basic subject studied by graph theory. The word "graph" was first used in this sense by J. J. Sylvester in 1878.[2][3]

6.1 Definitions

Definitions in graph theory vary. The following are some of the more basic ways of defining graphs and related mathematical structures.

6.1.1 Graph

In the most common sense of the term,[4] a *graph* is an ordered pair $G = (V, E)$ comprising a set V of *vertices, nodes* or *points* together with a set E of *edges, arcs* or *lines*, which are 2-element subsets of V (i.e., an edge is related with two vertices, and the relation is represented as an unordered pair of the vertices with respect to the particular edge). To avoid ambiguity, this type of graph may be described precisely as *undirected* and *simple*.

Other senses of *graph* stem from different conceptions of the edge set. In one more generalized notion,[5] E is a set together with a relation of *incidence* that associates with each edge two vertices. In another generalized notion, E is a multiset of unordered pairs of (not necessarily distinct) vertices. Many authors call this type of object a multigraph or pseudograph.

All of these variants and others are described more fully below.

The vertices belonging to an edge are called the *ends* or *end vertices* of the edge. A vertex may exist in a graph and not belong to an edge.

V and E are usually taken to be finite, and many of the well-known results are not true (or are rather different) for *infinite graphs* because many of the arguments fail in the infinite case. Moreover, V is often assumed to be non-empty, but E is allowed to be the empty set. The *order* of a graph is $|V|$, its number of vertices. The *size* of a graph is $|E|$, its number of edges. The *degree* or *valency* of a vertex is the number of edges that connect to it, where an edge that connects to the vertex at both ends (a loop) is counted twice.

For an edge $\{x, y\}$, graph theorists usually use the somewhat shorter notation xy.

6.1.2 Adjacency relation

The edges E of an undirected graph G induce a symmetric binary relation $\sim$ on V that is called the *adjacency relation* of G. Specifically, for each edge $\{x, y\}$, the vertices x and y are said to be *adjacent* to one another, which is denoted $x \sim y$.

6.2 Types of graphs

6.2.1 Distinction in terms of the main definition

As stated above, in different contexts it may be useful to refine the term *graph* with different degrees of generality. Whenever it is necessary to draw a strict distinction, the following terms are used. Most commonly, in modern texts in graph theory, unless stated otherwise, *graph* means "undirected simple finite graph" (see the definitions below).

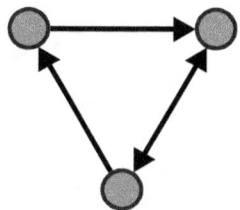

A directed graph.

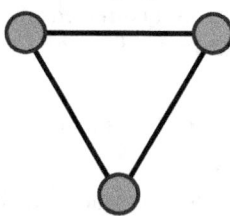

A simple undirected graph with three vertices and three edges. Each vertex has degree two, so this is also a regular graph.

Undirected graph

An *undirected graph* is a graph in which edges have no orientation. The edge (x, y) is identical to the edge (y, x), i.e., they are not ordered pairs, but sets $\{x, y\}$ (or 2-multisets) of vertices. The maximum number of edges in an undirected graph without a loop is $n(n-1)/2$.

Directed graph

Main article: Directed graph

A *directed graph* or *digraph* is a graph in which edges have orientations. It is written as an ordered pair $G = (V, A)$ (sometimes $G = (V, E)$) with

- V a set whose elements are called *vertices*, *nodes*, or *points*;

- A a set of ordered pairs of vertices, called *arrows*, *directed edges* (sometimes simply *edges* with the corresponding set named E instead of A), *directed arcs*, or *directed lines*.

An arrow (x, y) is considered to be directed *from x to y*; y is called the *head* and x is called the *tail* of the arrow; y is said to be a *direct successor* of x and x is said to be a *direct predecessor* of y. If a path leads from x to y, then y is said to be a *successor* of x and *reachable* from x, and x is said to be a *predecessor* of y. The arrow (y, x) is called the *inverted arrow* of (x, y).

A directed graph G is called *symmetric* if, for every arrow in G, the corresponding inverted arrow also belongs to G. A symmetric loopless directed graph $G = (V, A)$ is equivalent to a simple undirected graph $G' = (V, E)$, where the pairs of inverse arrows in A correspond one-to-one with the edges in E; thus the number of edges in G' is $|E| = |A|/2$, that is half the number of arrows in G.

An *oriented graph* is a directed graph in which at most one of (x, y) and (y, x) may be arrows of the graph.

Mixed graph

Main article: Mixed graph

A *mixed graph* is a graph in which some edges may be directed and some may be undirected. It is written as an ordered triple $G = (V, E, A)$ with V, E, and A defined as above. Directed and undirected graphs are special cases.

Multigraph

Main article: Multigraph

Multiple edges are two or more edges that connect the same two vertices. A *loop* is an edge (directed or undirected) that connects a vertex to itself; it may be permitted or not, according to the application. In this context, an edge with two different ends is called a *link*.

A *multigraph*, as opposed to a simple graph, is an undirected graph in which multiple edges (and sometimes loops) are allowed.

Where graphs are defined so as to *disallow* both multiple edges and loops, a multigraph is often defined to mean a graph which can have both multiple edges and loops,[6] although many use the term *pseudograph* for this meaning.[7] Where graphs are defined so as to *allow* both multiple edges and loops, a multigraph is often defined to mean a graph without loops.[8]

Simple graph

A *simple graph*, as opposed to a multigraph, is an undirected graph in which both multiple edges and loops are disallowed. In a simple graph the edges form a *set* (rather than a multiset) and each edge is a unordered pair of *distinct* vertices. In a simple graph with n vertices, the degree of every vertex is at most $n - 1$.

Quiver

Main article: Quiver (mathematics)

A *quiver* or *multidigraph* is a directed multigraph. A quiver may also have directed loops in it.

Weighted graph

A *weighted graph* is a graph in which a number (the weight) is assigned to each edge.[9] Such weights might represent for example costs, lengths or capacities, depending on the problem at hand. Some authors call such a graph a *network*.[10] Weighted correlation networks can be defined by soft-thresholding the pairwise correlations among variables (e.g. gene measurements). Such graphs arise in many contexts, for example in shortest path problems such as the traveling salesman problem.

Half-edges, loose edges

In certain situations it can be helpful to allow edges with only one end, called *half-edges*, or no ends, called *loose edges*; see the articles Signed graphs and Biased graphs.

6.2.2 Important classes of graph

Regular graph

Main article: Regular graph

A *regular graph* is a graph in which each vertex has the same number of neighbours, i.e., every vertex has the same degree. A regular graph with vertices of degree k is called a k-regular graph or regular graph of degree k.

Complete graph

Main article: Complete graph

 A *complete graph* is a graph in which each pair of vertices is joined by an edge. A complete graph contains all possible

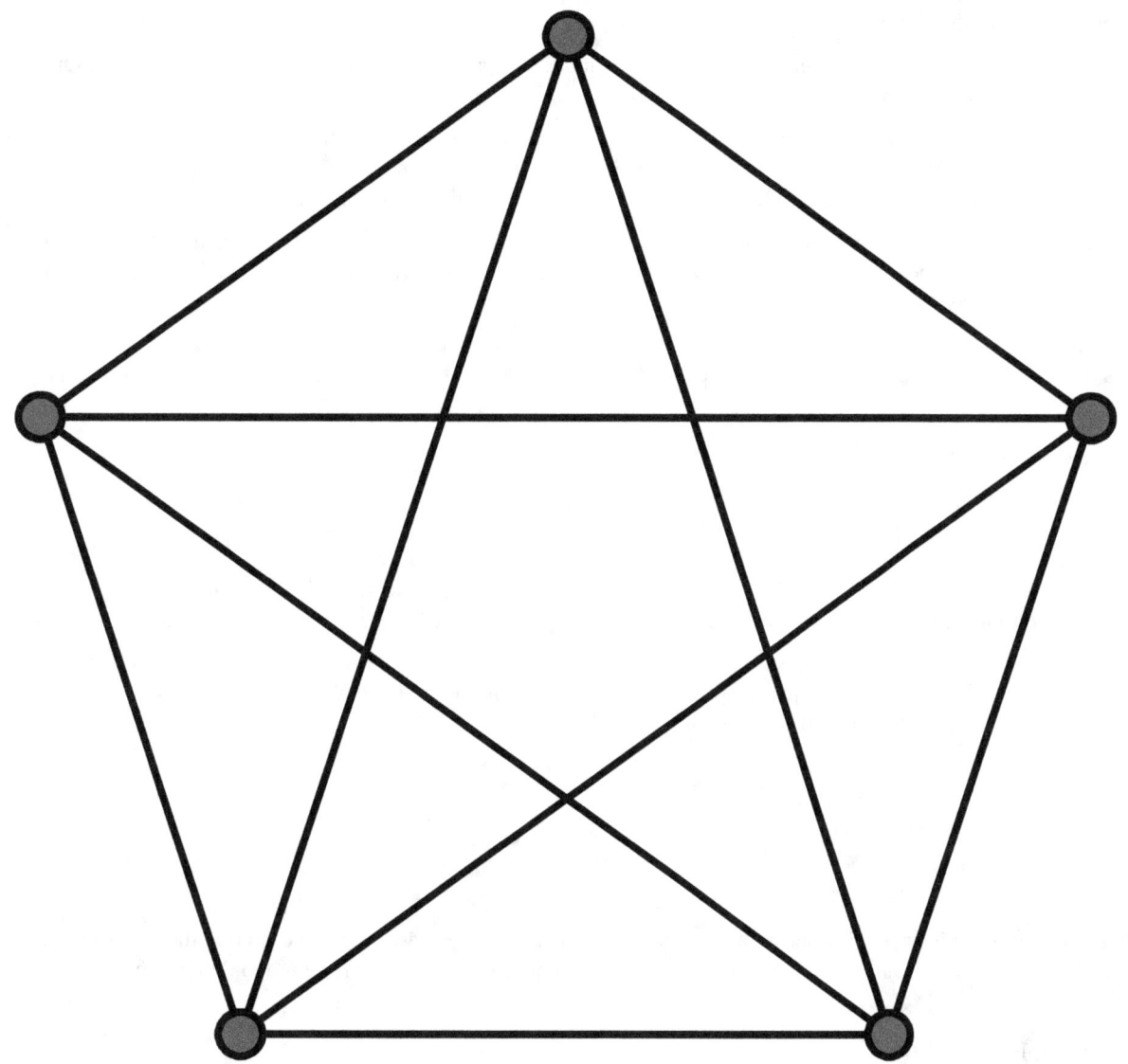

A complete graph with 5 vertices. Each vertex has an edge to every other vertex.

edges.

Finite graph

A *finite graph* is a graph in which the vertex set and the edge set are finite sets. Otherwise, it is called an *infinite graph*.

Most commonly in graph theory it is implied that the graphs discussed are finite. If the graphs are infinite, that is usually specifically stated.

Connected graph

Main article: Connectivity (graph theory)

In an undirected graph, an unordered pair of vertices $\{x, y\}$ is called *connected* if a path leads from x to y. Otherwise, the unordered pair is called *disconnected*.

A *connected graph* is an undirected graph in which every unordered pair of vertices in the graph is connected. Otherwise, it is called a *disconnected graph*.

In a directed graph, an ordered pair of vertices (x, y) is called *strongly connected* if a directed path leads from x to y. Otherwise, the ordered pair is called *weakly connected* if an undirected path leads from x to y after replacing all of its directed edges with undirected edges. Otherwise, the ordered pair is called *disconnected*.

A *strongly connected graph* is a directed graph in which every ordered pair of vertices in the graph is strongly connected. Otherwise, it is called a *weakly connected graph* if every ordered pair of vertices in the graph is weakly connected. Otherwise it is called a *disconnected graph*.

A *k-vertex-connected graph* or *k-edge-connected graph* is a graph in which no set of $k - 1$ vertices (respectively, edges) exists that, when removed, disconnects the graph. A k-vertex-connected graph is often called simply a *k-connected graph*.

Bipartite graph

Main article: Bipartite graph

A *bipartite graph* is a graph in which the vertex set can be partitioned into two sets, W and X, so that no two vertices in W share a common edge and no two vertices in X share a common edge. Alternatively, it is a graph with a chromatic number of 2.

In a complete bipartite graph, the vertex set is the union of two disjoint sets, W and X, so that every vertex in W is adjacent to every vertex in X but there are no edges within W or X.

Linear graph

Main article: Path graph

A *linear graph* or *path graph* of order n, the vertices can be listed in order, $v_0, v_1, \ldots, vn$, so that the edges are $\{vi_{-1}, vi\}$ for each $i = 1, 2, \ldots, n$. If a linear graph occurs as a subgraph of another graph, it is a path in that graph.

Planar graph

Main article: Planar graph

A *planar graph* is a graph whose vertices and edges can be drawn in a plane such that no two of the edges intersect.

Cycle graph

Main article: Cycle graph

A *cycle graph* of order $n \geq 3$, vertices can be named $v_1, \ldots, vn$ so that the edges are $\{vi_{-1}, vi\}$ for each $i = 2, \ldots, n$ in addition to $\{vn, v_1\}$. Cycle graphs can be characterized as connected 2-regular graphs. If a cycle graph occurs as a subgraph of another graph, it is a *cycle* or *circuit* in that graph.

Tree

Main article: Tree (graph theory)

A *tree* is a connected graph with no cycles.

A *forest* is a graph with no cycles, i.e. the disjoint union of one or more trees.

Advanced classes

More advanced kinds of graphs are:

- Petersen graph and its generalizations;

- perfect graphs;

- cographs;

- chordal graphs;

- other graphs with large automorphism groups: vertex-transitive, arc-transitive, and distance-transitive graphs;

- strongly regular graphs and their generalizations distance-regular graphs.

6.3   Properties of graphs

See also: Glossary of graph theory and Graph property

Two edges of a graph are called *adjacent* if they share a common vertex. Two arrows of a directed graph are called *consecutive* if the head of the first one is the tail of the second one. Similarly, two vertices are called *adjacent* if they share a common edge (*consecutive* if the first one is the tail and the second one is the head of an arrow), in which case the common edge is said to *join* the two vertices. An edge and a vertex on that edge are called *incident*.

The graph with only one vertex and no edges is called the *trivial graph*. A graph with only vertices and no edges is known as an *edgeless graph*. The graph with no vertices and no edges is sometimes called the *null graph* or *empty graph*, but the terminology is not consistent and not all mathematicians allow this object.

Normally, the vertices of a graph, by their nature as elements of a set, are distinguishable. This kind of graph may be called *vertex-labeled*. However, for many questions it is better to treat vertices as indistinguishable. (Of course, the vertices may be still distinguishable by the properties of the graph itself, e.g., by the numbers of incident edges.) The same remarks apply to edges, so graphs with labeled edges are called *edge-labeled*. Graphs with labels attached to edges or vertices are more generally designated as *labeled*. Consequently, graphs in which vertices are indistinguishable and edges are indistinguishable are called *unlabeled*. (Note that in the literature, the term *labeled* may apply to other kinds of labeling, besides that which serves only to distinguish different vertices or edges.)

The category of all graphs is the slice category Set $\downarrow D$ where D: Set $\rightarrow$ Set is the functor taking a set s to $s \times s$.

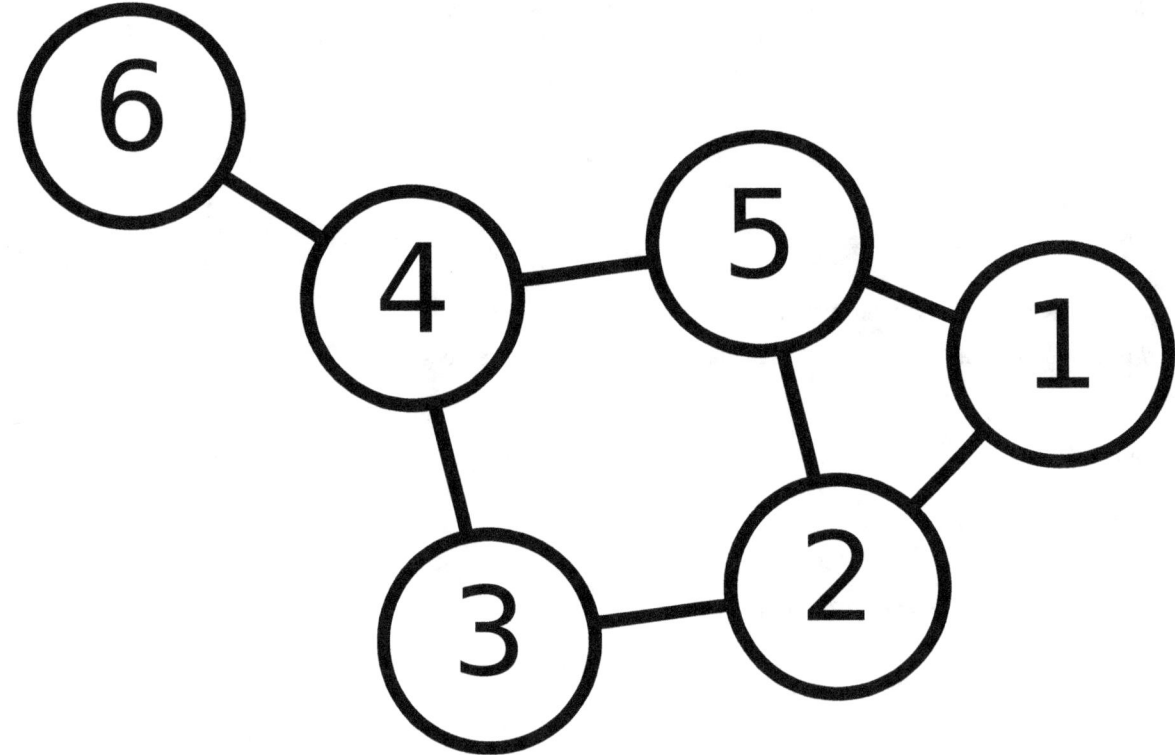

A graph with six nodes.

6.4 Examples

- The diagram at right is a graphic representation of the following graph:

    $V = \{1, 2, 3, 4, 5, 6\}$;
    $E = \{\{1, 2\}, \{1, 5\}, \{2, 3\}, \{2, 5\}, \{3, 4\}, \{4, 5\}, \{4, 6\}\}$.

- In category theory, a small category can be represented by a directed multigraph in which the objects of the category are represented as vertices and the morphisms as directed edges. Then, the functors between categories induce some, but not necessarily all, of the digraph morphisms of the graph.

- In computer science, directed graphs are used to represent knowledge (e.g., conceptual graph), finite state machines, and many other discrete structures.

- A binary relation R on a set X defines a directed graph. An element x of X is a direct predecessor of an element y of X if and only if xRy.

- A directed graph can model information networks such as Twitter, with one user following another.[11]

6.5 Graph operations

Main article: Graph operations

There are several operations that produce new graphs from initial ones, which might be classified into the following categories:

- *unary operations*, which create a new graph from an initial one, such as:
  - edge contraction,
  - line graph,
  - dual graph,
  - complement graph,
  - graph rewriting;
- *binary operations*, which create a new graph from two initial ones, such as:
  - disjoint union of graphs,
  - cartesian product of graphs,
  - tensor product of graphs,
  - strong product of graphs,
  - lexicographic product of graphs,
  - series-parallel graphs.

6.6 Generalizations

In a hypergraph, an edge can join more than two vertices.

An undirected graph can be seen as a simplicial complex consisting of 1-simplices (the edges) and 0-simplices (the vertices). As such, complexes are generalizations of graphs since they allow for higher-dimensional simplices.

Every graph gives rise to a matroid.

In model theory, a graph is just a structure. But in that case, there is no limitation on the number of edges: it can be any cardinal number, see continuous graph.

In computational biology, power graph analysis introduces power graphs as an alternative representation of undirected graphs.

In geographic information systems, geometric networks are closely modeled after graphs, and borrow many concepts from graph theory to perform spatial analysis on road networks or utility grids.

6.7 See also

- Conceptual graph
- Dual graph
- Glossary of graph theory
- Graph (data structure)
- Graph database
- Graph drawing
- Graph theory
- Hypergraph
- List of graph theory topics
- List of publications in graph theory
- Network theory

6.8 Notes

[1] Trudeau, Richard J. (1993). *Introduction to Graph Theory* (Corrected, enlarged republication. ed.). New York: Dover Pub. p. 19. ISBN 978-0-486-67870-2. Retrieved 8 August 2012. A graph is an object consisting of two sets called its *vertex set* and its *edge set*.

[2] See:

- J. J. Sylvester (February 7, 1878) "Chemistry and algebra," *Nature, 17* : 284. From page 284: "Every invariant and covariant thus becomes expressible by a *graph* precisely identical with a Kekuléan diagram or chemicograph."
- J. J. Sylvester (1878) "On an application of the new atomic theory to the graphical representation of the invariants and covariants of binary quantics, — with three appendices," *American Journal of Mathematics, Pure and Applied, 1* (1) : 64-90. The term "graph" first appears in this paper on page 65.

[3] Gross, Jonathan L.; Yellen, Jay (2004). *Handbook of graph theory.* CRC Press. p. 35. ISBN 978-1-58488-090-5.

[4] See, for instance, Iyanaga and Kawada, *69 J*, p. 234 or Biggs, p. 4.

[5] See, for instance, Graham et al., p. 5.

[6] For example, see. Bollobás, p. 7 and Diestel, p. 25.

[7] Gross (1998), p. 3, Gross (2003), p. 205, Harary, p.10, and Zwillinger, p. 220.

[8] For example, see Balakrishnan, p. 1, Gross (2003), p. 4, and Zwillinger, p. 220.

[9] Fletcher, Peter; Hoyle, Hughes; Patty, C. Wayne (1991). *Foundations of Discrete Mathematics* (International student ed.). Boston: PWS-KENT Pub. Co. p. 463. ISBN 0-53492-373-9. A *weighted graph* is a graph in which a number *w(e)*, called its *weight*, is assigned to each edge *e*.

[10] Strang, Gilbert (2005), *Linear Algebra and Its Applications* (4th ed.), Brooks Cole, ISBN 0-03-010567-6

[11] Pankaj Gupta, Ashish Goel, Jimmy Lin, Aneesh Sharma, Dong Wang, and Reza Bosagh Zadeh WTF: The who-to-follow system at Twitter, Proceedings of the 22nd international conference on World Wide Web

6.9 References

- Balakrishnan, V. K. (1997-02-01). *Graph Theory* (1st ed.). McGraw-Hill. ISBN 0-07-005489-4.

- Berge, Claude (1958). *Théorie des graphes et ses applications* (in French). Dunod, Paris: Collection Universitaire de Mathématiques, II. pp. viii+277. Translation: -. Dover, New York: Wiley. 2001 [1962].

- Biggs, Norman (1993). *Algebraic Graph Theory* (2nd ed.). Cambridge University Press. ISBN 0-521-45897-8.

- Bollobás, Béla (2002-08-12). *Modern Graph Theory* (1st ed.). Springer. ISBN 0-387-98488-7.

- Bang-Jensen, J.; Gutin, G. (2000). *Digraphs: Theory, Algorithms and Applications.* Springer.

- Diestel, Reinhard (2005). *Graph Theory* (3rd ed.). Berlin, New York: Springer-Verlag. ISBN 978-3-540-26183-4..

- Graham, R.L., Grötschel, M., and Lovász, L, ed. (1995). *Handbook of Combinatorics.* MIT Press. ISBN 0-262-07169-X.

- Gross, Jonathan L.; Yellen, Jay (1998-12-30). *Graph Theory and Its Applications.* CRC Press. ISBN 0-8493-3982-0.

- Gross, Jonathan L., & Yellen, Jay, ed. (2003-12-29). *Handbook of Graph Theory.* CRC. ISBN 1-58488-090-2.

- Harary, Frank (January 1995). *Graph Theory.* Addison Wesley Publishing Company. ISBN 0-201-41033-8.

- Iyanaga, Shôkichi; Kawada, Yukiyosi (1977). *Encyclopedic Dictionary of Mathematics*. MIT Press. ISBN 0-262-09016-3.

- Zwillinger, Daniel (2002-11-27). *CRC Standard Mathematical Tables and Formulae* (31st ed.). Chapman & Hall/CRC. ISBN 1-58488-291-3.

6.10 Further reading

- Trudeau, Richard J. (1993). *Introduction to Graph Theory* (Corrected, enlarged republication. ed.). New York: Dover Publications. ISBN 978-0-486-67870-2. Retrieved 8 August 2012.

6.11 External links

- Weisstein, Eric W., "Graph", *MathWorld*.

Chapter 7

Mathematical logic

For Quine's theory sometimes called "Mathematical Logic", see New Foundations.

Mathematical logic is a subfield of mathematics exploring the applications of formal logic to mathematics. It bears close connections to metamathematics, the foundations of mathematics, and theoretical computer science.[1] The unifying themes in mathematical logic include the study of the expressive power of formal systems and the deductive power of formal proof systems.

Mathematical logic is often divided into the fields of set theory, model theory, recursion theory, and proof theory. These areas share basic results on logic, particularly first-order logic, and definability. In computer science (particularly in the ACM Classification) mathematical logic encompasses additional topics not detailed in this article; see Logic in computer science for those.

Since its inception, mathematical logic has both contributed to, and has been motivated by, the study of foundations of mathematics. This study began in the late 19th century with the development of axiomatic frameworks for geometry, arithmetic, and analysis. In the early 20th century it was shaped by David Hilbert's program to prove the consistency of foundational theories. Results of Kurt Gödel, Gerhard Gentzen, and others provided partial resolution to the program, and clarified the issues involved in proving consistency. Work in set theory showed that almost all ordinary mathematics can be formalized in terms of sets, although there are some theorems that cannot be proven in common axiom systems for set theory. Contemporary work in the foundations of mathematics often focuses on establishing which parts of mathematics can be formalized in particular formal systems (as in reverse mathematics) rather than trying to find theories in which all of mathematics can be developed.

7.1 Subfields and scope

The *Handbook of Mathematical Logic* makes a rough division of contemporary mathematical logic into four areas:

1. set theory

2. model theory

3. recursion theory, and

4. proof theory and constructive mathematics (considered as parts of a single area).

Each area has a distinct focus, although many techniques and results are shared among multiple areas. The borderlines amongst these fields, and the lines separating mathematical logic and other fields of mathematics, are not always sharp. Gödel's incompleteness theorem marks not only a milestone in recursion theory and proof theory, but has also led to Löb's theorem in modal logic. The method of forcing is employed in set theory, model theory, and recursion theory, as well as in the study of intuitionistic mathematics.

The mathematical field of category theory uses many formal axiomatic methods, and includes the study of categorical logic, but category theory is not ordinarily considered a subfield of mathematical logic. Because of its applicability in diverse fields of mathematics, mathematicians including Saunders Mac Lane have proposed category theory as a foundational system for mathematics, independent of set theory. These foundations use toposes, which resemble generalized models of set theory that may employ classical or nonclassical logic.

7.2 History

Mathematical logic emerged in the mid-19th century as a subfield of mathematics independent of the traditional study of logic (Ferreirós 2001, p. 443). Before this emergence, logic was studied with rhetoric, through the syllogism, and with philosophy. The first half of the 20th century saw an explosion of fundamental results, accompanied by vigorous debate over the foundations of mathematics.

7.2.1 Early history

Further information: History of logic

Theories of logic were developed in many cultures in history, including China, India, Greece and the Islamic world. In 18th-century Europe, attempts to treat the operations of formal logic in a symbolic or algebraic way had been made by philosophical mathematicians including Leibniz and Lambert, but their labors remained isolated and little known.

7.2.2 19th century

In the middle of the nineteenth century, George Boole and then Augustus De Morgan presented systematic mathematical treatments of logic. Their work, building on work by algebraists such as George Peacock, extended the traditional Aristotelian doctrine of logic into a sufficient framework for the study of foundations of mathematics (Katz 1998, p. 686).

Charles Sanders Peirce built upon the work of Boole to develop a logical system for relations and quantifiers, which he published in several papers from 1870 to 1885. Gottlob Frege presented an independent development of logic with quantifiers in his *Begriffsschrift*, published in 1879, a work generally considered as marking a turning point in the history of logic. Frege's work remained obscure, however, until Bertrand Russell began to promote it near the turn of the century. The two-dimensional notation Frege developed was never widely adopted and is unused in contemporary texts.

From 1890 to 1905, Ernst Schröder published *Vorlesungen über die Algebra der Logik* in three volumes. This work summarized and extended the work of Boole, De Morgan, and Peirce, and was a comprehensive reference to symbolic logic as it was understood at the end of the 19th century.

Foundational theories

Concerns that mathematics had not been built on a proper foundation led to the development of axiomatic systems for fundamental areas of mathematics such as arithmetic, analysis, and geometry.

In logic, the term *arithmetic* refers to the theory of the natural numbers. Giuseppe Peano (1889) published a set of axioms for arithmetic that came to bear his name (Peano axioms), using a variation of the logical system of Boole and Schröder but adding quantifiers. Peano was unaware of Frege's work at the time. Around the same time Richard Dedekind showed that the natural numbers are uniquely characterized by their induction properties. Dedekind (1888) proposed a different characterization, which lacked the formal logical character of Peano's axioms. Dedekind's work, however, proved theorems inaccessible in Peano's system, including the uniqueness of the set of natural numbers (up to isomorphism) and the recursive definitions of addition and multiplication from the successor function and mathematical induction.

In the mid-19th century, flaws in Euclid's axioms for geometry became known (Katz 1998, p. 774). In addition to the independence of the parallel postulate, established by Nikolai Lobachevsky in 1826 (Lobachevsky 1840), mathematicians discovered that certain theorems taken for granted by Euclid were not in fact provable from his axioms. Among these is the theorem that a line contains at least two points, or that circles of the same radius whose centers are separated by that radius must intersect. Hilbert (1899) developed a complete set of axioms for geometry, building on previous work by Pasch (1882). The success in axiomatizing geometry motivated Hilbert to seek complete axiomatizations of other areas of mathematics, such as the natural numbers and the real line. This would prove to be a major area of research in the first half of the 20th century.

The 19th century saw great advances in the theory of real analysis, including theories of convergence of functions and Fourier series. Mathematicians such as Karl Weierstrass began to construct functions that stretched intuition, such as nowhere-differentiable continuous functions. Previous conceptions of a function as a rule for computation, or a smooth graph, were no longer adequate. Weierstrass began to advocate the arithmetization of analysis, which sought to axiomatize analysis using properties of the natural numbers. The modern (ε, δ)-definition of limit and continuous functions was already developed by Bolzano in 1817 (Felscher 2000), but remained relatively unknown. Cauchy in 1821 defined continuity in terms of infinitesimals (see Cours d'Analyse, page 34). In 1858, Dedekind proposed a definition of the real numbers in terms of Dedekind cuts of rational numbers (Dedekind 1872), a definition still employed in contemporary texts.

Georg Cantor developed the fundamental concepts of infinite set theory. His early results developed the theory of cardinality and proved that the reals and the natural numbers have different cardinalities (Cantor 1874). Over the next twenty years, Cantor developed a theory of transfinite numbers in a series of publications. In 1891, he published a new proof of the uncountability of the real numbers that introduced the diagonal argument, and used this method to prove Cantor's theorem that no set can have the same cardinality as its powerset. Cantor believed that every set could be well-ordered, but was unable to produce a proof for this result, leaving it as an open problem in 1895 (Katz 1998, p. 807).

7.2.3   20th century

In the early decades of the 20th century, the main areas of study were set theory and formal logic. The discovery of paradoxes in informal set theory caused some to wonder whether mathematics itself is inconsistent, and to look for proofs of consistency.

In 1900, Hilbert posed a famous list of 23 problems for the next century. The first two of these were to resolve the continuum hypothesis and prove the consistency of elementary arithmetic, respectively; the tenth was to produce a method that could decide whether a multivariate polynomial equation over the integers has a solution. Subsequent work to resolve these problems shaped the direction of mathematical logic, as did the effort to resolve Hilbert's *Entscheidungsproblem*, posed in 1928. This problem asked for a procedure that would decide, given a formalized mathematical statement, whether the statement is true or false.

Set theory and paradoxes

Ernst Zermelo (1904) gave a proof that every set could be well-ordered, a result Georg Cantor had been unable to obtain. To achieve the proof, Zermelo introduced the axiom of choice, which drew heated debate and research among mathematicians and the pioneers of set theory. The immediate criticism of the method led Zermelo to publish a second exposition of his result, directly addressing criticisms of his proof (Zermelo 1908a). This paper led to the general acceptance of the axiom of choice in the mathematics community.

Skepticism about the axiom of choice was reinforced by recently discovered paradoxes in naive set theory. Cesare Burali-Forti (1897) was the first to state a paradox: the Burali-Forti paradox shows that the collection of all ordinal numbers cannot form a set. Very soon thereafter, Bertrand Russell discovered Russell's paradox in 1901, and Jules Richard (1905) discovered Richard's paradox.

Zermelo (1908b) provided the first set of axioms for set theory. These axioms, together with the additional axiom of replacement proposed by Abraham Fraenkel, are now called Zermelo–Fraenkel set theory (ZF). Zermelo's axioms incorporated the principle of limitation of size to avoid Russell's paradox.

In 1910, the first volume of *Principia Mathematica* by Russell and Alfred North Whitehead was published. This seminal work developed the theory of functions and cardinality in a completely formal framework of type theory, which Russell and Whitehead developed in an effort to avoid the paradoxes. *Principia Mathematica* is considered one of the most influential works of the 20th century, although the framework of type theory did not prove popular as a foundational theory for mathematics (Ferreirós 2001, p. 445).

Fraenkel (1922) proved that the axiom of choice cannot be proved from the remaining axioms of Zermelo's set theory with urelements. Later work by Paul Cohen (1966) showed that the addition of urelements is not needed, and the axiom of choice is unprovable in ZF. Cohen's proof developed the method of forcing, which is now an important tool for establishing independence results in set theory.[2]

Symbolic logic

Leopold Löwenheim (1915) and Thoralf Skolem (1920) obtained the Löwenheim–Skolem theorem, which says that first-order logic cannot control the cardinalities of infinite structures. Skolem realized that this theorem would apply to first-order formalizations of set theory, and that it implies any such formalization has a countable model. This counterintuitive fact became known as Skolem's paradox.

In his doctoral thesis, Kurt Gödel (1929) proved the completeness theorem, which establishes a correspondence between syntax and semantics in first-order logic. Gödel used the completeness theorem to prove the compactness theorem, demonstrating the finitary nature of first-order logical consequence. These results helped establish first-order logic as the dominant logic used by mathematicians.

In 1931, Gödel published *On Formally Undecidable Propositions of Principia Mathematica and Related Systems*, which proved the incompleteness (in a different meaning of the word) of all sufficiently strong, effective first-order theories. This result, known as Gödel's incompleteness theorem, establishes severe limitations on axiomatic foundations for mathematics, striking a strong blow to Hilbert's program. It showed the impossibility of providing a consistency proof of arithmetic within any formal theory of arithmetic. Hilbert, however, did not acknowledge the importance of the incompleteness theorem for some time.

Gödel's theorem shows that a consistency proof of any sufficiently strong, effective axiom system cannot be obtained in the system itself, if the system is consistent, nor in any weaker system. This leaves open the possibility of consistency proofs that cannot be formalized within the system they consider. Gentzen (1936) proved the consistency of arithmetic using a finitistic system together with a principle of transfinite induction. Gentzen's result introduced the ideas of cut elimination and proof-theoretic ordinals, which became key tools in proof theory. Gödel (1958) gave a different consistency proof, which reduces the consistency of classical arithmetic to that of intutitionistic arithmetic in higher types.

Beginnings of the other branches

Alfred Tarski developed the basics of model theory.

Beginning in 1935, a group of prominent mathematicians collaborated under the pseudonym Nicolas Bourbaki to publish a series of encyclopedic mathematics texts. These texts, written in an austere and axiomatic style, emphasized rigorous presentation and set-theoretic foundations. Terminology coined by these texts, such as the words *bijection*, *injection*, and *surjection*, and the set-theoretic foundations the texts employed, were widely adopted throughout mathematics.

The study of computability came to be known as recursion theory, because early formalizations by Gödel and Kleene relied on recursive definitions of functions.[3] When these definitions were shown equivalent to Turing's formalization involving Turing machines, it became clear that a new concept – the computable function – had been discovered, and that this definition was robust enough to admit numerous independent characterizations. In his work on the incompleteness theorems in 1931, Gödel lacked a rigorous concept of an effective formal system; he immediately realized that the new definitions of computability could be used for this purpose, allowing him to state the incompleteness theorems in generality that could only be implied in the original paper.

Numerous results in recursion theory were obtained in the 1940s by Stephen Cole Kleene and Emil Leon Post. Kleene (1943) introduced the concepts of relative computability, foreshadowed by Turing (1939), and the arithmetical hierarchy. Kleene later generalized recursion theory to higher-order functionals. Kleene and Kreisel studied formal versions of intuitionistic mathematics, particularly in the context of proof theory.

7.3 Formal logical systems

At its core, mathematical logic deals with mathematical concepts expressed using formal logical systems. These systems, though they differ in many details, share the common property of considering only expressions in a fixed formal language. The systems of propositional logic and first-order logic are the most widely studied today, because of their applicability to foundations of mathematics and because of their desirable proof-theoretic properties.[4] Stronger classical logics such as second-order logic or infinitary logic are also studied, along with nonclassical logics such as intuitionistic logic.

7.3.1 First-order logic

Main article: First-order logic

First-order logic is a kind of formal system of logic designed to avoid Russell's Paradox. The well-formed formulas of first-order logic cannot be self-referential. That is, a sentence such as "This statement is true," cannot be expressed in a well-formed formula in first-order logic. The syntax of first-order logic involves only finite expressions, called well-formed formulas, while its semantics are characterized by the limitation of all quantifiers to a fixed domain of discourse.

Early results from formal logic established limitations of first-order logic. The Löwenheim–Skolem theorem (1919) showed that if a set of sentences in a countable first-order language has an infinite model then it has at least one model of each infinite cardinality. This shows that it is impossible for a set of first-order axioms to characterize (up to isomorphism) the natural numbers, the real numbers, or any other infinite structure. As the goal of early foundational studies was to produce axiomatic theories for all parts of mathematics, this limitation was particularly stark.

Gödel's completeness theorem (Gödel 1929) established the equivalence between semantic and syntactic definitions of logical consequence in first-order logic. It shows that if a particular sentence is true in every model that satisfies a particular set of axioms, then there must be a finite deduction of the sentence from the axioms. The compactness theorem first appeared as a lemma in Gödel's proof of the completeness theorem, and it took many years before logicians grasped its significance and began to apply it routinely. It says that a set of sentences has a model if and only if every finite subset has a model, or in other words that an inconsistent set of formulas must have a finite inconsistent subset. The completeness and compactness theorems allow for sophisticated analysis of logical consequence in first-order logic and the development of model theory, and they are a key reason for the prominence of first-order logic in mathematics.

Gödel's incompleteness theorems (Gödel 1931) establish additional limits on first-order axiomatizations. A logical system is effectively given if, given any formula in the language of the system, it is possible to decide whether or not the formula is an axiom. A logical system is sufficiently strong if it is possible to express the Peano axioms in the system. The **first incompleteness theorem** states that for any consistent, effectively given, sufficiently strong logical system, there exists a statement, called a Gödel sentence, which is true (in the sense that it holds for the natural numbers) but not provable within that logical system (and which indeed may fail in some non-standard models of arithmetic consistent with the logical system). When applied to first-order logic, the first incompleteness theorem implies that any consistent, effectively given, sufficiently strong first-order theory has models that are not elementarily equivalent, a stronger limitation than the one established by the Löwenheim–Skolem theorem. The **second incompleteness theorem** states that no consistent, sufficiently strong, effectively given axiom system for arithmetic can prove its own consistency. This has been interpreted to show that Hilbert's program cannot be completed.

7.3.2   Other classical logics

Many logics besides first-order logic are studied. These include infinitary logics, which allow for formulas to provide an infinite amount of information, and higher-order logics, which include a portion of set theory directly in their semantics.

The most well studied infinitary logic is $L_{\omega_1,\omega}$. In this logic, quantifiers may only be nested to finite depths, as in first-order logic, but formulas may have finite or countably infinite conjunctions and disjunctions within them. Thus, for example, it is possible to say that an object is a whole number using a formula of $L_{\omega_1,\omega}$ such as

$$(x = 0) \vee (x = 1) \vee (x = 2) \vee \cdots .$$

Higher-order logics allow for quantification not only of elements of the domain of discourse, but subsets of the domain of discourse, sets of such subsets, and other objects of higher type. The semantics are defined so that, rather than having a separate domain for each higher-type quantifier to range over, the quantifiers instead range over all objects of the appropriate type. The logics studied before the development of first-order logic, for example Frege's logic, had similar set-theoretic aspects. Although higher-order logics are more expressive, allowing complete axiomatizations of structures such as the natural numbers, they do not satisfy analogues of the completeness and compactness theorems from first-order logic, and are thus less amenable to proof-theoretic analysis.

Another type of logics are fixed-point logics that allow inductive definitions, like one writes for primitive recursive functions.

One can formally define an extension of first-order logic — a notion which encompasses all logics in this section because they behave like first-order logic in certain fundamental ways, but does not encompass all logics in general, e.g. it does not encompass intuitionistic, modal or fuzzy logic. Lindström's theorem implies that the only extension of first-order logic satisfying both the compactness theorem and the Downward Löwenheim–Skolem theorem is first-order logic.

7.3.3   Nonclassical and modal logic

Modal logics include additional modal operators, such as an operator which states that a particular formula is not only true, but necessarily true. Although modal logic is not often used to axiomatize mathematics, it has been used to study the properties of first-order provability (Solovay 1976) and set-theoretic forcing (Hamkins and Löwe 2007).

Intuitionistic logic was developed by Heyting to study Brouwer's program of intuitionism, in which Brouwer himself avoided formalization. Intuitionistic logic specifically does not include the law of the excluded middle, which states that each sentence is either true or its negation is true. Kleene's work with the proof theory of intuitionistic logic showed that constructive information can be recovered from intuitionistic proofs. For example, any provably total function in intuitionistic arithmetic is computable; this is not true in classical theories of arithmetic such as Peano arithmetic.

7.3.4   Algebraic logic

Algebraic logic uses the methods of abstract algebra to study the semantics of formal logics. A fundamental example is the use of Boolean algebras to represent truth values in classical propositional logic, and the use of Heyting algebras to represent truth values in intuitionistic propositional logic. Stronger logics, such as first-order logic and higher-order logic, are studied using more complicated algebraic structures such as cylindric algebras.

7.4   Set theory

Main article: Set theory

Set theory is the study of sets, which are abstract collections of objects. Many of the basic notions, such as ordinal and cardinal numbers, were developed informally by Cantor before formal axiomatizations of set theory were developed. The first such axiomatization, due to Zermelo (1908b), was extended slightly to become Zermelo–Fraenkel set theory (ZF), which is now the most widely used foundational theory for mathematics.

Other formalizations of set theory have been proposed, including von Neumann–Bernays–Gödel set theory (NBG), Morse–Kelley set theory (MK), and New Foundations (NF). Of these, ZF, NBG, and MK are similar in describing a cumulative hierarchy of sets. New Foundations takes a different approach; it allows objects such as the set of all sets at the cost of restrictions on its set-existence axioms. The system of Kripke–Platek set theory is closely related to generalized recursion theory.

Two famous statements in set theory are the axiom of choice and the continuum hypothesis. The axiom of choice, first stated by Zermelo (1904), was proved independent of ZF by Fraenkel (1922), but has come to be widely accepted by mathematicians. It states that given a collection of nonempty sets there is a single set C that contains exactly one element from each set in the collection. The set C is said to "choose" one element from each set in the collection. While the ability to make such a choice is considered obvious by some, since each set in the collection is nonempty, the lack of a general, concrete rule by which the choice can be made renders the axiom nonconstructive. Stefan Banach and Alfred Tarski (1924) showed that the axiom of choice can be used to decompose a solid ball into a finite number of pieces which can then be rearranged, with no scaling, to make two solid balls of the original size. This theorem, known as the Banach–Tarski paradox, is one of many counterintuitive results of the axiom of choice.

The continuum hypothesis, first proposed as a conjecture by Cantor, was listed by David Hilbert as one of his 23 problems in 1900. Gödel showed that the continuum hypothesis cannot be disproven from the axioms of Zermelo–Fraenkel set theory (with or without the axiom of choice), by developing the constructible universe of set theory in which the continuum hypothesis must hold. In 1963, Paul Cohen showed that the continuum hypothesis cannot be proven from the axioms of Zermelo–Fraenkel set theory (Cohen 1966). This independence result did not completely settle Hilbert's question, however, as it is possible that new axioms for set theory could resolve the hypothesis. Recent work along these lines has been conducted by W. Hugh Woodin, although its importance is not yet clear (Woodin 2001).

Contemporary research in set theory includes the study of large cardinals and determinacy. Large cardinals are cardinal numbers with particular properties so strong that the existence of such cardinals cannot be proved in ZFC. The existence of the smallest large cardinal typically studied, an inaccessible cardinal, already implies the consistency of ZFC. Despite the fact that large cardinals have extremely high cardinality, their existence has many ramifications for the structure of the real line. *Determinacy* refers to the possible existence of winning strategies for certain two-player games (the games are said to be *determined*). The existence of these strategies implies structural properties of the real line and other Polish spaces.

7.5 Model theory

Main article: Model theory

Model theory studies the models of various formal theories. Here a theory is a set of formulas in a particular formal logic and signature, while a model is a structure that gives a concrete interpretation of the theory. Model theory is closely related to universal algebra and algebraic geometry, although the methods of model theory focus more on logical considerations than those fields.

The set of all models of a particular theory is called an elementary class; classical model theory seeks to determine the properties of models in a particular elementary class, or determine whether certain classes of structures form elementary classes.

The method of quantifier elimination can be used to show that definable sets in particular theories cannot be too complicated. Tarski (1948) established quantifier elimination for real-closed fields, a result which also shows the theory of the field of real numbers is decidable. (He also noted that his methods were equally applicable to algebraically closed fields of arbitrary characteristic.) A modern subfield developing from this is concerned with o-minimal structures.

Morley's categoricity theorem, proved by Michael D. Morley (1965), states that if a first-order theory in a countable language is categorical in some uncountable cardinality, i.e. all models of this cardinality are isomorphic, then it is categorical in all uncountable cardinalities.

A trivial consequence of the continuum hypothesis is that a complete theory with less than continuum many nonisomorphic countable models can have only countably many. Vaught's conjecture, named after Robert Lawson Vaught, says that this is true even independently of the continuum hypothesis. Many special cases of this conjecture have been established.

7.6   Recursion theory

Main article: Recursion theory

Recursion theory, also called **computability theory**, studies the properties of computable functions and the Turing degrees, which divide the uncomputable functions into sets that have the same level of uncomputability. Recursion theory also includes the study of generalized computability and definability. Recursion theory grew from the work of Alonzo Church and Alan Turing in the 1930s, which was greatly extended by Kleene and Post in the 1940s.

Classical recursion theory focuses on the computability of functions from the natural numbers to the natural numbers. The fundamental results establish a robust, canonical class of computable functions with numerous independent, equivalent characterizations using Turing machines, λ calculus, and other systems. More advanced results concern the structure of the Turing degrees and the lattice of recursively enumerable sets.

Generalized recursion theory extends the ideas of recursion theory to computations that are no longer necessarily finite. It includes the study of computability in higher types as well as areas such as hyperarithmetical theory and α-recursion theory.

Contemporary research in recursion theory includes the study of applications such as algorithmic randomness, computable model theory, and reverse mathematics, as well as new results in pure recursion theory.

7.6.1   Algorithmically unsolvable problems

An important subfield of recursion theory studies algorithmic unsolvability; a decision problem or function problem is **algorithmically unsolvable** if there is no possible computable algorithm that returns the correct answer for all legal inputs to the problem. The first results about unsolvability, obtained independently by Church and Turing in 1936, showed that the Entscheidungsproblem is algorithmically unsolvable. Turing proved this by establishing the unsolvability of the halting problem, a result with far-ranging implications in both recursion theory and computer science.

There are many known examples of undecidable problems from ordinary mathematics. The word problem for groups was proved algorithmically unsolvable by Pyotr Novikov in 1955 and independently by W. Boone in 1959. The busy beaver problem, developed by Tibor Radó in 1962, is another well-known example.

Hilbert's tenth problem asked for an algorithm to determine whether a multivariate polynomial equation with integer coefficients has a solution in the integers. Partial progress was made by Julia Robinson, Martin Davis and Hilary Putnam. The algorithmic unsolvability of the problem was proved by Yuri Matiyasevich in 1970 (Davis 1973).

7.7   Proof theory and constructive mathematics

Main article: Proof theory

Proof theory is the study of formal proofs in various logical deduction systems. These proofs are represented as formal mathematical objects, facilitating their analysis by mathematical techniques. Several deduction systems are commonly considered, including Hilbert-style deduction systems, systems of natural deduction, and the sequent calculus developed by Gentzen.

The study of **constructive mathematics**, in the context of mathematical logic, includes the study of systems in non-classical logic such as intuitionistic logic, as well as the study of predicative systems. An early proponent of predicativism was Hermann Weyl, who showed it is possible to develop a large part of real analysis using only predicative methods (Weyl 1918).

Because proofs are entirely finitary, whereas truth in a structure is not, it is common for work in constructive mathematics to emphasize provability. The relationship between provability in classical (or nonconstructive) systems and provability in intuitionistic (or constructive, respectively) systems is of particular interest. Results such as the Gödel–Gentzen negative translation show that it is possible to embed (or *translate*) classical logic into intuitionistic logic, allowing some properties about intuitionistic proofs to be transferred back to classical proofs.

Recent developments in proof theory include the study of proof mining by Ulrich Kohlenbach and the study of proof-theoretic ordinals by Michael Rathjen.

7.8   Connections with computer science

Main article: Logic in computer science

The study of computability theory in computer science is closely related to the study of computability in mathematical logic. There is a difference of emphasis, however. Computer scientists often focus on concrete programming languages and feasible computability, while researchers in mathematical logic often focus on computability as a theoretical concept and on noncomputability.

The theory of semantics of programming languages is related to model theory, as is program verification (in particular, model checking). The Curry–Howard isomorphism between proofs and programs relates to proof theory, especially intuitionistic logic. Formal calculi such as the lambda calculus and combinatory logic are now studied as idealized programming languages.

Computer science also contributes to mathematics by developing techniques for the automatic checking or even finding of proofs, such as automated theorem proving and logic programming.

Descriptive complexity theory relates logics to computational complexity. The first significant result in this area, Fagin's theorem (1974) established that NP is precisely the set of languages expressible by sentences of existential second-order logic.

7.9   Foundations of mathematics

Main article: Foundations of mathematics

In the 19th century, mathematicians became aware of logical gaps and inconsistencies in their field. It was shown that Euclid's axioms for geometry, which had been taught for centuries as an example of the axiomatic method, were incomplete. The use of infinitesimals, and the very definition of function, came into question in analysis, as pathological examples such as Weierstrass' nowhere-differentiable continuous function were discovered.

Cantor's study of arbitrary infinite sets also drew criticism. Leopold Kronecker famously stated "God made the integers; all else is the work of man," endorsing a return to the study of finite, concrete objects in mathematics. Although Kronecker's argument was carried forward by constructivists in the 20th century, the mathematical community as a whole rejected them. David Hilbert argued in favor of the study of the infinite, saying "No one shall expel us from the Paradise that Cantor has created."

Mathematicians began to search for axiom systems that could be used to formalize large parts of mathematics. In addition to removing ambiguity from previously naive terms such as function, it was hoped that this axiomatization would allow for consistency proofs. In the 19th century, the main method of proving the consistency of a set of axioms was to provide a model for it. Thus, for example, non-Euclidean geometry can be proved consistent by defining *point* to mean a point on a fixed sphere and *line* to mean a great circle on the sphere. The resulting structure, a model of elliptic geometry, satisfies the axioms of plane geometry except the parallel postulate.

With the development of formal logic, Hilbert asked whether it would be possible to prove that an axiom system is consistent by analyzing the structure of possible proofs in the system, and showing through this analysis that it is impossible to prove a contradiction. This idea led to the study of proof theory. Moreover, Hilbert proposed that the analysis should be entirely concrete, using the term *finitary* to refer to the methods he would allow but not precisely defining them. This project, known as Hilbert's program, was seriously affected by Gödel's incompleteness theorems, which show that the consistency of formal theories of arithmetic cannot be established using methods formalizable in those theories. Gentzen showed that it is possible to produce a proof of the consistency of arithmetic in a finitary system augmented with axioms of transfinite induction, and the techniques he developed to do so were seminal in proof theory.

A second thread in the history of foundations of mathematics involves nonclassical logics and constructive mathematics. The study of constructive mathematics includes many different programs with various definitions of *constructive*. At the most accommodating end, proofs in ZF set theory that do not use the axiom of choice are called constructive by many mathematicians. More limited versions of constructivism limit themselves to natural numbers, number-theoretic functions, and sets of natural numbers (which can be used to represent real numbers, facilitating the study of mathematical analysis). A common idea is that a concrete means of computing the values of the function must be known before the function itself can be said to exist.

In the early 20th century, Luitzen Egbertus Jan Brouwer founded intuitionism as a philosophy of mathematics. This philosophy, poorly understood at first, stated that in order for a mathematical statement to be true to a mathematician, that person must be able to *intuit* the statement, to not only believe its truth but understand the reason for its truth. A consequence of this definition of truth was the rejection of the law of the excluded middle, for there are statements that, according to Brouwer, could not be claimed to be true while their negations also could not be claimed true. Brouwer's philosophy was influential, and the cause of bitter disputes among prominent mathematicians. Later, Kleene and Kreisel would study formalized versions of intuitionistic logic (Brouwer rejected formalization, and presented his work in unformalized natural language). With the advent of the BHK interpretation and Kripke models, intuitionism became easier to reconcile with classical mathematics.

7.10   See also

- Knowledge representation and reasoning

- List of computability and complexity topics

- List of first-order theories

- List of logic symbols

- List of mathematical logic topics

- List of set theory topics

- Metalogic

7.11   Notes

[1] Undergraduate texts include Boolos, Burgess, and Jeffrey (2002), Enderton (2001), and Mendelson (1997). A classic graduate text by Shoenfield (2001) first appeared in 1967.

[2] See also Cohen 2008.

[3] A detailed study of this terminology is given by Soare (1996).

[4] Ferreirós (2001) surveys the rise of first-order logic over other formal logics in the early 20th century.

7.12 References

7.12.1 Undergraduate texts

- Walicki, Michał (2011), *Introduction to Mathematical Logic*, Singapore: World Scientific Publishing, ISBN 978-981-4343-87-9.

- Boolos, George; Burgess, John; Jeffrey, Richard (2002), *Computability and Logic* (4th ed.), Cambridge: Cambridge University Press, ISBN 978-0-521-00758-0.

- Crossley, J.N.; Ash, C.J.; Brickhill, C.J.; Stillwell, J.C.; Williams, N.H. (1972), *What is mathematical logic?*, London-Oxford-New York: Oxford University Press, ISBN 0-19-888087-1, Zbl 0251.02001.

- Enderton, Herbert (2001), *A mathematical introduction to logic* (2nd ed.), Boston, MA: Academic Press, ISBN 978-0-12-238452-3.

- Hamilton, A.G. (1988), *Logic for Mathematicians* (2nd ed.), Cambridge: Cambridge University Press, ISBN 978-0-521-36865-0.

- Ebbinghaus, H.-D.; Flum, J.; Thomas, W. (1994), *Mathematical Logic* (2nd ed.), New York: Springer, ISBN 0-387-94258-0.

- Katz, Robert (1964), *Axiomatic Analysis*, Boston, MA: D. C. Heath and Company.

- Mendelson, Elliott (1997), *Introduction to Mathematical Logic* (4th ed.), London: Chapman & Hall, ISBN 978-0-412-80830-2.

- Rautenberg, Wolfgang (2010), *A Concise Introduction to Mathematical Logic* (3rd ed.), New York: Springer Science+Business Media, doi:10.1007/978-1-4419-1221-3, ISBN 978-1-4419-1220-6.

- Schwichtenberg, Helmut (2003–2004), *Mathematical Logic* (PDF), Munich, Germany: Mathematisches Institut der Universität München.

- Shawn Hedman, *A first course in logic: an introduction to model theory, proof theory, computability, and complexity*, Oxford University Press, 2004, ISBN 0-19-852981-3. Covers logics in close relation with computability theory and complexity theory

7.12.2 Graduate texts

- Andrews, Peter B. (2002), *An Introduction to Mathematical Logic and Type Theory: To Truth Through Proof* (2nd ed.), Boston: Kluwer Academic Publishers, ISBN 978-1-4020-0763-7.

- Barwise, Jon, ed. (1989), *Handbook of Mathematical Logic*, Studies in Logic and the Foundations of Mathematics, North Holland, ISBN 978-0-444-86388-1.

- Hodges, Wilfrid (1997), *A shorter model theory*, Cambridge: Cambridge University Press, ISBN 978-0-521-58713-6.

- Jech, Thomas (2003), *Set Theory: Millennium Edition*, Springer Monographs in Mathematics, Berlin, New York: Springer-Verlag, ISBN 978-3-540-44085-7.

- Shoenfield, Joseph R. (2001) [1967], *Mathematical Logic* (2nd ed.), A K Peters, ISBN 978-1-56881-135-2.

- Troelstra, Anne Sjerp; Schwichtenberg, Helmut (2000), *Basic Proof Theory*, Cambridge Tracts in Theoretical Computer Science (2nd ed.), Cambridge: Cambridge University Press, ISBN 978-0-521-77911-1.

7.12.3 Research papers, monographs, texts, and surveys

- Cohen, P. J. (1966), *Set Theory and the Continuum Hypothesis*, Menlo Park, CA: W. A. Benjamin.

- Cohen, Paul Joseph (2008) [1966]. *Set theory and the continuum hypothesis*. Mineola, New York: Dover Publications. ISBN 978-0-486-46921-8..

- J.D. Sneed, *The Logical Structure of Mathematical Physics*. Reidel, Dordrecht, 1971 (revised edition 1979).

- Davis, Martin (1973), "Hilbert's tenth problem is unsolvable", *The American Mathematical Monthly* (The American Mathematical Monthly, Vol. 80, No. 3) **80** (3): 233–269, doi:10.2307/2318447, JSTOR 2318447, reprinted as an appendix in Martin Davis, Computability and Unsolvability, Dover reprint 1982. JStor

- Felscher, Walter (2000), "Bolzano, Cauchy, Epsilon, Delta", *The American Mathematical Monthly* (The American Mathematical Monthly, Vol. 107, No. 9) **107** (9): 844–862, doi:10.2307/2695743, JSTOR 2695743. JSTOR

- Ferreirós, José (2001), "The Road to Modern Logic-An Interpretation", *Bulletin of Symbolic Logic* (The Bulletin of Symbolic Logic, Vol. 7, No. 4) **7** (4): 441–484, doi:10.2307/2687794, JSTOR 2687794. JStor

- Hamkins, Joel David; Benedikt Löwe, "The modal logic of forcing", *Transactions of the American Mathematical Society*. arXiv 0509616

- Katz, Victor J. (1998), *A History of Mathematics*, Addison–Wesley, ISBN 0-321-01618-1.

- Morley, Michael (1965), "Categoricity in Power", *Transactions of the American Mathematical Society* (Transactions of the American Mathematical Society, Vol. 114, No. 2) **114** (2): 514–538, doi:10.2307/1994188, JSTOR 1994188.

- Soare, Robert I. (1996), "Computability and recursion", *Bulletin of Symbolic Logic* (The Bulletin of Symbolic Logic, Vol. 2, No. 3) **2** (3): 284–321, doi:10.2307/420992, JSTOR 420992.

- Solovay, Robert M. (1976), "Provability Interpretations of Modal Logic", *Israel Journal of Mathematics* **25** (3–4): 287–304, doi:10.1007/BF02757006.

- Woodin, W. Hugh (2001), "The Continuum Hypothesis, Part I", *Notices of the American Mathematical Society* **48** (6). PDF

7.12.4 Classical papers, texts, and collections

- Burali-Forti, Cesare (1897), *A question on transfinite numbers*, reprinted in van Heijenoort 1976, pp. 104–111.

- Dedekind, Richard (1872), *Stetigkeit und irrationale Zahlen*. English translation of title: "Consistency and irrational numbers".

- Dedekind, Richard (1888), *Was sind und was sollen die Zahlen? Two English translations:*

  - 1963 (1901). *Essays on the Theory of Numbers*. Beman, W. W., ed. and trans. Dover.

  - 1996. In *From Kant to Hilbert: A Source Book in the Foundations of Mathematics*, 2 vols, Ewald, William B., ed., Oxford University Press: 787–832.

- Fraenkel, Abraham A. (1922), "Der Begriff 'definit' und die Unabhängigkeit des Auswahlsaxioms", *Sitzungsberichte der Preussischen Akademie der Wissenschaften, Physikalisch-mathematische Klasse*, pp. 253–257 (German), reprinted in English translation as "The notion of 'definite' and the independence of the axiom of choice", van Heijenoort 1976, pp. 284–289.

- Frege Gottlob (1879), *Begriffsschrift, eine der arithmetischen nachgebildete Formelsprache des reinen Denkens*. Halle a. S.: Louis Nebert. Translation: *Concept Script, a formal language of pure thought modelled upon that of arithmetic*, by S. Bauer-Mengelberg in Jean Van Heijenoort, ed., 1967. *From Frege to Gödel: A Source Book in Mathematical Logic, 1879–1931*. Harvard University Press.

- Frege Gottlob (1884), *Die Grundlagen der Arithmetik: eine logisch-mathematische Untersuchung über den Begriff der Zahl*. Breslau: W. Koebner. Translation: J. L. Austin, 1974. *The Foundations of Arithmetic: A logico-mathematical enquiry into the concept of number*, 2nd ed. Blackwell.

- Gentzen, Gerhard (1936), "Die Widerspruchsfreiheit der reinen Zahlentheorie", *Mathematische Annalen* **112**: 132–213, doi:10.1007/BF01565428, reprinted in English translation in Gentzen's *Collected works*, M. E. Szabo, ed., North-Holland, Amsterdam, 1969.

- Gödel, Kurt (1929), *Über die Vollständigkeit des Logikkalküls*, doctoral dissertation, University Of Vienna. English translation of title: "Completeness of the logical calculus".

- Gödel, Kurt (1930), "Die Vollständigkeit der Axiome des logischen Funktionen-kalküls", *Monatshefte für Mathematik und Physik* **37**: 349–360, doi:10.1007/BF01696781. English translation of title: "The completeness of the axioms of the calculus of logical functions".

- Gödel, Kurt (1931), "Über formal unentscheidbare Sätze der Principia Mathematica und verwandter Systeme I", *Monatshefte für Mathematik und Physik* **38** (1): 173–198, doi:10.1007/BF01700692, see On Formally Undecidable Propositions of Principia Mathematica and Related Systems for details on English translations.

- Gödel, Kurt (1958), "Über eine bisher noch nicht benützte Erweiterung des finiten Standpunktes", *Dialectica. International Journal of Philosophy* **12** (3–4): 280–287, doi:10.1111/j.1746-8361.1958.tb01464.x, reprinted in English translation in Gödel's *Collected Works*, vol II, Soloman Feferman et al., eds. Oxford University Press, 1990.

- van Heijenoort, Jean, ed. (1967, 1976 3rd printing with corrections), *From Frege to Gödel: A Source Book in Mathematical Logic, 1879–1931* (3rd ed.), Cambridge, Mass: Harvard University Press, ISBN 0-674-32449-8, (pbk.) Check date values in: |date= (help)

- Hilbert, David (1899), *Grundlagen der Geometrie*, Leipzig: Teubner, English 1902 edition (*The Foundations of Geometry*) republished 1980, Open Court, Chicago.

- David, Hilbert (1929),"Probleme der Grundlegung der Mathematik",*Mathematische Annalen***102**: 1–9,doi:10. Lecture given at the International Congress of Mathematicians, 3 September 1928. Published in English translation as "The Grounding of Elementary Number Theory", in Mancosu 1998, pp. 266–273.

- Kleene, Stephen Cole (1943), "Recursive Predicates and Quantifiers", *American Mathematical Society Transactions* (Transactions of the American Mathematical Society, Vol. 53, No. 1) **54** (1): 41–73, doi:10.2307/1990131, JSTOR 1990131.

- Lobachevsky, Nikolai (1840), *Geometrishe Untersuchungen zur Theorie der Parellellinien* (German). Reprinted in English translation as "Geometric Investigations on the Theory of Parallel Lines" in *Non-Euclidean Geometry*, Robert Bonola (ed.), Dover, 1955. ISBN 0-486-60027-0

- Löwenheim, Leopold (1915), "Über Möglichkeiten im Relativkalkül", *Mathematische Annalen* **76** (4): 447–470, doi:10.1007/BF01458217, ISSN 0025-5831 (German). Translated as "On possibilities in the calculus of relatives" in Jean van Heijenoort, 1967. *A Source Book in Mathematical Logic, 1879–1931*. Harvard Univ. Press: 228–251.

- Mancosu, Paolo, ed. (1998), *From Brouwer to Hilbert. The Debate on the Foundations of Mathematics in the 1920s*, Oxford: Oxford University Press.

- Pasch, Moritz (1882), *Vorlesungen über neuere Geometrie*.

- Peano, Giuseppe (1889), *Arithmetices principia, nova methodo exposita* (Latin), excerpt reprinted in English stranslation as "The principles of arithmetic, presented by a new method", van Heijenoort 1976, pp. 83 97.

- Richard, Jules (1905), "Les principes des mathématiques et le problème des ensembles", *Revue générale des sciences pures et appliquées* **16**: 541 (French), reprinted in English translation as "The principles of mathematics and the problems of sets", van Heijenoort 1976, pp. 142–144.

- Skolem, Thoralf (1920), "Logisch-kombinatorische Untersuchungen über die Erfüllbarkeit oder Beweisbarkeit mathematischer Sätze nebst einem Theoreme über dichte Mengen", *Videnskapsselskapet Skrifter, I. Matematisk-naturvidenskabelig Klasse* **6**: 1–36.

- Tarski, Alfred (1948), *A decision method for elementary algebra and geometry*, Santa Monica, California: RAND Corporation

- Turing, Alan M. (1939), "Systems of Logic Based on Ordinals", *Proceedings of the London Mathematical Society* **45** (2): 161–228, doi:10.1112/plms/s2-45.1.161

- Zermelo, Ernst (1904), "Beweis, daß jede Menge wohlgeordnet werden kann", *Mathematische Annalen* **59** (4): 514–516, doi:10.1007/BF01445300 (German), reprinted in English translation as "Proof that every set can be well-ordered", van Heijenoort 1976, pp. 139–141.

- Zermelo, Ernst (1908a), "Neuer Beweis für die Möglichkeit einer Wohlordnung", *Mathematische Annalen* **65**: 107–128, doi:10.1007/BF01450054, ISSN 0025-5831 (German), reprinted in English translation as "A new proof of the possibility of a well-ordering", van Heijenoort 1976, pp. 183–198.

- Zermelo, Ernst (1908b), "Untersuchungen über die Grundlagen der Mengenlehre", *Mathematische Annalen* **65** (2): 261–281, doi:10.1007/BF01449999.

7.13   External links

- Hazewinkel, Michiel, ed. (2001), "Mathematical logic", *Encyclopedia of Mathematics*, Springer, ISBN 978-1-55608-010-4

- Polyvalued logic and Quantity Relation Logic

- *forall x: an introduction to formal logic*, a free textbook by P. D. Magnus.

- *A Problem Course in Mathematical Logic*, a free textbook by Stefan Bilaniuk.

- Detlovs, Vilnis, and Podnieks, Karlis (University of Latvia), *Introduction to Mathematical Logic.* (hyper-textbook).

- In the Stanford Encyclopedia of Philosophy:

    Classical Logic by Stewart Shapiro.
    First-order Model Theory by Wilfrid Hodges.

- In the London Philosophy Study Guide:

    Mathematical Logic
    Set Theory & Further Logic
    Philosophy of Mathematics

Chapter 8

Semantics

Semantics (from Ancient Greek: σημαντικός *sēmantikós*, "significant")[1][2] is the study of meaning. It focuses on the relation between *signifiers*, like words, phrases, signs, and symbols, and what they stand for; their denotation. Linguistic semantics is the study of meaning that is used for understanding human expression through language. Other forms of semantics include the semantics of programming languages, formal logics, and semiotics. In international scientific vocabulary semantics is also called *semasiology*.

The word *semantics* itself denotes a range of ideas—from the popular to the highly technical. It is often used in ordinary language for denoting a problem of understanding that comes down to word selection or connotation. This problem of understanding has been the subject of many formal enquiries, over a long period of time, especially in the field of formal semantics. In linguistics, it is the study of the interpretation of signs or symbols used in agents or communities within particular circumstances and contexts.[3] Within this view, sounds, facial expressions, body language, and proxemics have semantic (meaningful) content, and each comprises several branches of study. In written language, things like paragraph structure and punctuation bear semantic content; other forms of language bear other semantic content.[3]

The formal study of semantics intersects with many other fields of inquiry, including lexicology, syntax, pragmatics, etymology and others. Independently, semantics is also a well-defined field in its own right, often with synthetic properties. In the philosophy of language, semantics and reference are closely connected. Further related fields include philology, communication, and semiotics. The formal study of semantics can therefore be manifold and complex.

Semantics contrasts with syntax, the study of the combinatorics of units of a language (without reference to their meaning), and pragmatics, the study of the relationships between the symbols of a language, their meaning, and the users of the language.[5] Semantics as a field of study also has significant ties to various representational theories of meaning including truth theories of meaning, coherence theories of meaning, and correspondence theories of meaning. Each of these is related to the general philosophical study of reality and the representation of meaning.

8.1 Linguistics

In linguistics, **semantics** is the subfield that is devoted to the study of meaning, as inherent at the levels of words, phrases, sentences, and larger units of discourse (termed *texts*, or *narratives*). The study of semantics is also closely linked to the subjects of representation, reference and denotation. The basic study of semantics is oriented to the examination of the meaning of signs, and the study of relations between different linguistic units and compounds: homonymy, synonymy, antonymy, hypernymy, hyponymy, meronymy, metonymy, holonymy, paronyms. A key concern is how meaning attaches to larger chunks of text, possibly as a result of the composition from smaller units of meaning. Traditionally, semantics has included the study of *sense* and denotative *reference*, truth conditions, argument structure, thematic roles, discourse analysis, and the linkage of all of these to syntax.

8.2   Montague grammar

In the late 1960s, Richard Montague proposed a system for defining semantic entries in the lexicon in terms of the lambda calculus. In these terms, the syntactic parse of the sentence *John ate every bagel* would consist of a subject (*John*) and a predicate (*ate every bagel*); Montague demonstrated that the meaning of the sentence altogether could be decomposed into the meanings of its parts and in relatively few rules of combination. The logical predicate thus obtained would be elaborated further, e.g. using truth theory models, which ultimately relate meanings to a set of Tarskiian universals, which may lie outside the logic. The notion of such meaning atoms or primitives is basic to the language of thought hypothesis from the 1970s.

Despite its elegance, Montague grammar was limited by the context-dependent variability in word sense, and led to several attempts at incorporating context, such as:

- Situation semantics (1980s): truth-values are incomplete, they get assigned based on context

- Generative lexicon (1990s): categories (types) are incomplete, and get assigned based on context

8.3   Dynamic turn in semantics

In Chomskyan linguistics there was no mechanism for the learning of semantic relations, and the nativist view considered all semantic notions as inborn. Thus, even novel concepts were proposed to have been dormant in some sense. This view was also thought unable to address many issues such as metaphor or associative meanings, and semantic change, where meanings within a linguistic community change over time, and qualia or subjective experience. Another issue not addressed by the nativist model was how perceptual cues are combined in thought, e.g. in mental rotation.[6]

This view of semantics, as an innate finite meaning inherent in a lexical unit that can be composed to generate meanings for larger chunks of discourse, is now being fiercely debated in the emerging domain of cognitive linguistics[7] and also in the non-Fodorian camp in philosophy of language.[8] The challenge is motivated by:

- factors internal to language, such as the problem of resolving indexical or anaphora (e.g. *this x*, *him*, *last week*). In these situations *context* serves as the input, but the interpreted utterance also modifies the context, so it is also the output. Thus, the interpretation is necessarily dynamic and the meaning of sentences is viewed as contexts changing potentials instead of propositions.

- factors external to language, i.e. language is not a set of labels stuck on things, but "a toolbox, the importance of whose elements lie in the way they function rather than their attachments to things."[8] This view reflects the position of the later Wittgenstein and his famous *game* example, and is related to the positions of Quine, Davidson, and others.

A concrete example of the latter phenomenon is semantic underspecification – meanings are not complete without some elements of context. To take an example of one word, *red*, its meaning in a phrase such as *red book* is similar to many other usages, and can be viewed as compositional.[9] However, the colours implied in phrases such as *red wine* (very dark), and *red hair* (coppery), or *red soil*, or *red skin* are very different. Indeed, these colours by themselves would not be called *red* by native speakers. These instances are contrastive, so *red wine* is so called only in comparison with the other kind of wine (which also is not *white* for the same reasons). This view goes back to de Saussure:

> Each of a set of synonyms like *redouter* ('to dread'), *craindre* ('to fear'), *avoir peur* ('to be afraid') has its particular value only because they stand in contrast with one another. No word has a value that can be identified independently of what else is in its vicinity.[10]

and may go back to earlier Indian views on language, especially the Nyaya view of words as indicators and not carriers of meaning.[11]

An attempt to defend a system based on propositional meaning for semantic underspecification can be found in the generative lexicon model of James Pustejovsky, who extends contextual operations (based on type shifting) into the lexicon. Thus meanings are generated "on the fly" (as you go), based on finite context.

8.4 Prototype theory

Another set of concepts related to fuzziness in semantics is based on prototypes. The work of Eleanor Rosch in the 1970s led to a view that natural categories are not characterizable in terms of necessary and sufficient conditions, but are graded (fuzzy at their boundaries) and inconsistent as to the status of their constituent members. One may compare it with Jung's archetype, though the concept of archetype sticks to static concept. Some post-structuralists are against the fixed or static meaning of the words. Derrida, following Nietzsche, talked about slippages in fixed meanings.

Systems of categories are not objectively *out there* in the world but are rooted in people's experience. These categories evolve as learned concepts of the world – meaning is not an objective truth, but a subjective construct, learned from experience, and language arises out of the "grounding of our conceptual systems in shared embodiment and bodily experience".[12] A corollary of this is that the conceptual categories (i.e. the lexicon) will not be identical for different cultures, or indeed, for every individual in the same culture. This leads to another debate (see the Sapir–Whorf hypothesis or Eskimo words for snow).

8.5 Theories in semantics

8.5.1 Model theoretic semantics

Main article: formal semantics (linguistics)

Originates from Montague's work (see above). A highly formalized theory of natural language semantics in which expressions are assigned denotations (meanings) such as individuals, truth values, or functions from one of these to another. The truth of a sentence, and more interestingly, its logical relation to other sentences, is then evaluated relative to a model.

8.5.2 Formal (or truth-conditional) semantics

Main article: truth-conditional semantics

Pioneered by the philosopher Donald Davidson, another formalized theory, which aims to associate each natural language sentence with a meta-language description of the conditions under which it is true, for example: 'Snow is white' is true if and only if snow is white. The challenge is to arrive at the truth conditions for any sentences from fixed meanings assigned to the individual words and fixed rules for how to combine them. In practice, truth-conditional semantics is similar to model-theoretic semantics; conceptually, however, they differ in that truth-conditional semantics seeks to connect language with statements about the real world (in the form of meta-language statements), rather than with abstract models.

8.5.3 Lexical and conceptual semantics

Main article: conceptual semantics

This theory is an effort to explain properties of argument structure. The assumption behind this theory is that syntactic properties of phrases reflect the meanings of the words that head them.[13] With this theory, linguists can better deal with the fact that subtle differences in word meaning correlate with other differences in the syntactic structure that the word appears in.[13] The way this is gone about is by looking at the internal structure of words.[14] These small parts that make up the internal structure of words are termed *semantic primitives*.[14]

8.5.4   Lexical semantics

Main article: lexical semantics

A linguistic theory that investigates word meaning. This theory understands that the meaning of a word is fully reflected by its context. Here, the meaning of a word is constituted by its contextual relations.[15] Therefore, a distinction between degrees of participation as well as modes of participation are made.[15] In order to accomplish this distinction any part of a sentence that bears a meaning and combines with the meanings of other constituents is labeled as a semantic constituent. Semantic constituents that cannot be broken down into more elementary constituents are labeled minimal semantic constituents.[15]

8.5.5   Computational semantics

Main article: computational semantics

Computational semantics is focused on the processing of linguistic meaning. In order to do this concrete algorithms and architectures are described. Within this framework the algorithms and architectures are also analyzed in terms of decidability, time/space complexity, data structures they require and communication protocols.[16]

8.6   Computer science

Main article: Semantics (computer science)

In computer science, the term *semantics* refers to the meaning of languages, as opposed to their form (syntax). According to Euzenat, semantics "provides the rules for interpreting the syntax which do not provide the meaning directly but constrains the possible interpretations of what is declared."[17] In other words, semantics is about interpretation of an expression. Additionally, the term is applied to certain types of data structures specifically designed and used for representing information content.

8.6.1   Programming languages

The semantics of programming languages and other languages is an important issue and area of study in computer science. Like the syntax of a language, its semantics can be defined exactly.

For instance, the following statements use different syntaxes, but cause the same instructions to be executed:

Generally these operations would all perform an arithmetical addition of 'y' to 'x' and store the result in a variable called 'x'.

Various ways have been developed to describe the semantics of programming languages formally, building on mathematical logic:[18]

- Operational semantics: The meaning of a construct is specified by the computation it induces when it is executed on a machine. In particular, it is of interest *how* the effect of a computation is produced.

- Denotational semantics: Meanings are modelled by mathematical objects that represent the effect of executing the constructs. Thus *only* the effect is of interest, not how it is obtained.

- Axiomatic semantics: Specific properties of the effect of executing the constructs are expressed as *assertions*. Thus there may be aspects of the executions that are ignored.

8.6.2 Semantic models

Terms such as *semantic network* and *semantic data model* are used to describe particular types of data model characterized by the use of directed graphs in which the vertices denote concepts or entities in the world, and the arcs denote relationships between them.

The Semantic Web refers to the extension of the World Wide Web via embedding added semantic metadata, using semantic data modelling techniques such as Resource Description Framework (RDF) and Web Ontology Language (OWL).

8.7 Psychology

In psychology, *semantic memory* is memory for meaning – in other words, the aspect of memory that preserves only the *gist*, the general significance, of remembered experience – while episodic memory is memory for the ephemeral details – the individual features, or the unique particulars of experience. The term 'episodic memory' was introduced by Tulving and Schacter in the context of 'declarative memory' which involved simple association of factual or objective information concerning its object. Word meaning is measured by the company they keep, i.e. the relationships among words themselves in a semantic network. The memories may be transferred intergenerationally or isolated in one generation due to a cultural disruption. Different generations may have different experiences at similar points in their own time-lines. This may then create a vertically heterogeneous semantic net for certain words in an otherwise homogeneous culture.[19] In a network created by people analyzing their understanding of the word (such as Wordnet) the links and decomposition structures of the network are few in number and kind, and include *part of*, *kind of*, and similar links. In automated ontologies the links are computed vectors without explicit meaning. Various automated technologies are being developed to compute the meaning of words: latent semantic indexing and support vector machines as well as natural language processing, neural networks and predicate calculus techniques.

Ideasthesia is a psychological phenomenon in which activation of concepts evokes sensory experiences. For example, in synesthesia, activation of a concept of a letter (e.g., that of the letter *A*) evokes sensory-like experiences (e.g., of red color).

8.8 See also

8.8.1 Linguistics and semiotics

- Asemic writing

- Cognitive semantics

- Colorless green ideas sleep furiously

- Computational semantics

- Discourse representation theory

- General semantics

- Generative semantics

- Hermeneutics

- Natural semantic metalanguage

- Onomasiology

- Phono-semantic matching

- Pragmatic maxim

- Pragmaticism

- Pragmatism

- Problem of universals

- Semantic change or progression

- Semantic class

- Semantic feature

- Semantic field

- Semantic lexicon

- Semantic primes

- Semantic property

- Sememe

- Semiosis

- Semiotics

- SPL notation

8.8.2   Logic and mathematics

- Formal logic

- Game semantics

- Model theory

- Gödel's incompleteness theorems

- Proof-theoretic semantics

- Semantic consequence

- Semantic theory of truth

- Semantics of logic

- Truth-value semantics

8.8.3   Computer science

- Formal semantics of programming languages

- Knowledge representation

- Semantic networks

- Semantic transversal

- Semantic analysis

- Semantic compression

- Semantic HTML

- Semantic integration

- Semantic interpretation

- Semantic link

- Semantic reasoner

- Semantic service oriented architecture

- Semantic spectrum

- Semantic unification

- Semantic Web

8.8.4 Psychology

- Ideasthesia

8.9 References

[1] σημαντικός. Liddell, Henry George; Scott, Robert; *A Greek–English Lexicon* at the Perseus Project

[2] The word is derived from the Ancient Greek word σημαντικός (*semantikos*), "related to meaning, significant", from σημαίνω *semaino*, "to signify, to indicate", which is from σῆμα *sema*, "sign, mark, token". The plural is used in analogy with words similar to *physics*, which was in the neuter plural in Ancient Greek and meant "things relating to nature".

[3] Neurath, Otto; Carnap, Rudolf; Morris, Charles F. W. (Editors) (1955). *International Encyclopedia of Unified Science*. Chicago, IL: University of Chicago Press.

[4] Cruse, Alan; *Meaning and Language: An introduction to Semantics and Pragmatics*, Chapter 1, Oxford Textbooks in Linguistics, 2004; Kearns, Kate; *Semantics*, Palgrave MacMillan 2000; Cruse, D. A.; *Lexical Semantics*, Cambridge, MA, 1986.

[5] Kitcher, Philip; Salmon, Wesley C. (1989). *Scientific Explanation*. Minneapolis, MN: University of Minnesota Press. p. 35.

[6] Barsalou, L.; *Perceptual Symbol Systems*, Behavioral and Brain Sciences, 22(4), 1999

[7] Langacker, Ronald W. (1999). *Grammar and Conceptualization*. Berlin/New York: Mouton de Gruyer. ISBN 3-11-016603-8.

[8] Peregrin, Jaroslav (2003). *Meaning: The Dynamic Turn. Current Research in the Semantics/Pragmatics Interface*. London: Elsevier.

[9] Gärdenfors, Peter (2000). *Conceptual Spaces: The Geometry of Thought*. MIT Press/Bradford Books. ISBN 978-0-585-22837-2.

[10] de Saussure, Ferdinand (1916). *The Course of General Linguistics (Cours de linguistique générale)*.

[11] Matilal, Bimal Krishna (1990). *The Word and the World: India's Contribution to the Study of Language*. Oxford. The Nyaya and Mimamsa schools in Indian vyākaraṇa tradition conducted a centuries-long debate on whether sentence meaning arises through composition on word meanings, which are primary; or whether word meanings are obtained through analysis of sentences where they appear. (Chapter 8).

[12] Lakoff, George; Johnson, Mark (1999). *Philosophy in the Flesh: The embodied mind and its challenge to Western thought. Chapter 1*. New York, NY: Basic Books. OCLC 93961754.

[13] Levin, Beth; Pinker, Steven; *Lexical & Conceptual Semantics*, Blackwell, Cambridge, MA, 1991

[14] Jackendoff, Ray; *Semantic Structures*, MIT Press, Cambridge, MA, 1990

[15] Cruse, D.; *Lexical Semantics*, Cambridge University Press, Cambridge, MA, 1986

[16] Nerbonne, J.; *The Handbook of Contemporary Semantic Theory* (ed. Lappin, S.), Blackwell Publishing, Cambridge, MA, 1996

[17] Euzenat, Jerome. *Ontology Matching.* Springer-Verlag Berlin Heidelberg, 2007, p. 36

[18] Nielson, Hanne Riis; Nielson, Flemming (1995). *Semantics with Applications, A Formal Introduction* (1st ed.). Chicester, England: John Wiley & Sons. ISBN 0-471-92980-8.

[19] Giannini, A. J.; *Semiotic and Semantic Implications of "Authenticity"*, Psychological Reports, 106(2):611–612, 2010

8.10   External links

- semanticsarchive.net

- Teaching page for A-level semantics

- Chomsky, Noam; *On Referring*, Harvard University, 30 October 2007 (video)

- Jackendoff, Ray; *Conceptual Semantics*, Harvard University, 13 November 2007 (video)

- *Semantics: an interview with Jerry Fodor* (ReVEL, vol. 5, no. 8 (2007))

Chapter 9

Syntax

For other uses, see Syntax (disambiguation). Not to be confused with Sin tax. See also Syntaxis.
"Sentence structure" redirects here. For sentence types in traditional grammar, see Sentence clause structure.

In linguistics, **syntax** is the set of rules, principles, and processes that govern the structure of sentences in a given language. The term *syntax* is also used to refer to the study of such principles and processes.[1] The goal of many syntacticians is to discover the syntactic rules common to all languages.

In mathematics, *syntax* refers to the rules governing the behavior of mathematical systems, such as formal languages used in logic. (See logical syntax.)

9.1 Etymology

From Ancient Greek: σύνταξις "coordination" from σύν *syn*, "together," and τάξις *táxis*, "an ordering".

9.2 Early history

Works on grammar were written long before modern syntax came about; the *Aṣṭādhyāyī* of Pāṇini (c. 4th century BC) is often cited as an example of a premodern work that approaches the sophistication of a modern syntactic theory.[2] In the West, the school of thought that came to be known as "traditional grammar" began with the work of Dionysius Thrax.

For centuries, work in syntax was dominated by a framework known as *grammaire générale*, first expounded in 1660 by Antoine Arnauld in a book of the same title. This system took as its basic premise the assumption that language is a direct reflection of thought processes and therefore there is a single, most natural way to express a thought.

However, in the 19th century, with the development of historical-comparative linguistics, linguists began to realize the sheer diversity of human language and to question fundamental assumptions about the relationship between language and logic. It became apparent that there was no such thing as the most natural way to express a thought, and therefore logic could no longer be relied upon as a basis for studying the structure of language.

The Port-Royal grammar modeled the study of syntax upon that of logic. (Indeed, large parts of the Port-Royal Logic were copied or adapted from the *Grammaire générale*.[3]) Syntactic categories were identified with logical ones, and all sentences were analyzed in terms of "Subject – Copula – Predicate." Initially, this view was adopted even by the early comparative linguists such as Franz Bopp.

The central role of syntax within theoretical linguistics became clear only in the 20th century, which could reasonably be called the "century of syntactic theory" as far as linguistics is concerned. (For a detailed and critical survey of the history of syntax in the last two centuries, see the monumental work by Giorgio Graffi (2001).)[4]

9.3   Modern theories

There are a number of theoretical approaches to the discipline of syntax. One school of thought, founded in the works of Derek Bickerton,[5] sees syntax as a branch of biology, since it conceives of syntax as the study of linguistic knowledge as embodied in the human mind. Other linguists (e.g., Gerald Gazdar) take a more Platonistic view, since they regard syntax to be the study of an abstract formal system.[6] Yet others (e.g., Joseph Greenberg) consider syntax a taxonomical device to reach broad generalizations across languages.

9.3.1   Generative grammar

Main article: Generative grammar

The hypothesis of generative grammar is that language is a structure of the human mind. The goal of generative grammar is to make a complete model of this inner language (known as *i-language*). This model could be used to describe all human language and to predict the grammaticality of any given utterance (that is, to predict whether the utterance would sound correct to native speakers of the language). This approach to language was pioneered by Noam Chomsky. Most generative theories (although not all of them) assume that syntax is based upon the constituent structure of sentences. Generative grammars are among the theories that focus primarily on the form of a sentence, rather than its communicative function.

Among the many generative theories of linguistics, the Chomskyan theories are:

- Transformational grammar (TG) (Original theory of generative syntax laid out by Chomsky in *Syntactic Structures* in 1957)[7]

- Government and binding theory (GB) (revised theory in the tradition of TG developed mainly by Chomsky in the 1970s and 1980s)[8]

- Minimalist program (MP) (a reworking of the theory out of the GB framework published by Chomsky in 1995)[9]

Other theories that find their origin in the generative paradigm are:

- Arc pair grammar

- Generalized phrase structure grammar (GPSG; now largely out of date)

- Generative semantics (now largely out of date)

- Head-driven phrase structure grammar (HPSG)

- Lexical functional grammar (LFG)

- Nanosyntax

- Relational grammar (RG) (now largely out of date)

9.3.2   Categorial grammar

Main article: Categorial grammar

Categorial grammar is an approach that attributes the syntactic structure not to rules of grammar, but to the properties of the syntactic categories themselves. For example, rather than asserting that sentences are constructed by a rule that combines a noun phrase (NP) and a verb phrase (VP) (e.g., the phrase structure rule S → NP VP), in categorial grammar, such principles are embedded in the category of the head word itself. So the syntactic category for an intransitive verb is a complex formula representing the fact that the verb acts as a function word requiring an NP as an input and produces a sentence level structure as an output. This complex category is notated as (NP\S) instead of V. NP\S is read as "a category that searches to the left (indicated by \) for an NP (the element on the left) and outputs a sentence (the element on the right)." The category of transitive verb is defined as an element that requires two NPs (its subject and its direct object) to form a sentence. This is notated as (NP/(NP\S)) which means "a category that searches to the right (indicated by /) for an NP (the object), and generates a function (equivalent to the VP) which is (NP\S), which in turn represents a function that searches to the left for an NP and produces a sentence."

Tree-adjoining grammar is a categorial grammar that adds in partial tree structures to the categories.

9.3.3 Dependency grammar

Main article: Dependency grammar
Dependency grammar is an approach to sentence structure where syntactic units are arranged according to the dependency

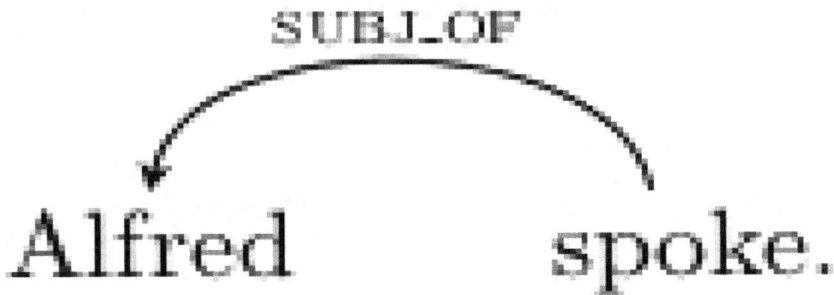

A syntactic parse of "Alfred spoke" under the dependency formalism

relation, as opposed to the constituency relation of phrase structure grammars. Dependencies are directed links between words. The (finite) verb is seen as the root of all clause structure and all the other words in the clause are either directly or indirectly dependent on this root. Some prominent dependency-based theories of syntax are:

- Recursive categorical syntax, or Algebraic syntax
- Functional generative description
- Meaning–text theory
- Operator grammar
- Word grammar

Lucien Tesnière (1893–1954) is widely seen as the father of modern dependency-based theories of syntax and grammar. He argued vehemently against the binary division of the clause into subject and predicate that is associated with the grammars of his day (S → NP VP) and which remains at the core of most phrase structure grammars. In the place of this division, he positioned the verb as the root of all clause structure.[10]

9.3.4 Stochastic/probabilistic grammars/network theories

Theoretical approaches to syntax that are based upon probability theory are known as stochastic grammars. One common implementation of such an approach makes use of a neural network or connectionism.

9.3.5   Functionalist grammars

Main article: Functional theories of grammar

Functionalist theories, although focused upon form, are driven by explanation based upon the function of a sentence (i.e. its communicative function). Some typical functionalist theories include:

- Cognitive grammar

- Construction grammar (CxG)

- Emergent grammar

- Functional discourse grammar (Dik)

- Prague linguistic circle

- Role and reference grammar (RRG)

- Systemic functional grammar

9.4   See also

9.4.1   Syntactic terms

- Adjective

- Adjective phrase

- Adjunct

- Adpositional phrase

- Adverb

- Anaphora

- Answer ellipsis

- Antecedent

- Antecedent-contained deletion

- Appositive

- Argument

- Article

- Aspect

- Attributive adjective and predicative adjective

- Auxiliary verb

- Binding

- Branching

- c-command

- Case

- Category

- Catena

- Clause

- Closed class word

- Comparative

- Complement

- Compound noun and adjective

- Conjugation

- Conjunction

- Constituent

- Coordination

- Coreference

- Crossover

- Dangling modifier

- Declension

- Dependency grammar

- Dependent marking

- Determiner

- Discontinuity

- Do-support

- Dual (form for two)

- Ellipsis

- Endocentric

- Exceptional case-marking

- Expletive

- Extraposition

- Finite verb

- Function word

- Gapping

- Gender

- Gerund

- Government

- Head

- Head marking

- Infinitive

- Inverse copular construction

- Inversion

- Lexical item

- m-command

- Measure word (classifier)

- Merge

- Modal particle

- Modal verb

- Modifier

- Mood

- Movement

- Movement paradox

- Nanosyntax

- Negative inversion

- Non-configurational language

- Non-finite verb

- Noun

- Noun ellipsis

- Noun phrase

- Number

- Object

- Open class word

- Parasitic gap

- Part of speech

- Particle

- Periphrasis

- Person

- Personal pronoun

- Pied-piping

- Phrasal verb

- Phrase

- Phrase structure grammar

- Plural

- Predicate

- Predicative expression

- Preposition and postposition

- Pronoun

- Pseudogapping

- Raising

- **Relation** (Grammatical relation)

- Restrictiveness

- Right node raising

- Sandhi

- Scrambling

- Selection

- Sentence

- Separable verb

- Shifting

- Singular

- Sluicing

- Small clause

- Stripping

- Subcategorization

- Subject

- Subject-auxiliary inversion

- Subject-verb inversion

- Subordination

- Superlative

- Tense

- Topicalization

- Tough movement

- Uninflected word

- V2 word order

- Valency

- Verb

- Verb phrase

- Verb phrase ellipsis

- Voice

- Wh-movement

- Word order

- X-bar theory

9.5 Notes

[1] Chomsky, Noam (2002) [1957]. *Syntactic Structures*. p. 11.

[2] Fortson IV, Benjamin W. (2004). *Indo-European Language and Culture: An Introduction*. Blackwell. p. 186. ISBN 978-1405188968. [The *Aṣṭādhyāyī*] is a highly precise and thorough description of the structure of Sanskrit somewhat resembling modern generative grammar...[it] remained the most advanced linguistic analysis of any kind until the twentieth century.

[3] Arnauld, Antoine (1683). *La logique* (5th ed.). Paris: G. Desprez. p. 137. *Nous avons emprunté...ce que nous avons dit...d'un petit Livre...sous le titre de Grammaire générale.*

[4] Giorgio, Graffi (2001). *200 Years of Syntax: A Critical Survey* (googlebook preview). John Benjamins Publishing.

[5] See Bickerton, Derek (1990). *Language and Species*. University of Chicago Press. ISBN 0-226-04610-9. and, for more recent advances, Derek Bickerton; Eörs Szathmáry, ed. (2009). *Biological foundations and origin of syntax*. MIT Press. ISBN 978-0-262-01356-7.

[6] Ted Briscoe, 2 May 2001, Interview with Gerald Gazdar. Retrieved 2008-06-04.

[7] Chomsky, Noam. 1957. *Syntactic Structures*. The Hague/Paris: Mouton, p. 15.

[8] Chomsky, Noam (1981/1993). Lectures on Government and Binding: The Pisa Lectures. Mouton de Gruyter.

[9] Chomsky, Noam (1995). The Minimalist Program. MIT Press.

[10] Concerning Tesnière's rejection of the binary division of the clause into subject and predicate and in favor of the verb as the root of all structure, see Tesnière (1969:103–105).

9.6 References

- Brown, Keith; Jim Miller (eds.) (1996). *Concise Encyclopedia of Syntactic Theories*. New York: Elsevier Science. ISBN 0-08-042711-1.

- Carnie, Andrew (2006). *Syntax: A Generative Introduction* (2nd ed.). Oxford: Wiley-Blackwell. ISBN 1-4051-3384-8.

- Freidin, Robert; Howard Lasnik (eds.) (2006). *Syntax*. Critical Concepts in Linguistics. New York: Routledge. ISBN 0-415-24672-5.

- Graffi, Giorgio (2001). *200 Years of Syntax. A Critical Survey*. Studies in the History of the Language Sciences 98. Amsterdam: Benjamins. ISBN 90-272-4587-8.

- Mieszko Talasiewicz (2009). *Philosophy of Syntax—Foundational Topics*. Springer. ISBN 978-90-481-3287-4. An interdisciplinary essay on the interplay between logic and linguistics on syntactic theories.

- Tesnière, Lucien 1969. Éleménts de syntaxe structurale. 2nd edition. Paris: Klincksieck.

9.7 Further reading

- Martin Everaert, Henk Van Riemsdijk, Rob Goedemans and Bart Hollebrandse, ed. (2006). *The Blackwell companion to syntax*. Blackwell. ISBN 978-1-4051-1485-1. 5 Volumes; 77 case studies of syntactic phenomena.

- Brian Roark; Richard William Sproat (2007). *Computational approaches to morphology and syntax*. Oxford University Press. ISBN 978-0-19-927477-2. part II: Computational approaches to syntax.

- Isac, Daniela; Charles Reiss (2013). *I-language: An Introduction to Linguistics as Cognitive Science, 2nd edition*. Oxford University Press. ISBN 978-0199660179.

- Edith A. Moravcsik (2006). *An introduction to syntax: fundamentals of syntactic analysis*. Continuum International Publishing Group. ISBN 978-0-8264-8945-6. Attempts to be a theory-neutral introduction. The companion Edith A. Moravcsik (2006). *An introduction to syntactic theory*. Continuum International Publishing Group. ISBN 978-0-8264-8943-2. surveys the major theories. Jointly reviewed in *The Canadian Journal of Linguistics* 54(1), March 2009, pp. 172–175

9.8 External links

- The syntax of natural language: An online introduction using the Trees program—Beatrice Santorini & Anthony Kroch, University of Pennsylvania, 2007

Chapter 10

Universal algebra

Universal algebra (sometimes called **general algebra**) is the field of mathematics that studies algebraic structures themselves, not examples ("models") of algebraic structures. For instance, rather than take particular groups as the object of study, in universal algebra one takes "the theory of groups" as an object of study.

10.1 Basic idea

In universal algebra, an **algebra** (or **algebraic structure**) is a set A together with a collection of operations on A. An **n-ary operation** on A is a function that takes n elements of A and returns a single element of A. Thus, a 0-ary operation (or *nullary operation*) can be represented simply as an element of A, or a *constant*, often denoted by a letter like a. A 1-ary operation (or *unary operation*) is simply a function from A to A, often denoted by a symbol placed in front of its argument, like $\sim x$. A 2-ary operation (or *binary operation*) is often denoted by a symbol placed between its arguments, like $x * y$. Operations of higher or unspecified *arity* are usually denoted by function symbols, with the arguments placed in parentheses and separated by commas, like $f(x,y,z)$ or $f(x_1,...,xn)$. Some researchers allow infinitary operations, such as $\bigwedge_{\alpha \in J} x_\alpha$ where J is an infinite index set, thus leading into the algebraic theory of complete lattices. One way of talking about an algebra, then, is by referring to it as an algebra of a certain type Ω, where Ω is an ordered sequence of natural numbers representing the arity of the operations of the algebra.

10.1.1 Equations

After the operations have been specified, the nature of the algebra can be further limited by axioms, which in universal algebra often take the form of identities, or **equational laws.** An example is the associative axiom for a binary operation, which is given by the equation $x * (y * z) = (x * y) * z$. The axiom is intended to hold for all elements x, y, and z of the set A.

10.2 Varieties

Main article: Variety (universal algebra)

An algebraic structure that can be defined by identities is called a **variety,** and these are sufficiently important that some authors consider varieties the only object of study in universal algebra, while others consider them an object.

Restricting one's study to varieties rules out:

- Predicate logic, notably quantification, including universal quantification ($\forall$), except before an equation, and existential quantification ($\exists$)

- All relations except equality, in particular inequalities, both $a \neq b$ and order relations

In this narrower definition, universal algebra can be seen as a special branch of model theory, typically dealing with structures having operations only (i.e. the type can have symbols for functions but not for relations other than equality), and in which the language used to talk about these structures uses equations only.

Not all algebraic structures in a wider sense fall into this scope. For example ordered groups are not studied in mainstream universal algebra because they involve an ordering relation.

A more fundamental restriction is that universal algebra cannot study the class of fields, because there is no type (a.k.a. signature) in which all field laws can be written as equations (inverses of elements are defined for all *non-zero* elements in a field, so inversion cannot simply be added to the type).

One advantage of this restriction is that the structures studied in universal algebra can be defined in any category that has *finite products*. For example, a topological group is just a group in the category of topological spaces.

10.2.1 Examples

Most of the usual algebraic systems of mathematics are examples of varieties, but not always in an obvious way – the usual definitions often involve quantification or inequalities.

Groups

To see how this works, let's consider the definition of a group. Normally a group is defined in terms of a single binary operation $*$, subject to these axioms:

- Associativity (as in the previous section): $x * (y * z) = (x * y) * z$; formally: $\forall x,y,z.\ x*(y*z)=(x*y)*z$.

- Identity element: There exists an element e such that for each element x, $e * x = x = x * e$; formally: $\exists e\, \forall x.\ e*x=x=x*e$.

- Inverse element: It can easily be seen that the identity element is unique. If this unique identity element is denoted by e then for each x, there exists an element i such that $x * i = e = i * x$; formally: $\forall x\, \exists i.\ x*i=e=i*x$.

(Some authors also use an axiom called "closure", stating that $x * y$ belongs to the set A whenever x and y do. But from a universal algebraist's point of view, that is already implied by calling $*$ a binary operation.)

This definition of a group is problematic from the point of view of universal algebra. The reason is that the axioms of the identity element and inversion are not stated purely in terms of equational laws but also have clauses involving the phrase "there exists ... such that ...". This is inconvenient; the list of group properties can be simplified to universally quantified equations by adding a nullary operation e and a unary operation $\sim$ in addition to the binary operation $*$. Then list the axioms for these three operations as follows:

- Associativity: $x * (y * z) = (x * y) * z$.

- Identity element: $e * x = x = x * e$; formally: $\forall x.\ e*x=x=x*e$.

- Inverse element: $x * (\sim x) = e = (\sim x) * x$ formally: $\forall x.\ x*\sim x=e=\sim x*x$.

(Of course, we usually write "x^{-1}" instead of "$\sim x$", which shows that the notation for operations of low arity is not *always* as given in the second paragraph.)

What has changed is that in the usual definition there are:

- a single binary operation (signature (2))

- 1 equational law (associativity)

- 2 quantified laws (identity and inverse)

...while in the universal algebra definition there are

- 3 operations: one binary, one unary, and one nullary (signature (2,1,0))

- 3 equational laws (associativity, identity, and inverse)

- no quantified laws (except for outermost universal quantifiers which are allowed in varieties)

It is important to check that this really does capture the definition of a group. The reason that it might not is that specifying one of these universal groups might give more information than specifying one of the usual kind of group. After all, nothing in the usual definition said that the identity element e was *unique*; if there is another identity element e', then it is ambiguous which one should be the value of the nullary operator e. Proving that it is unique is a common beginning exercise in classical group theory textbooks. The same thing is true of inverse elements. So, the universal algebraist's definition of a group is equivalent to the usual definition.

At first glance this is simply a technical difference, replacing quantified laws with equational laws. However, it has immediate practical consequences – when defining a group object in category theory, where the object in question may not be a set, one must use equational laws (which make sense in general categories), and cannot use quantified laws (which do not make sense, as objects in general categories do not have elements). Further, the perspective of universal algebra insists not only that the inverse and identity exist, but that they be maps in the category. The basic example is of a topological group – not only must the inverse exist element-wise, but the inverse map must be continuous (some authors also require the identity map to be a closed inclusion, hence cofibration, again referring to properties of the map).

10.3   Basic constructions

We assume that the type, Ω, has been fixed. Then there are three basic constructions in universal algebra: homomorphic image, subalgebra, and product.

A homomorphism between two algebras A and B is a function $h: A \rightarrow B$ from the set A to the set B such that, for every operation fA of A and corresponding fB of B (of arity, say, n), $h(fA(x_1,...,xn)) = fB(h(x_1),...,h(xn))$. (Sometimes the subscripts on f are taken off when it is clear from context which algebra your function is from.) For example, if e is a constant (nullary operation), then $h(eA) = eB$. If $\sim$ is a unary operation, then $h(\sim x) = \sim h(x)$. If $*$ is a binary operation, then $h(x * y) = h(x) * h(y)$. And so on. A few of the things that can be done with homomorphisms, as well as definitions of certain special kinds of homomorphisms, are listed under the entry Homomorphism. In particular, we can take the homomorphic image of an algebra, $h(A)$.

A subalgebra of A is a subset of A that is closed under all the operations of A. A product of some set of algebraic structures is the cartesian product of the sets with the operations defined coordinatewise.

10.4   Some basic theorems

- The isomorphism theorems, which encompass the isomorphism theorems of groups, rings, modules, etc.

- Birkhoff's HSP Theorem, which states that a class of algebras is a variety if and only if it is closed under homomorphic images, subalgebras, and arbitrary direct products.

10.5   Motivations and applications

In addition to its unifying approach, universal algebra also gives deep theorems and important examples and counterexamples. It provides a useful framework for those who intend to start the study of new classes of algebras. It can enable the use of methods invented for some particular classes of algebras to other classes of algebras, by recasting the methods in terms of universal algebra (if possible), and then interpreting these as applied to other classes. It has also provided conceptual clarification; as J.D.H. Smith puts it, *"What looks messy and complicated in a particular framework may turn out to be simple and obvious in the proper general one."*

In particular, universal algebra can be applied to the study of monoids, rings, and lattices. Before universal algebra came along, many theorems (most notably the isomorphism theorems) were proved separately in all of these fields, but with universal algebra, they can be proven once and for all for every kind of algebraic system.

The 1956 paper by Higgins referenced below has been well followed up for its framework for a range of particular algebraic systems, while his 1963 paper is notable for its discussion of algebras with operations which are only partially defined, typical examples for this being categories and groupoids. This leads on to the subject of higher-dimensional algebra which can be defined as the study of algebraic theories with partial operations whose domains are defined under geometric conditions. Notable examples of these are various forms of higher-dimensional categories and groupoids.

10.6 Generalizations

Further information: Category theory, Operad theory and Partial algebra

A more generalised programme along these lines is carried out by category theory. Given a list of operations and axioms in universal algebra, the corresponding algebras and homomorphisms are the objects and morphisms of a category. Category theory applies to many situations where universal algebra does not, extending the reach of the theorems. Conversely, many theorems that hold in universal algebra do not generalise all the way to category theory. Thus both fields of study are useful.

A more recent development in category theory that generalizes operations is operad theory – an operad is a set of operations, similar to a universal algebra.

Another development is partial algebra where the operators can be partial functions.

An important generalization of universal algebra theory is model theory, which is sometimes described as "universal theory + logic".

10.7 History

In Alfred North Whitehead's book *A Treatise on Universal Algebra,* published in 1898, the term *universal algebra* had essentially the same meaning that it has today. Whitehead credits William Rowan Hamilton and Augustus De Morgan as originators of the subject matter, and James Joseph Sylvester with coining the term itself.[1]

At the time structures such as Lie algebras and hyperbolic quaternions drew attention to the need to expand algebraic structures beyond the associatively multiplicative class. In a review Alexander Macfarlane wrote: "The main idea of the work is not unification of the several methods, nor generalization of ordinary algebra so as to include them, but rather the comparative study of their several structures." At the time George Boole's algebra of logic made a strong counterpoint to ordinary number algebra, so the term "universal" served to calm strained sensibilities.

Whitehead's early work sought to unify quaternions (due to Hamilton), Grassmann's Ausdehnungslehre, and Boole's algebra of logic. Whitehead wrote in his book:

> *"Such algebras have an intrinsic value for separate detailed study; also they are worthy of comparative study, for the sake of the light thereby thrown on the general theory of symbolic reasoning, and on algebraic symbolism in particular. The comparative study necessarily presupposes some previous separate study, comparison being impossible without knowledge."*[2]

Whitehead, however, had no results of a general nature. Work on the subject was minimal until the early 1930s, when Garrett Birkhoff and Øystein Ore began publishing on universal algebras. Developments in metamathematics and category theory in the 1940s and 1950s furthered the field, particularly the work of Abraham Robinson, Alfred Tarski, Andrzej Mostowski, and their students (Brainerd 1967).

In the period between 1935 and 1950, most papers were written along the lines suggested by Birkhoff's papers, dealing with free algebras, congruence and subalgebra lattices, and homomorphism theorems. Although the development of mathematical logic had made applications to algebra possible, they came about slowly; results published by Anatoly Maltsev in the 1940s went unnoticed because of the war. Tarski's lecture at the 1950 International Congress of Mathematicians in Cambridge ushered in a new period in which model-theoretic aspects were developed, mainly by Tarski himself, as well as C.C. Chang, Leon Henkin, Bjarni Jónsson, Roger Lyndon, and others.

In the late 1950s, Edward Marczewski[3] emphasized the importance of free algebras, leading to the publication of more than 50 papers on the algebraic theory of free algebras by Marczewski himself, together with Jan Mycielski, Władysław Narkiewicz, Witold Nitka, J. Płonka, S. Świerczkowski, K. Urbanik, and others.

10.8   See also

- Graph algebra

- Homomorphism

- Lattice theory

- Signature

- Term algebra

- Variety

- Clone

- Universal algebraic geometry

- Model theory

10.9   Footnotes

[1] Grätzer, George. **Universal Algebra,** Van Nostrand Co., Inc., 1968, p. *v.*

[2] Quoted in Grätzer, George. **Universal Algebra,** Van Nostrand Co., Inc., 1968.

[3] Marczewski, E. "A general scheme of the notions of independence in mathematics." Bull. Acad. Polon. Sci. Ser. Sci. Math. Astronom. Phys. **6** (1958), 731–736.

10.10   References

- Bergman, George M., 1998. *An Invitation to General Algebra and Universal Constructions* (pub. Henry Helson, 15 the Crescent, Berkeley CA, 94708) 398 pp. ISBN 0-9655211-4-1.

- Birkhoff, Garrett, 1946. Universal algebra. *Comptes Rendus du Premier Congrès Canadien de Mathématiques,* University of Toronto Press, Toronto, pp. 310–326.

- Brainerd, Barron, Aug–Sep 1967. Review of *Universal Algebra* by P. M. Cohn. *American Mathematical Monthly,* 74(7): 878–880.

- Burris, Stanley N., and H.P. Sankappanavar, 1981. *A Course in Universal Algebra* Springer-Verlag. ISBN 3-540-90578-2 *Free online edition.*

- Cohn, Paul Moritz, 1981. *Universal Algebra*. Dordrecht, Netherlands: D.Reidel Publishing. ISBN 90-277-1213-1 *(First published in 1965 by Harper & Row)*

- Freese, Ralph, and Ralph McKenzie, 1987. *Commutator Theory for Congruence Modular Varieties, 1st ed. London Mathematical Society Lecture Note Series, 125. Cambridge Univ. Press. ISBN 0-521-34832-3. Free online second edition.*

- Grätzer, George, 1968. *Universal Algebra* D. Van Nostrand Company, Inc.

- Higgins, P. J. Groups with multiple operators. Proc. London Math. Soc. (3) 6 (1956), 366–416.

- Higgins, P.J., Algebras with a scheme of operators. *Mathematische Nachrichten* (27) (1963) 115–132.

- Hobby, David, and Ralph McKenzie, 1988. *The Structure of Finite Algebras* American Mathematical Society. ISBN 0-8218-3400-2. *Free online edition.*

- Jipsen, Peter, and Henry Rose, 1992. *Varieties of Lattices*, Lecture Notes in Mathematics 1533. Springer Verlag. ISBN 0-387-56314-8. *Free online edition.*

- Pigozzi, Don. *General Theory of Algebras.*

- Smith, J.D.H., 1976. *Mal'cev Varieties*, Springer-Verlag.

- Whitehead, Alfred North, 1898. *A Treatise on Universal Algebra*, Cambridge. (*Mainly of historical interest.*)

10.11 External links

- *Algebra Universalis*—a journal dedicated to Universal Algebra.

Chapter 11

Algebraic geometry

Not to be confused with Geometric algebra, an application of Clifford algebra to geometry.

For the book by Robin Hartshorne, see Algebraic Geometry (book).

Algebraic geometry is a branch of mathematics, classically studying zeros of multivariate polynomials. Modern algebraic geometry is based on the use of abstract algebraic techniques, mainly from commutative algebra, for solving geometrical problems about these sets of zeros.

The fundamental objects of study in algebraic geometry are algebraic varieties, which are geometric manifestations of solutions of systems of polynomial equations. Examples of the most studied classes of algebraic varieties are: plane algebraic curves, which include lines, circles, parabolas, ellipses, hyperbolas, cubic curves like elliptic curves and quartic curves like lemniscates, and Cassini ovals. A point of the plane belongs to an algebraic curve if its coordinates satisfy a given polynomial equation. Basic questions involve the study of the points of special interest like the singular points, the inflection points and the points at infinity. More advanced questions involve the topology of the curve and relations between the curves given by different equations.

Algebraic geometry occupies a central place in modern mathematics and has multiple conceptual connections with such diverse fields as complex analysis, topology and number theory. Initially a study of systems of polynomial equations in several variables, the subject of algebraic geometry starts where equation solving leaves off, and it becomes even more important to understand the intrinsic properties of the totality of solutions of a system of equations, than to find a specific solution; this leads into some of the deepest areas in all of mathematics, both conceptually and in terms of technique.

In the 20th century, algebraic geometry has split into several subareas.

- The main stream of algebraic geometry is devoted to the study of the complex points of the algebraic varieties and more generally to the points with coordinates in an algebraically closed field.

- The study of the points of an algebraic variety with coordinates in the field of the rational numbers or in a number field became arithmetic geometry (or more classically Diophantine geometry), a subfield of algebraic number theory.

- The study of the real points of an algebraic variety is the subject of real algebraic geometry.

- A large part of singularity theory is devoted to the singularities of algebraic varieties.

- With the rise of the computers, a computational algebraic geometry area has emerged, which lies at the intersection of algebraic geometry and computer algebra. It consists essentially in developing algorithms and software for studying and finding the properties of explicitly given algebraic varieties.

Much of the development of the main stream of algebraic geometry in the 20th century occurred within an abstract algebraic framework, with increasing emphasis being placed on "intrinsic" properties of algebraic varieties not dependent on any particular way of embedding the variety in an ambient coordinate space; this parallels developments in topology, differential and complex geometry. One key achievement of this abstract algebraic geometry is Grothendieck's scheme

This Togliatti surface is an algebraic surface of degree five. The picture represents a portion of its real locus.

theory which allows one to use sheaf theory to study algebraic varieties in a way which is very similar to its use in the study of differential and analytic manifolds. This is obtained by extending the notion of point: In classical algebraic geometry, a point of an affine variety may be identified, through Hilbert's Nullstellensatz, with a maximal ideal of the coordinate ring, while the points of the corresponding affine scheme are all prime ideals of this ring. This means that a point of such a scheme may be either a usual point or a subvariety. This approach also enables a unification of the language and the tools of classical algebraic geometry, mainly concerned with complex points, and of algebraic number theory. Wiles's proof of the longstanding conjecture called Fermat's last theorem is an example of the power of this approach.

11.1 Basic notions

Further information: Algebraic variety

11.1.1   Zeros of simultaneous polynomials

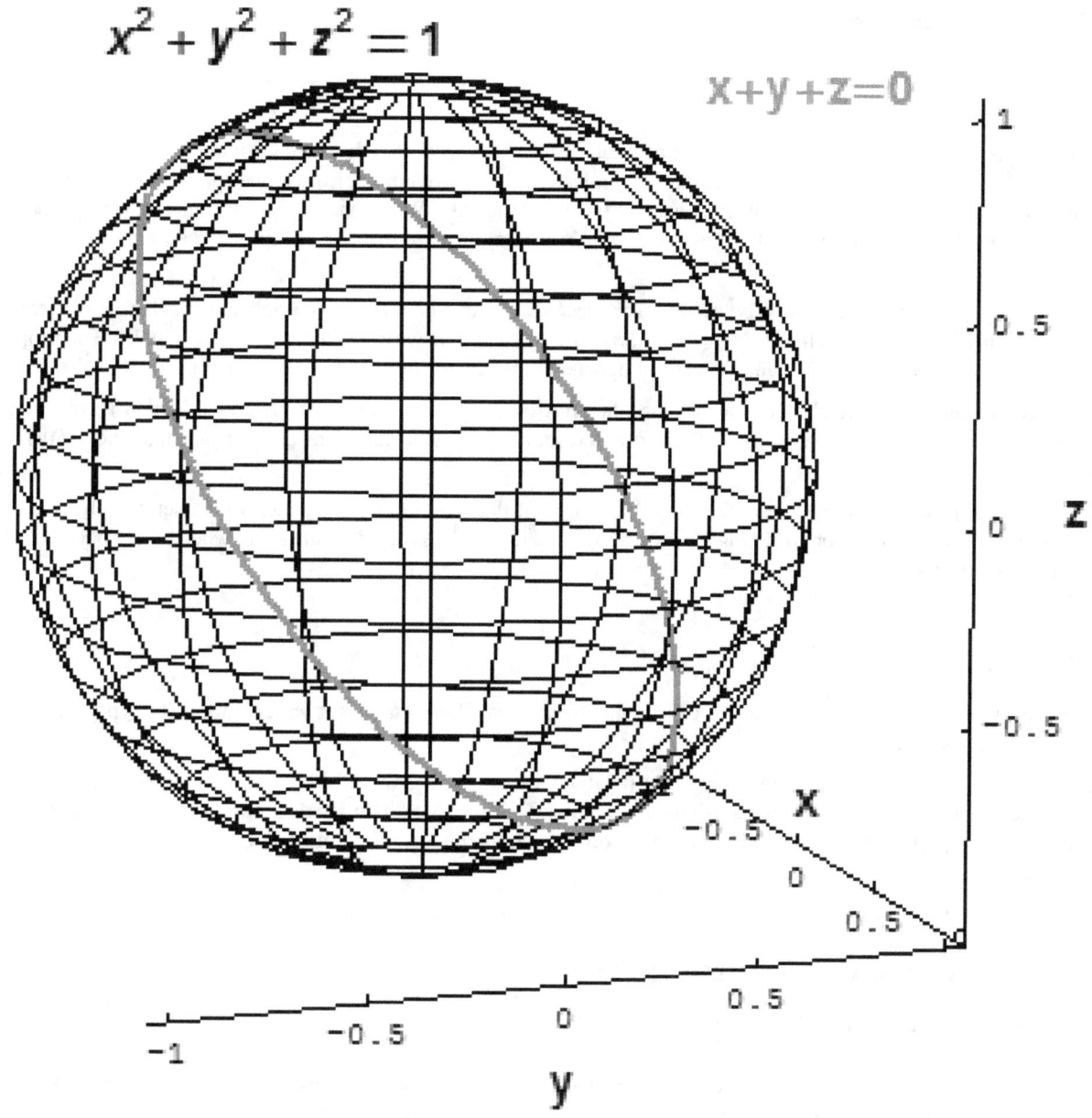

Sphere and slanted circle

In classical algebraic geometry, the main objects of interest are the vanishing sets of collections of polynomials, meaning the set of all points that simultaneously satisfy one or more polynomial equations. For instance, the two-dimensional sphere in three-dimensional Euclidean space $\mathbf{R}^3$ could be defined as the set of all points (x,y,z) with

$$x^2 + y^2 + z^2 - 1 = 0.$$

A "slanted" circle in $\mathbf{R}^3$ can be defined as the set of all points (x,y,z) which satisfy the two polynomial equations

$$x^2 + y^2 + z^2 - 1 = 0,$$

$x + y + z = 0$.

11.1.2 Affine varieties

Main article: Affine variety

First we start with a field k. In classical algebraic geometry, this field was always the complex numbers **C**, but many of the same results are true if we assume only that k is algebraically closed. We consider the affine space of dimension n over k, denoted $\mathbf{A}^n(k)$ (or more simply $\mathbf{A}^n$, when k is clear from the context). When one fixes a coordinates system, one may identify $\mathbf{A}^n(k)$ with k^n. The purpose of not working with k^n is to emphasize that one "forgets" the vector space structure that k^n carries.

A function $f : \mathbf{A}^n \to \mathbf{A}^1$ is said to be *polynomial* (or *regular*) if it can be written as a polynomial, that is, if there is a polynomial p in $k[x_1,...,xn]$ such that $f(M) = p(t_1,...,tn)$ for every point M with coordinates $(t_1,...,tn)$ in $\mathbf{A}^n$. The property of a function to be polynomial (or regular) does not depend on the choice of a coordinate system in $\mathbf{A}^n$.

When a coordinate system is chosen, the regular functions on the affine n-space may be identified with the ring of polynomial functions in n variables over k. Therefore, the set of the regular functions on $\mathbf{A}^n$ is a ring, which is denoted $k[\mathbf{A}^n]$.

We say that a polynomial *vanishes* at a point if evaluating it at that point gives zero. Let S be a set of polynomials in $k[\mathbf{A}^n]$. The *vanishing set of S* (or *vanishing locus* or *zero set*) is the set $V(S)$ of all points in $\mathbf{A}^n$ where every polynomial in S vanishes. In other words,

$$V(S) = \{(t_1, \ldots, t_n) | \forall p \in S, p(t_1, \ldots, t_n) = 0\}.$$

A subset of $\mathbf{A}^n$ which is $V(S)$, for some S, is called an *algebraic set*. The V stands for *variety* (a specific type of algebraic set to be defined below).

Given a subset U of $\mathbf{A}^n$, can one recover the set of polynomials which generate it? If U is *any* subset of $\mathbf{A}^n$, define $I(U)$ to be the set of all polynomials whose vanishing set contains U. The I stands for ideal: if two polynomials f and g both vanish on U, then $f+g$ vanishes on U, and if h is any polynomial, then hf vanishes on U, so $I(U)$ is always an ideal of the polynomial ring $k[\mathbf{A}^n]$.

Two natural questions to ask are:

- Given a subset U of $\mathbf{A}^n$, when is $U = V(I(U))$?

- Given a set S of polynomials, when is $S = I(V(S))$?

The answer to the first question is provided by introducing the Zariski topology, a topology on $\mathbf{A}^n$ whose closed sets are the algebraic sets, and which directly reflects the algebraic structure of $k[\mathbf{A}^n]$. Then $U = V(I(U))$ if and only if U is an algebraic set or equivalently a Zariski-closed set. The answer to the second question is given by Hilbert's Nullstellensatz. In one of its forms, it says that $I(V(S))$ is the radical of the ideal generated by S. In more abstract language, there is a Galois connection, giving rise to two closure operators; they can be identified, and naturally play a basic role in the theory; the example is elaborated at Galois connection.

For various reasons we may not always want to work with the entire ideal corresponding to an algebraic set U. Hilbert's basis theorem implies that ideals in $k[\mathbf{A}^n]$ are always finitely generated.

An algebraic set is called *irreducible* if it cannot be written as the union of two smaller algebraic sets. Any algebraic set is a finite union of irreducible algebraic sets and this decomposition is unique. Thus its elements are called the *irreducible components* of the algebraic set. An irreducible algebraic set is also called a *variety*. It turns out that an algebraic set is a variety if and only if it may be defined as the vanishing set of a prime ideal of the polynomial ring.

Some authors do not make a clear distinction between algebraic sets and varieties and use *irreducible variety* to make the distinction when needed.

11.1.3   Regular functions

Main article: Regular function

Just as continuous functions are the natural maps on topological spaces and smooth functions are the natural maps on differentiable manifolds, there is a natural class of functions on an algebraic set, called *regular functions* or *polynomial functions*. A regular function on an algebraic set V contained in $\mathbf{A}^n$ is the restriction to V of a regular function on $\mathbf{A}^n$. For an algebraic set defined on the field of the complex numbers, the regular functions are smooth and even analytic.

It may seem unnaturally restrictive to require that a regular function always extend to the ambient space, but it is very similar to the situation in a normal topological space, where the Tietze extension theorem guarantees that a continuous function on a closed subset always extends to the ambient topological space.

Just as with the regular functions on affine space, the regular functions on V form a ring, which we denote by $k[V]$. This ring is called the *coordinate ring of V*.

Since regular functions on V come from regular functions on $\mathbf{A}^n$, there is a relationship between the coordinate rings. Specifically, if a regular function on V is the restriction of two functions f and g in $k[\mathbf{A}^n]$, then $f - g$ is a polynomial function which is null on V and thus belongs to $I(V)$. Thus $k[V]$ may be identified with $k[\mathbf{A}^n]/I(V)$.

11.1.4   Morphism of affine varieties

Using regular functions from an affine variety to $\mathbf{A}^1$, we can define regular maps from one affine variety to another. First we will define a regular map from a variety into affine space: Let V be a variety contained in $\mathbf{A}^n$. Choose m regular functions on V, and call them $f_1, ..., fm$. We define a *regular map f* from V to $\mathbf{A}^m$ by letting $f = (f_1, ..., fm)$. In other words, each fi determines one coordinate of the range of f.

If V' is a variety contained in $\mathbf{A}^m$, we say that f is a *regular map* from V to V' if the range of f is contained in V'.

The definition of the regular maps apply also to algebraic sets. The regular maps are also called *morphisms*, as they make the collection of all affine algebraic sets into a category, where the objects are the affine algebraic sets and the morphisms are the regular maps. The affine varieties is a subcategory of the category of the algebraic sets.

Given a regular map g from V to V' and a regular function f of $k[V']$, then $f \circ g \in k[V]$. The map $f \rightarrow f \circ g$ is a ring homomorphism from $k[V']$ to $k[V]$. Conversely, every ring homomorphism from $k[V']$ to $k[V]$ defines a regular map from V to V'. This defines an equivalence of categories between the category of algebraic sets and the opposite category of the finitely generated reduced k-algebras. This equivalence is one of the starting points of scheme theory.

11.1.5   Rational function and birational equivalence

Main article: Rational mapping

Contrarily to the preceding ones, this section concerns only varieties and not algebraic sets. On the other hand, the definitions extend naturally to projective varieties (next section), as an affine variety and its projective completion have the same field of functions.

If V is an affine variety, its coordinate ring is an integral domain and has thus a field of fractions which is denoted $k(V)$ and called the *field of the rational functions* on V or, shortly, the *function field* of V. Its elements are the restrictions to V of the rational functions over the affine space containing V. The domain of a rational function f is not V but the complement of the subvariety (a hypersurface) where the denominator of f vanishes.

Like for regular maps, one may define a *rational map* from a variety V to a variety V'. Like for the regular maps, the rational maps from V to V' may be identified to the field homomorphisms from $k(V')$ to $k(V)$.

Two affine varieties are *birationally equivalent* if there are two rational functions between them which are inverse one to the other in the regions where both are defined. Equivalently, they are birationally equivalent if their function fields are isomorphic.

An affine variety is a *rational variety* if it is birationally equivalent to an affine space. This means that the variety admits a rational parameterization. For example, the circle of equation $x^2 + y^2 - 1 = 0$ is a rational curve, as it has the parameterization

$$x = \frac{2t}{1 + t^2}$$

$$y = \frac{1 - t^2}{1 + t^2},$$

which may also be viewed as a rational map from the line to the circle.

The problem of resolution of singularities is to know if every algebraic variety is birationally equivalent to a variety whose projective completion is nonsingular (see also smooth completion). It has been positively solved in characteristic 0 by Heisuke Hironaka in 1964 and is yet unsolved in finite characteristic.

11.1.6   Projective variety

Main article: Algebraic geometry of projective spaces
Just as the formulas for the roots of 2nd, 3rd and 4th degree polynomials suggest extending real numbers to the more

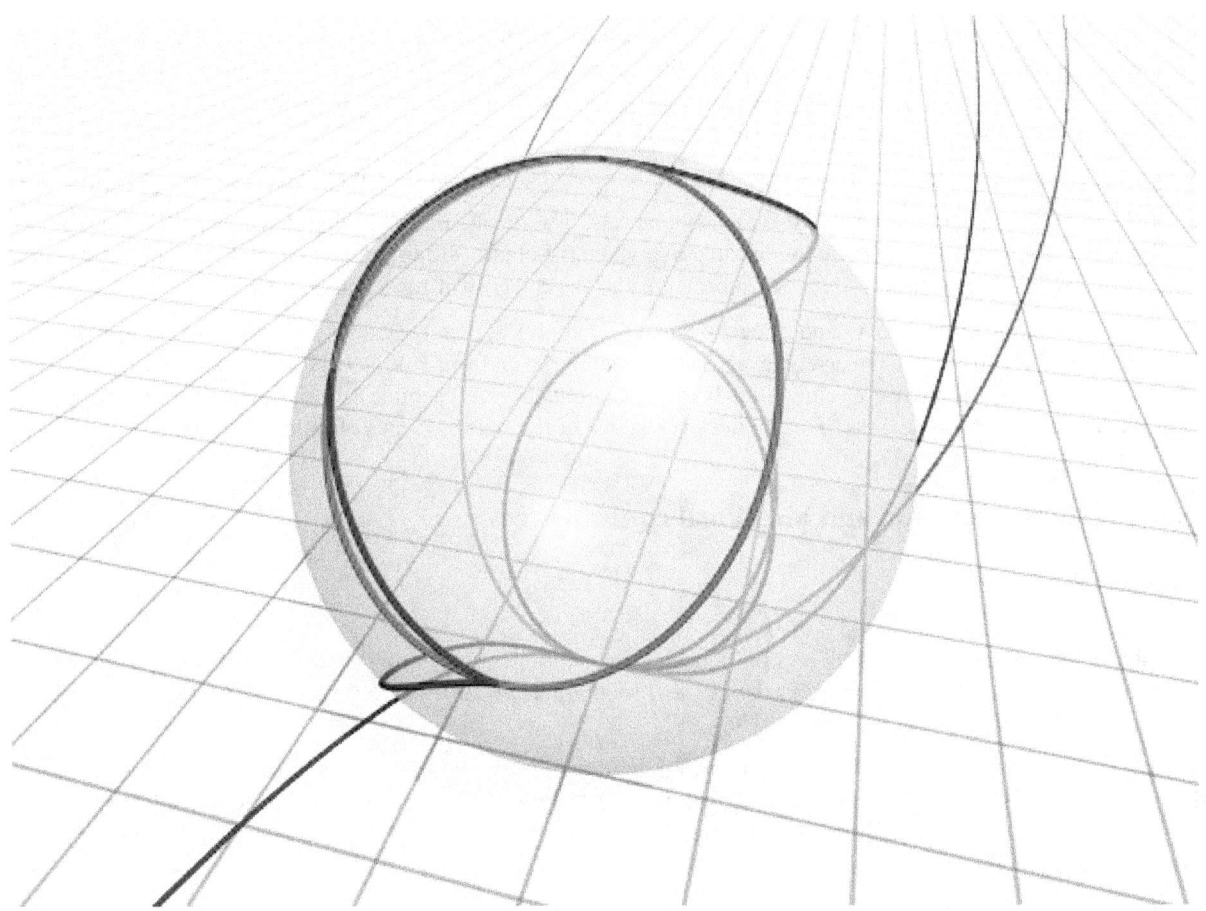

parabola (y = x^2, red) and cubic (y = x^3, blue) in projective space

algebraically complete setting of the complex numbers, many properties of algebraic varieties suggest extending affine space to a more geometrically complete projective space. Whereas the complex numbers are obtained by adding the number i, a root of the polynomial x^2 + 1, projective space is obtained by adding in appropriate points "at infinity", points where parallel lines may meet.

To see how this might come about, consider the variety $V(y - x^2)$. If we draw it, we get a parabola. As x goes to positive infinity, the slope of the line from the origin to the point (x, x^2) also goes to positive infinity. As x goes to negative infinity, the slope of the same line goes to negative infinity.

Compare this to the variety $V(y - x^3)$. This is a cubic curve. As x goes to positive infinity, the slope of the line from the origin to the point (x, x^3) goes to positive infinity just as before. But unlike before, as x goes to negative infinity, the slope of the same line goes to positive infinity as well; the exact opposite of the parabola. So the behavior "at infinity" of $V(y - x^3)$ is different from the behavior "at infinity" of $V(y - x^2)$.

The consideration of the *projective completion* of the two curves, which is their prolongation "at infinity" in the projective plane, allows to quantify this difference: the point at infinity of the parabola is a regular point, whose tangent is the line at infinity, while the point at infinity of the cubic curve is a cusp. Also, both curves are rational, as they are parameterized by x, and Riemann-Roch theorem implies that the cubic curve must have a singularity, which must be at infinity, as all its points in the affine space are regular.

Thus many of the properties of algebraic varieties, including birational equivalence and all the topological properties, depend on the behavior "at infinity" and so it is natural to study the varieties in projective space. Furthermore, the introduction of projective techniques made many theorems in algebraic geometry simpler and sharper: For example, Bézout's theorem on the number of intersection points between two varieties can be stated in its sharpest form only in projective space. For these reasons, projective space plays a fundamental role in algebraic geometry.

Nowadays, the *projective space* $\mathbf{P}^n$ of dimension n is usually defined as the set of the lines passing through a point, considered as the origin, in the affine space of dimension $n+1$, or equivalently to the set of the vector lines in a vector space of dimension $n+1$. When a coordinate system has been chosen in the space of dimension $n+1$, all the points of a line have the same set of coordinates, up to the multiplication by an element of k. This defines the homogeneous coordinates of a point of $\mathbf{P}^n$ as a sequence of $n+1$ elements of the base field k, defined up to the multiplication by a nonzero element of k (the same for the whole sequence).

Given a polynomial in $n+1$ variables, it vanishes at all the point of a line passing through the origin if and only if it is homogeneous. In this case, one says that the polynomial *vanishes* at the corresponding point of $\mathbf{P}^n$. This allows to define a *projective algebraic set* in $\mathbf{P}^n$ as the set $V(f_1, ..., fk)$ where a finite set of homogeneous polynomials $\{f_1, ..., fk\}$ vanishes. Like for affine algebraic sets, there is a bijection between the projective algebraic sets and the reduced homogeneous ideals which define them. The *projective varieties* are the projective algebraic sets whose defining ideal is prime. In other words, a projective variety is a projective algebraic set, whose homogeneous coordinate ring is an integral domain, the *projective coordinates ring* being defined as the quotient of the graded ring or the polynomials in $n+1$ variables by the homogeneous (reduced) ideal defining the variety. Every projective algebraic set may be uniquely decomposed into a finite union of projective varieties.

The only regular functions which may be defined properly on a projective variety are the constant functions. Thus this notion is not used in projective situations. On the other hand, the *field of the rational functions* or *function field* is a useful notion, which, similarly as in the affine case, is defined as the set of the quotients of two homogeneous elements of the same degree in the homogeneous coordinate ring.

11.2   Real algebraic geometry

Main article: Real algebraic geometry

The real algebraic geometry is the study of the real points of the algebraic geometry.

The fact that the field of the reals number is an ordered field may not be occulted in such a study. For example, the curve of equation $x^2 + y^2 - a = 0$ is a circle if $a > 0$, but does not have any real point if $a < 0$. It follows that real algebraic geometry is not only the study of the real algebraic varieties, but has been generalized to the study of the *semi-algebraic sets*, which are the solutions of systems of polynomial equations and polynomial inequalities. For example, a branch of the hyperbola of equation $xy - 1 = 0$ is not an algebraic variety, but is a semi-algebraic set defined by $xy - 1 = 0$ and

$x > 0$ or by $xy - 1 = 0$ and $x + y > 0$.

One of the challenging problems of real algebraic geometry is the unsolved Hilbert's sixteenth problem: Decide which respective positions are possible for the ovals of a nonsingular plane curve of degree 8.

11.3 Computational algebraic geometry

One may date the origin of computational algebraic geometry to meeting EUROSAM'79 (International Symposium on Symbolic and Algebraic Manipulation) held at Marseille, France in June 1979. At this meeting,

- Dennis S. Arnon showed that George E. Collins's Cylindrical algebraic decomposition (CAD) allows the computation of the topology of semi-algebraic sets,

- Bruno Buchberger presented the Gröbner bases and his algorithm to compute them,

- Daniel Lazard presented a new algorithm for solving systems of homogeneous polynomial equations with a computational complexity which is essentially polynomial in the expected number of solutions and thus simply exponential in the number of the unknowns. This algorithm is strongly related with Macaulay's multivariate resultant.

Since then, most results in this area are related to one or several of these items either by using or improving one of these algorithms, or by finding algorithms whose complexity is simply exponential in the number of the variables.

11.3.1 Gröbner basis

Main article: Gröbner basis

A Gröbner basis is a system of generators of a polynomial ideal whose computation allows the deduction of many properties of the affine algebraic variety defined by the ideal.

Given an ideal I defining an algebraic set V:

- V is empty (over an algebraically closed extension of the basis field), if and only if the Gröbner basis for any monomial ordering is reduced to $\{1\}$.

- By mean of the Hilbert series one may compute the dimension and the degree of V from any Gröbner basis of I for a monomial ordering refining the total degree.

- If the dimension of V is 0, one may compute the points (finite in number) of V from any Gröbner basis of I (see systems of polynomial equations).

- A Gröbner basis computation allows to remove from V all irreducible components which are contained in a given hyper surface.

- A Gröbner basis computation allows to compute the Zariski closure of the image of V by the projection on the k first coordinates, and the subset of the image where the projection is not proper.

- More generally Gröbner basis computations allows to compute the Zariski closure of the image and the critical points of a rational function of V into another affine variety.

Gröbner basis computations do not allow to compute directly the primary decomposition of I nor the prime ideals defining the irreducible components of V, but most algorithms for this involve Gröbner basis computation. The algorithms which are not based on Gröbner bases use regular chains but may need Gröbner bases in some exceptional situations.

Gröbner base are deemed to be difficult to compute. In fact they may contain, in the worst case, polynomials whose degree is doubly exponential in the number of variables and a number of polynomials which is also doubly exponential.

However, this is only a worst case complexity, and the complexity bound of Lazard's algorithm of 1979 may frequently apply. Faugère's F4 and F5 algorithms realize this complexity, as F5 algorithm may be viewed as an improvement of Lazard's 1979 algorithm. It follows that the best implementations allow to compute almost routinely with algebraic sets of degree more than 100. This means that, presently, the difficulty of computing a Gröbner basis is strongly related to the intrinsic difficulty of the problem.

11.3.2 Cylindrical Algebraic Decomposition (CAD)

CAD is an algorithm which had been introduced in 1973 by G. Collins to implement with an acceptable complexity the Tarski–Seidenberg theorem on quantifier elimination over the real numbers.

This theorem concerns the formulas of the first-order logic whose atomic formulas are polynomial equalities or inequalities between polynomials with real coefficients. These formulas are thus the formulas which may be constructed from the atomic formulas by the logical operators *and* ($\wedge$), *or* ($\vee$), *not* ($\neg$), *for all* ($\forall$) and *exists* ($\exists$). Tarski's theorem asserts that, from such a formula, one may compute an equivalent formula without quantifier ($\forall$, $\exists$).

The complexity of CAD is doubly exponential in the number of variables. This means that CAD allow, in theory, to solve every problem of real algebraic geometry which may be expressed by such a formula, that is almost every problem concerning explicitly given varieties and semi-algebraic sets.

While Gröbner basis computation has doubly exponential complexity only in rare cases, CAD has almost always this high complexity. This implies that, unless if most polynomials appearing in the input are linear, it may not solve problems with more than four variables.

Since 1973, most of the research on this subject is devoted either to improve CAD or to find alternate algorithms in special cases of general interest.

As an example of the state of art, there are efficient algorithms to find at least a point in every connected component of a semi-algebraic set, and thus to test if a semi-algebraic set is empty. On the other hand, CAD is yet, in practice, the best algorithm to count the number of connected components.

11.3.3 Asymptotic complexity vs. practical efficiency

The basic general algorithms of computational geometry have a double exponential worst case complexity. More precisely, if d is the maximal degree of the input polynomials and n the number of variables, their complexity is at most $d^{2^{cn}}$ for some constant c, and, for some inputs, the complexity is at least $d^{2^{c'n}}$ for another constant c'.

During the last 20 years of 20th century, various algorithms have been introduced to solve specific subproblems with a better complexity. Most of these algorithms have a complexity $d^{O(n^2)}$.

Among these algorithms which solve a sub problem of the problems solved by Gröbner bases, one may cite *testing if an affine variety is empty* and *solving nonhomogeneous polynomial systems which have a finite number of solutions*. Such algorithms are rarely implemented because, on most entries Faugère's F4 and F5 algorithms have a better practical efficiency and probably a similar or better complexity (*probably* because the evaluation of the complexity of Gröbner basis algorithms on a particular class of entries is a difficult task which has be done only in few special cases).

The main algorithms of real algebraic geometry which solve a problem solved by CAD are related to the topology of semi-algebraic sets. One may cite *counting the number of connected components, testing if two points are in the same components* or *computing a Whitney stratification of a real algebraic set*. They have a complexity of $d^{O(n^2)}$, but the constant involved by O notation is so high that using them to solve any nontrivial problem effectively solved by CAD, is impossible even if one could use all the existing computing power in the world. Therefore, these algorithms have never been implemented and this is an active research area to search for algorithms with have together a good asymptotic complexity and a good practical efficiency.

11.4 Abstract modern viewpoint

The modern approaches to algebraic geometry redefine and effectively extend the range of basic objects in various levels of generality to schemes, formal schemes, ind-schemes, algebraic spaces, algebraic stacks and so on. The need for this arises already from the useful ideas within theory of varieties, e.g. the formal functions of Zariski can be accommodated by introducing nilpotent elements in structure rings; considering spaces of loops and arcs, constructing quotients by group actions and developing formal grounds for natural intersection theory and deformation theory lead to some of the further extensions.

Most remarkably, in late 1950s, algebraic varieties were subsumed into Alexander Grothendieck's concept of a scheme. Their local objects are affine schemes or prime spectra which are locally ringed spaces which form a category which is antiequivalent to the category of commutative unital rings, extending the duality between the category of affine algebraic varieties over a field k, and the category of finitely generated reduced k-algebras. The gluing is along Zariski topology; one can glue within the category of locally ringed spaces, but also, using the Yoneda embedding, within the more abstract category of presheaves of sets over the category of affine schemes. The Zariski topology in the set theoretic sense is then replaced by a Grothendieck topology. Grothendieck introduced Grothendieck topologies having in mind more exotic but geometrically finer and more sensitive examples than the crude Zariski topology, namely the étale topology, and the two flat Grothendieck topologies: fppf and fpqc; nowadays some other examples became prominent including Nisnevich topology. Sheaves can be furthermore generalized to stacks in the sense of Grothendieck, usually with some additional representability conditions leading to Artin stacks and, even finer, Deligne-Mumford stacks, both often called algebraic stacks.

Sometimes other algebraic sites replace the category of affine schemes. For example, Nikolai Durov has introduced commutative algebraic monads as a generalization of local objects in a generalized algebraic geometry. Versions of a tropical geometry, of an absolute geometry over a field of one element and an algebraic analogue of Arakelov's geometry were realized in this setup.

Another formal generalization is possible to Universal algebraic geometry in which every variety of algebras has its own algebraic geometry. The term *variety of algebras* should not be confused with *algebraic variety*.

The language of schemes, stacks and generalizations has proved to be a valuable way of dealing with geometric concepts and became cornerstones of modern algebraic geometry.

Algebraic stacks can be further generalized and for many practical questions like deformation theory and intersection theory, this is often the most natural approach. One can extend the Grothendieck site of affine schemes to a higher categorical site of derived affine schemes, by replacing the commutative rings with an infinity category of differential graded commutative algebras, or of simplicial commutative rings or a similar category with an appropriate variant of a Grothendieck topology. One can also replace presheaves of sets by presheaves of simplicial sets (or of infinity groupoids). Then, in presence of an appropriate homotopic machinery one can develop a notion of derived stack as such a presheaf on the infinity category of derived affine schemes, which is satisfying certain infinite categorical version of a sheaf axiom (and to be algebraic, inductively a sequence of representability conditions). Quillen model categories, Segal categories and quasicategories are some of the most often used tools to formalize this yielding the *derived algebraic geometry*, introduced by the school of Carlos Simpson, including Andre Hirschowitz, Bertrand Toën, Gabrielle Vezzosi, Michel Vaquié and others; and developed further by Jacob Lurie, Bertrand Toën, and Gabrielle Vezzosi. Another (noncommutative) version of derived algebraic geometry, using A-infinity categories has been developed from early 1990-s by Maxim Kontsevich and followers.

11.5 History

11.5.1 Prehistory: before the 16th century

Some of the roots of algebraic geometry date back to the work of the Hellenistic Greeks from the 5th century BC. The Delian problem, for instance, was to construct a length x so that the cube of side x contained the same volume as the rectangular box a^2b for given sides a and b. Menaechmus (circa 350 BC) considered the problem geometrically by intersecting the pair of plane conics $ay = x^2$ and $xy = ab$.[1] The later work, in the 3rd century BC, of Archimedes and Apollonius studied more systematically problems on conic sections,[2] and also involved the use of coordinates.[1] The Arab mathematicians were able to solve by purely algebraic means certain cubic equations, and then to interpret the results geometrically. This was done, for instance, by Ibn al-Haytham in the 10th century AD.[3] Subsequently, Persian mathematician Omar Khayyám (born 1048 A.D.) discovered the general method of solving cubic equations by intersecting a parabola with a circle.[4] Each of these early developments in algebraic geometry dealt with questions of finding and describing the intersections of algebraic curves.

11.5.2   Renaissance

Such techniques of applying geometrical constructions to algebraic problems were also adopted by a number of Renaissance mathematicians such as Gerolamo Cardano and Niccolò Fontana "Tartaglia" on their studies of the cubic equation. The geometrical approach to construction problems, rather than the algebraic one, was favored by most 16th and 17th century mathematicians, notably Blaise Pascal who argued against the use of algebraic and analytical methods in geometry.[5] The French mathematicians Franciscus Vieta and later René Descartes and Pierre de Fermat revolutionized the conventional way of thinking about construction problems through the introduction of coordinate geometry. They were interested primarily in the properties of *algebraic curves*, such as those defined by Diophantine equations (in the case of Fermat), and the algebraic reformulation of the classical Greek works on conics and cubics (in the case of Descartes).

During the same period, Blaise Pascal and Gérard Desargues approached geometry from a different perspective, developing the synthetic notions of projective geometry. Pascal and Desargues also studied curves, but from the purely geometrical point of view: the analog of the Greek *ruler and compass construction*. Ultimately, the analytic geometry of Descartes and Fermat won out, for it supplied the 18th century mathematicians with concrete quantitative tools needed to study physical problems using the new calculus of Newton and Leibniz. However, by the end of the 18th century, most of the algebraic character of coordinate geometry was subsumed by the *calculus of infinitesimals* of Lagrange and Euler.

11.5.3   19th and early 20th century

It took the simultaneous 19th century developments of non-Euclidean geometry and Abelian integrals in order to bring the old algebraic ideas back into the geometrical fold. The first of these new developments was seized up by Edmond Laguerre and Arthur Cayley, who attempted to ascertain the generalized metric properties of projective space. Cayley introduced the idea of *homogeneous polynomial forms*, and more specifically quadratic forms, on projective space. Subsequently, Felix Klein studied projective geometry (along with other types of geometry) from the viewpoint that the geometry on a space is encoded in a certain class of transformations on the space. By the end of the 19th century, projective geometers were studying more general kinds of transformations on figures in projective space. Rather than the projective linear transformations which were normally regarded as giving the fundamental Kleinian geometry on projective space, they concerned themselves also with the higher degree birational transformations. This weaker notion of congruence would later lead members of the 20th century Italian school of algebraic geometry to classify algebraic surfaces up to birational isomorphism.

The second early 19th century development, that of Abelian integrals, would lead Bernhard Riemann to the development of Riemann surfaces.

In the same period began the algebraization of the algebraic geometry through commutative algebra. The prominent results in this direction are Hilbert's basis theorem and Hilbert's Nullstellensatz, which are the basis of the connexion between algebraic geometry and commutative algebra, and Macaulay's multivariate resultant, which is the basis of elimination theory. Probably because of the size of the computation which is implied by multivariate resultants, elimination theory was forgotten during the middle of the 20th century until it was renewed by singularity theory and computational algebraic geometry.[6]

11.5.4   20th century

B. L. van der Waerden, Oscar Zariski and André Weil developed a foundation for algebraic geometry based on contemporary commutative algebra, including valuation theory and the theory of ideals. One of the goals was to give a rigorous framework for proving the results of Italian school of algebraic geometry. In particular, this school used systematically the notion of generic point without any precise definition, which was first given by these authors during the 1930s.

In the 1950s and 1960s Jean-Pierre Serre and Alexander Grothendieck recast the foundations making use of sheaf theory. Later, from about 1960, and largely led by Grothendieck, the idea of schemes was worked out, in conjunction with a very refined apparatus of homological techniques. After a decade of rapid development the field stabilized in the 1970s, and new applications were made, both to number theory and to more classical geometric questions on algebraic varieties, singularities and moduli.

An important class of varieties, not easily understood directly from their defining equations, are the abelian varieties, which are the projective varieties whose points form an abelian group. The prototypical examples are the elliptic curves, which have a rich theory. They were instrumental in the proof of Fermat's last theorem and are also used in elliptic curve cryptography.

In parallel with the abstract trend of the algebraic geometry, which is concerned with general statements about varieties, methods for effective computation with concretely-given varieties have also been developed, which lead to the new area of computational algebraic geometry. One of the founding methods of this area is the theory of Gröbner bases, introduced by Bruno Buchberger in 1965. Another founding method, more specially devoted to real algebraic geometry, is the cylindrical algebraic decomposition, introduced by George E. Collins in 1973.

11.6 Analytic geometry

An **analytic variety** is defined locally as the set of common solutions of several equations involving analytic functions. It is analogous to the included concept of real or complex algebraic variety. Any complex manifold is an analytic variety. Since analytic varieties may have singular points, not all analytic varieties are manifolds.

Modern analytic geometry is essentially equivalent to real and complex algebraic geometry, as has been shown by Jean-Pierre Serre in his paper *GAGA*, the name of which is French for *Algebraic geometry and analytic geometry*. Nevertheless, the two fields remain distinct, as the methods of proof are quite different and algebraic geometry includes also geometry in finite characteristic.

11.7 Applications

Algebraic geometry now finds applications in statistics,[7] control theory,[8][9] robotics,[10] error-correcting codes,[11] phylogenetics[12] and geometric modelling.[13] There are also connections to string theory,[14] game theory,[15] graph matchings,[16] solitons[17] and integer programming.[18]

11.8 See also

- Algebraic statistics

- Differential geometry

- Geometric algebra

- Glossary of classical algebraic geometry

- Intersection theory

- Important publications in algebraic geometry

- List of algebraic surfaces

- Noncommutative algebraic geometry

- Differential algebraic geometry

- Real algebraic geometry

11.9   Notes

[1] Dieudonné, Jean (1972). "The historical development of algebraic geometry". *The American Mathematical Monthly* **79** (8): 827–866. doi:10.2307/2317664. JSTOR 2317664.

[2] Kline, M. (1972) *Mathematical Thought from Ancient to Modern Times* (Volume 1). Oxford University Press. pp. 108, 90.

[3] Kline, M. (1972) *Mathematical Thought from Ancient to Modern Times* (Volume 1). Oxford University Press. p. 193.

[4] Kline, M. (1972) *Mathematical Thought from Ancient to Modern Times* (Volume 1). Oxford University Press. pp. 193–195.

[5] Kline, M. (1972) *Mathematical Thought from Ancient to Modern Times* (Volume 1). Oxford University Press. p. 279.

[6] A witness of this oblivion is the fact that Van der Waerden removed the chapter on elimination theory from the third edition (and all the subsequent ones) of his treatise *Moderne algebra* (in German).

[7] Drton, Mathias; Sturmfels, Bernd; Sullivant, Seth (2009). *Lectures on Algebraic Statistics*. Springer. ISBN 978-3-7643-8904-8.

[8] Falb, Peter (1990). *Methods of Algebraic Geometry in Control Theory Part II Multivariable Linear Systems and Projective Algebraic Geometry*. Springer. ISBN 978-0-8176-4113-9.

[9] Allen Tannenbaum (1982), Invariance and Systems Theory: Algebraic and Geometric Aspects, Lecture Notes in Mathematics, volume 845, Springer-Verlag, ISBN 9783540105657

[10] Selig, J.M. (2005). *Geometric Fundamentals of Robotics*. Springer. ISBN 978-0-387-20874-9.

[11] Tsfasman, Michael A.; Vlăduţ, Serge G.; Nogin, Dmitry (1990). *Algebraic Geometric Codes Basic Notions*. American Mathematical Soc. ISBN 978-0-8218-7520-9.

[12] Barry A. Cipra (2007), Algebraic Geometers See Ideal Approach to Biology, SIAM News, Volume 40, Number 6

[13] Jüttler, Bert; Piene, Ragni (2007). *Geometric Modeling and Algebraic Geometry*. Springer. ISBN 978-3-540-72185-7.

[14] Cox, David A.; Katz, Sheldon (1999). *Mirror Symmetry and Algebraic Geometry*. American Mathematical Soc. ISBN 978-0-8218-2127-5.

[15] Blume, L. E.; Zame, W. R. (1994). "The algebraic geometry of perfect and sequential equilibrium" (PDF). *Econometrica* **62** (4): 783–794. JSTOR 2951732.

[16] Kenyon, Richard; Okounkov, Andrei; Sheffield, Scott (2003). "Dimers and Amoebae". arXiv:math-ph/0311005 [math-ph].

[17] Fordy, Allan P. (1990). *Soliton Theory A Survey of Results*. Manchester University Press. ISBN 978-0-7190-1491-8.

[18] Cox, David A.; Sturmfels, Bernd. Manocha, Dinesh N., ed. *Applications of Computational Algebraic Geometry*. American Mathematical Soc. ISBN 978-0-8218-6758-7.

11.10   Further reading

Some classic textbooks that predate schemes

- van der Waerden, B. L. (1945). *Einfuehrung in die algebraische Geometrie*. Dover.

- Hodge, W. V. D.; Pedoe, Daniel (1994). *Methods of Algebraic Geometry Volume 1*. Cambridge University Press. ISBN 0-521-46900-7. Zbl 0796.14001.

- Hodge, W. V. D.; Pedoe, Daniel (1994). *Methods of Algebraic Geometry Volume 2*. Cambridge University Press. ISBN 0-521-46901-5. Zbl 0796.14002.

- Hodge, W. V. D.; Pedoe, Daniel (1994). *Methods of Algebraic Geometry Volume 3*. Cambridge University Press. ISBN 0-521-46775-6. Zbl 0796.14003.

Modern textbooks that do not use the language of schemes

- Garrity, Thomas; et al. (2013). *Algebraic Geometry A Problem Solving Approach*. American Mathematical Society. ISBN 0-821-89396-3.

- Griffiths, Phillip; Harris, Joe (1994). *Principles of Algebraic Geometry*. Wiley-Interscience. ISBN 0-471-05059-8. Zbl 0836.14001.

- Harris, Joe (1995). *Algebraic Geometry A First Course*. Springer-Verlag. ISBN 0-387-97716-3. Zbl 0779.14001.

- Mumford, David (1995). *Algebraic Geometry I Complex Projective Varieties* (2nd ed.). Springer-Verlag. ISBN 3-540-58657-1. Zbl 0821.14001.

- Reid, Miles (1988). *Undergraduate Algebraic Geometry*. Cambridge University Press. ISBN 0-521-35662-8. Zbl 0701.14001.

- Shafarevich, Igor (1995). *Basic Algebraic Geometry I Varieties in Projective Space* (2nd ed.). Springer-Verlag. ISBN 0-387-54812-2. Zbl 0797.14001.

Textbooks in computational algebraic geometry

- Cox, David A.; Little, John; O'Shea, Donal (1997). *Ideals, Varieties, and Algorithms* (2nd ed.). Springer-Verlag. ISBN 0-387-94680-2. Zbl 0861.13012.

- Basu, Saugata; Pollack, Richard; Roy, Marie-Françoise (2006). *Algorithms in real algebraic geometry*. Springer-Verlag.

- González-Vega, Laureano; Recio, Tómas (1996). *Algorithms in algebraic geometry and applications*. Birkhaüser.

- Elkadi, Mohamed; Mourrain, Bernard; Piene, Ragni, eds. (2006). *Algebraic geometry and geometric modeling*. Springer-Verlag.

- Dickenstein, Alicia; Schreyer, Frank-Olaf; Sommese, Andrew J., eds. (2008). *Algorithms in Algebraic Geometry*. The IMA Volumes in Mathematics and its Applications **146**. Springer. ISBN 9780387751559. LCCN 2007938208.

- Cox, David A.; Little, John B.; O'Shea, Donal (1998). *Using algebraic geometry*. Springer-Verlag.

- Caviness, Bob F.; Johnson, Jeremy R. (1998). *Quantifier elimination and cylindrical algebraic decomposition*. Springer-Verlag.

Textbooks and references for schemes

- Eisenbud, David; Harris, Joe (1998). *The Geometry of Schemes*. Springer-Verlag. ISBN 0-387-98637-5. Zbl 0960.14002.

- Grothendieck, Alexander (1960). *Éléments de géométrie algébrique*. Publications Mathématiques de l'IHÉS. Zbl 0118.36206.

- Grothendieck, Alexander; Dieudonné, Jean Alexandre (1971). *Éléments de géométrie algébrique* **1** (2nd ed.). Springer-Verlag. ISBN 3-540-05113-9. Zbl 0203.23301.

- Hartshorne, Robin (1977). *Algebraic Geometry*. Springer-Verlag. ISBN 0-387-90244-9. Zbl 0367.14001.

- Mumford, David (1999). *The Red Book of Varieties and Schemes Includes the Michigan Lectures on Curves and Their Jacobians* (2nd ed.). Springer-Verlag. ISBN 3-540-63293-X. Zbl 0945.14001.

- Shafarevich, Igor (1995). *Basic Algebraic Geometry II Schemes and complex manifolds* (2nd ed.). Springer-Verlag. ISBN 3-540-57554-5. Zbl 0797.14002.

11.11　External links

- *Foundations of Algebraic Geometry* by Ravi Vakil, 764 pp.

- *Algebraic geometry* entry on PlanetMath

- English translation of the van der Waerden textbook

- The History of Algebraic Geometry (1.425 Gigabyte MOV file), a 1972 talk by Jean Dieudonné at the Department of Mathematics of the University of Wisconsin-Milwaukee

- The Stacks Project, an open source textbook and reference work on algebraic stacks and algebraic geometry

Chapter 12

Finite model theory

Finite Model Theory (FMT) is a subarea of model theory (MT). MT is the branch of mathematical logic which deals with the relation between a formal language (syntax) and its interpretations (semantics). FMT is a restriction of MT to interpretations of finite structures, which have a finite universe.

- Since many central theorems of MT do not hold when restricted to finite structures, FMT is quite different from MT in its methods of proof. Central results of classical model theory that fail for finite structures include the compactness theorem, Gödel's completeness theorem, and the method of ultraproducts for first-order logic (FO).

- As MT is closely related to mathematical algebra, FMT became an "unusually effective"[1] instrument in computer science. In other words: "In the history of mathematical logic most interest has concentrated on infinite structures....Yet, the objects computers have and hold are always finite. To study computation we need a theory of finite structures."[2] Thus the main application areas of FMT are: descriptive complexity theory, database theory and formal language theory.

- FMT is mainly about discrimination of structures. The usual motivating question is whether a given class of structures can be described (up to isomorphism) in a given language. For instance, can all cyclic graphs be discriminated (from the non-cyclic ones) by a sentence of the first-order logic of graphs? This can also be phrased as: is the property "cyclic" FO expressible?

12.1 Basic Challenges

A single structure can always be axiomatized in first-order logic, where axiomatized in a language L means described uniquely up to isomorphism by a single L-sentence. Some finite sets of structures can also be axiomatized in FO. However, FO is not sufficient to axiomatize any set containing infinite structures.

12.1.1 Characterisation of a Single Structure

Is a language L expressive enough to axiomatize a single finite structure S?

Problem

A structure like (1) in the figure can be described by FO sentences in the logic of graphs like

1. Every node has an edge to another node: $\forall_x \exists_y G(x, y)$.

2. No node has an edge to itself: $\forall_{x,y}(G(x, y) \Rightarrow x \neq y)$.

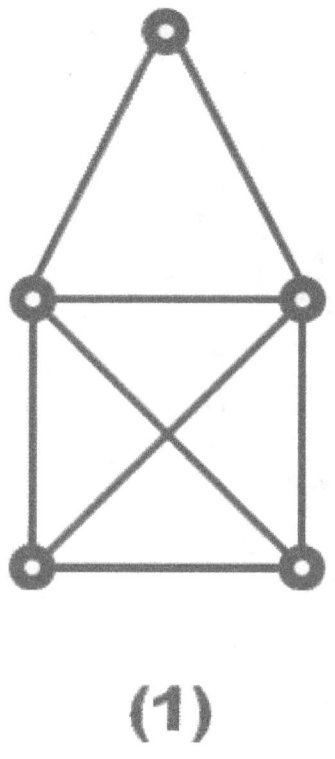

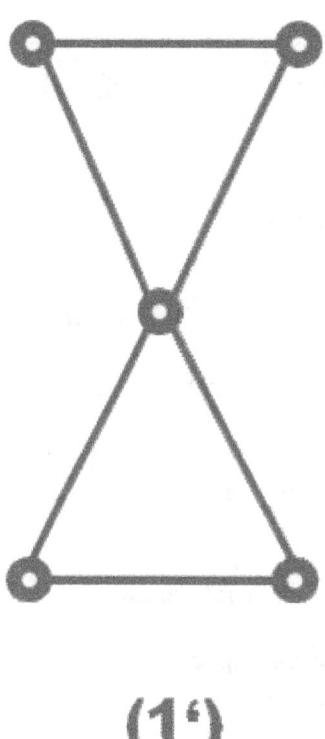

Single graphs (1) and (1') having common properties.

3. There is at least one node that is connected to all others: $\exists_x \forall_y (x \neq y \Rightarrow G(x,y))$.

However, these properties do not axiomatize the structure, since for structure (1') the above properties hold as well, yet structures (1) and (1') are not isomorphic.

Informally the question is whether by adding enough properties, these properties together describe exactly (1) and are valid (all together) for no other structure (up to isomorphism).

Approach

For a single finite structure it is always possible to precisely describe the structure by a single FO sentence. The principle is illustrated here for a structure with one binary relation R and without constants:

1. say that there are at least n elements: $\varphi_1 = \bigwedge_{i \neq j} \neg(x_i = x_j)$

2. say that there are at most n elements: $\varphi_2 = \forall_y \bigvee_i (x_i = y)$

3. state every element of the relation R: $\varphi_3 = \bigwedge_{(a_i,a_j) \in R} R(x_i, x_j)$

4. state every non-element of the relation R: $\varphi_4 = \bigwedge_{(a_i,a_j) \notin R} \neg R(x_i, x_j)$

all for the same tuple $x_1..x_n$, yielding the FO sentence $\exists_{x_1} \ldots \exists_{x_n} (\varphi_1 \wedge \varphi_2 \wedge \varphi_3 \wedge \varphi_4)$.

Extension to a fixed Number of Structures

The method of describing a single structure by means of a first-order sentence can easily be extended for any fixed number of structures. A unique description can be obtained by the disjunction of the descriptions for each structure. For instance, for 2 structures this would be

$$\exists_{x_1}...\exists_{x_n}(\varphi_1 \wedge \varphi_2 \wedge \varphi_3 \wedge \varphi_4) \vee \exists_{x_1}...\exists_{x_n}(\varrho_1 \wedge \varrho_2 \wedge \varrho_3 \wedge \varrho_4).$$

Extension to an infinite Structure

By definition, a set containing an infinite structure falls outside the area that FMT deals with. Note that infinite structures can never be discriminated in FO because of the compactness theorem of classical MT: for every infinite model a non-isomorphic one can be found, but which has exactly the same FO properties.

The most famous example is probably Skolem's theorem, that there is a countable non-standard model of arithmetic.

12.1.2 Characterisation of a Class of Structures

Is a language L expressive enough to describe exactly those finite structures that have certain property P in common (up to isomorphism)?

Problem

The descriptions given so far all specify the number of elements of the universe. Unfortunately most interesting sets of structures are not restricted to a certain size, like all graphs that are trees, are connected or are acyclic. Thus to discriminate a finite number of structures is of special importance.

Approach

Instead of a general statement, the following is a sketch of a methodology to differentiate between structures that can and cannot be discriminated.

1. The core idea is that whenever one wants to see if a Property P can be expressed in FO, one chooses structures A and B, where A does have P and B doesn't. If for A and B the same FO sentences hold, then P cannot be expressed in FO (else it can). In short:

$A \in P, B \notin P$ and $A \equiv B$

where $A \equiv B$ is shorthand for $A \models \alpha \Leftrightarrow B \models \alpha$ for all FO-sentences α, and P represents the class of structures with property P.

2. The methodology considers countably many subsets of the language, the union of which forms the language itself. For instance, for FO consider classes FO[m] for each m. For each m the above core idea then has to be shown. That is:

$A \in P, B \notin P$ and $A \equiv_m B$

with a pair A, B for each m and α (in $\equiv$) from FO[m]. It may be appropriate to choose the classes FO[m] to form a partition of the language.

3. One common way to define FO[m] is by means of the quantifier rank qr(α) of a FO formula α, which expresses the depth of quantifier nesting. For example for a formula in prenex normal form, qr is simply the total number of its quantifiers. Then FO[m] can be defined as all FO formulas α with qr(α) $\leq$ m (or, if a partition is desired, as those FO formulas with quantifier rank equal to m).

4. Thus it all comes down to showing $A \models \alpha \Leftrightarrow B \models \alpha$ on the subsets FO[m]. The main approach here is to use the algebraic characterization provided by Ehrenfeucht–Fraïssé games. Informally, these take a single partial isomorphism on A and B and extend it m times, in order to either prove or disprove $A \equiv_m B$, dependent on who wins the game.

Example

We want to show that the property that the size of an orderered structure **A**=(A, $\leq$) is even, can not be expressed in FO.

1. The idea is to pick **A** $\in$ EVEN and **B** $\notin$ EVEN, where EVEN is the class of all structures of even size.

2. We start with 2 ordered structures **A**$_2$ and **B**$_2$ with universes $A_2 = \{1, 2, 3, 4\}$ and $B_2 = \{1, 2, 3, 4, 5\}$. Obviously **A**$_2$ $\in$ EVEN and **B**$_2 \notin$ EVEN.

3. For m = 2, we can now show* that in a 2-move Ehrenfeucht–Fraïssé game on **A**$_2$ and **B**$_2$ the duplicator always wins, and thus **A**$_2$ and **B**$_2$ cannot be discriminated in FO[2], i.e. **A**$_2 \models \alpha \Leftrightarrow$ **B**$_2 \models \alpha$ for every $\alpha \in$ FO[2].

4. Next we have to scale the structures up by increasing m. For example, for m = 3 we must find an **A**$_3$ and **B**$_3$ such that the duplicator always wins the 3-move game. This can be achieved by $A_3 = \{1, ..., 8\}$ and $B_3 = \{1, ..., 9\}$. More generally, we can choose $A_m = \{1, ..., 2^m\}$ and $B_m = \{1, ..., 2^m+1\}$; for any m the duplicator always wins the m-move game for this pair of structures*.

5. Thus EVEN on finite ordered structures cannot be expressed in FO.

(*) Note that the proof of the result of the Ehrenfeucht–Fraïssé game has been omitted, since it is not the main focus here.

12.2 Applications

12.2.1 Database Theory

A substantial fragment of SQL (namely that which is effectively relational algebra) is based on first-order logic (more precisely can be translated in domain relational calculus by means of Codd's theorem), as the following example illustrates: Think of a database table "GIRLS" with the columns "FIRST_NAME" and "LAST_NAME". This corresponds to a binary relation, say G(f, l) on FIRST_NAME X LAST_NAME. The FO query {l : G(**'Judy'**, l)}, which returns all the last names where the first name is 'Judy', would look in SQL like this:

select LAST_NAME from GIRLS where FIRST_NAME = 'Judy'

Notice, we assume here, that all last names appear only once (or we should use SELECT DISTINCT since we assume that relations and answers are sets, not bags).

Next we want to make a more complex statement. Therefore in addition to the "GIRLS" table we have a table "BOYS" also with the columns "FIRST_NAME" and "LAST_NAME". Now we want to query the last names of all the girls that have the same last name as at least one of the boys. The FO query is {(**f,l**) : $\exists$**h** (**G(f, l)** $\wedge$ **B(h, l)**)}, and the corresponding SQL statement is:

select FIRST_NAME, LAST_NAME from GIRLS where LAST_NAME IN (select LAST_NAME from BOYS);

Notice that in order to express the "$\wedge$" we introduced the new language element "IN" with a subsequent select statement. This makes the language more expressive for the price of higher difficulty to learn and implement. This is a common trade-off in formal language design. The way shown above ("IN") is by far not the only one to extend the language. An alternative way is e.g. to introduce a "JOIN" operator, that is:

select distinct g.FIRST_NAME, g.LAST_NAME from GIRLS g, BOYS b where g.LAST_NAME=b.LAST_NAME;

First-order logic is too restrictive for some database applications, for instance because of its inability to express transitive closure. This has led to more powerful constructs being added to database query languages, such as recursive WITH in SQL:1999. More expressive logics, like fixpoint logics, have therefore been studied in finite model theory because of their relevance to database theory and applications.

12.2.2 Querying & Search

Narrative data contains no defined relations. Thus the logical structure of text search queries can be expressed in Propositional Logic, like in:

("Java" AND NOT "island") OR ("C#" AND NOT "music")

Note that the challenges in full text search are different from database querying, like ranking of results.

12.3 History

1. Trakhtenbrot 1950: failure of completeness theorem in FO,

2. Scholz 1952: characterisation of spectra in FO,

3. Fagin 1974: the set of all properties expressible in existential second-order logic is precisely the complexity class NP,

4. Chandra, Harel 1979/ 80: fixed-point FO extension for db query languages capable of expressing transitive closure -> queries as central objects of FMT.

5. Immerman, Vardi 1982: fixed point logic over ordered structures captures PTIME -> descriptive complexity (... Immerman–Szelepcsényi theorem)

6. Ebbinghaus, Flum 1995: First comprehensive book "Finite Model Theory"

7. Abiteboul, Hull, Vianu 1995: Book "Foundations of Databases"

8. Immerman 1999: Book "Descriptive Complexity"

9. Kuper, Libkin, Paredaens 2000: Book "Constraint Databases"

10. Darmstadt 2005/ Aachen2006: first international workshops on "Algorithmic Model Theory"

12.4 Citations

[1] Fagin, Ronald (1993). "Finite-model theory – a personal perspective" (PDF). *Theoretical Computer Science* **116**: 3–31. doi:10.1016/0304-3975(93)90218-I.

[2] Immerman, Neil (1999). *Descriptive Complexity*. New York: Springer-Verlag. p. 6. ISBN 0-387-98600-6.

12.5 References

- Ebbinghaus, Heinz-Dieter; Flum, Jörg (1995). *Finite Model Theory*. Springer. ISBN 978-3-540-60149-4.

- Libkin, Leonid (2004). *Elements of Finite Model Theory*. Springer. ISBN 3-540-21202-7.

- Abiteboul, Serge; Hull, Richard; Vianu, Victor (1995). *Foundations of Databases*. Addison-Wesley. ISBN 0-201-53771-0.

- Immerman, Neil (1999). *Descriptive Complexity*. New York: Springer. ISBN 0-387-98600-6.

12.6   Further reading

- Grädel, Erich; Kolaitis, Phokion G.; Libkin, Leonid; Maarten, Marx; Spencer, Joel; Vardi, Moshe Y.; Venema, Yde; Weinstein, Scott (2007). *Finite model theory and its applications*. Texts in Theoretical Computer Science. An EATCS Series. Berlin: Springer-Verlag. ISBN 978-3-540-00428-8. Zbl 1133.03001.

12.7   External links

- Libkin, Leonid (2009). "The finite model theory toolbox of a database theoretician". *PODS 2009: Proceedings of the twenty-eighth ACM SIGACT–SIGMOD symposium on Principles of database systems*. pp. 65–76. doi:10.1145 Also suitable as a general introduction and overview.

- Leonid Libkin. Introductory chapter of "Elements of Finite Model Theory". Motivates three main application areas: databases, complexity and formal languages.

- Jouko Väänänen. A Short Course on Finite Model Theory. Department of Mathematics, University of Helsinki. Based on lectures from 1993-1994.

- Anuj Dawar. Infinite and Finite Model Theory, slides, University of Cambridge, 2002.

- "Algorithmic Model Theory". RWTH Aachen. Retrieved 7 November 2013. Includes a list of open FMT problems.

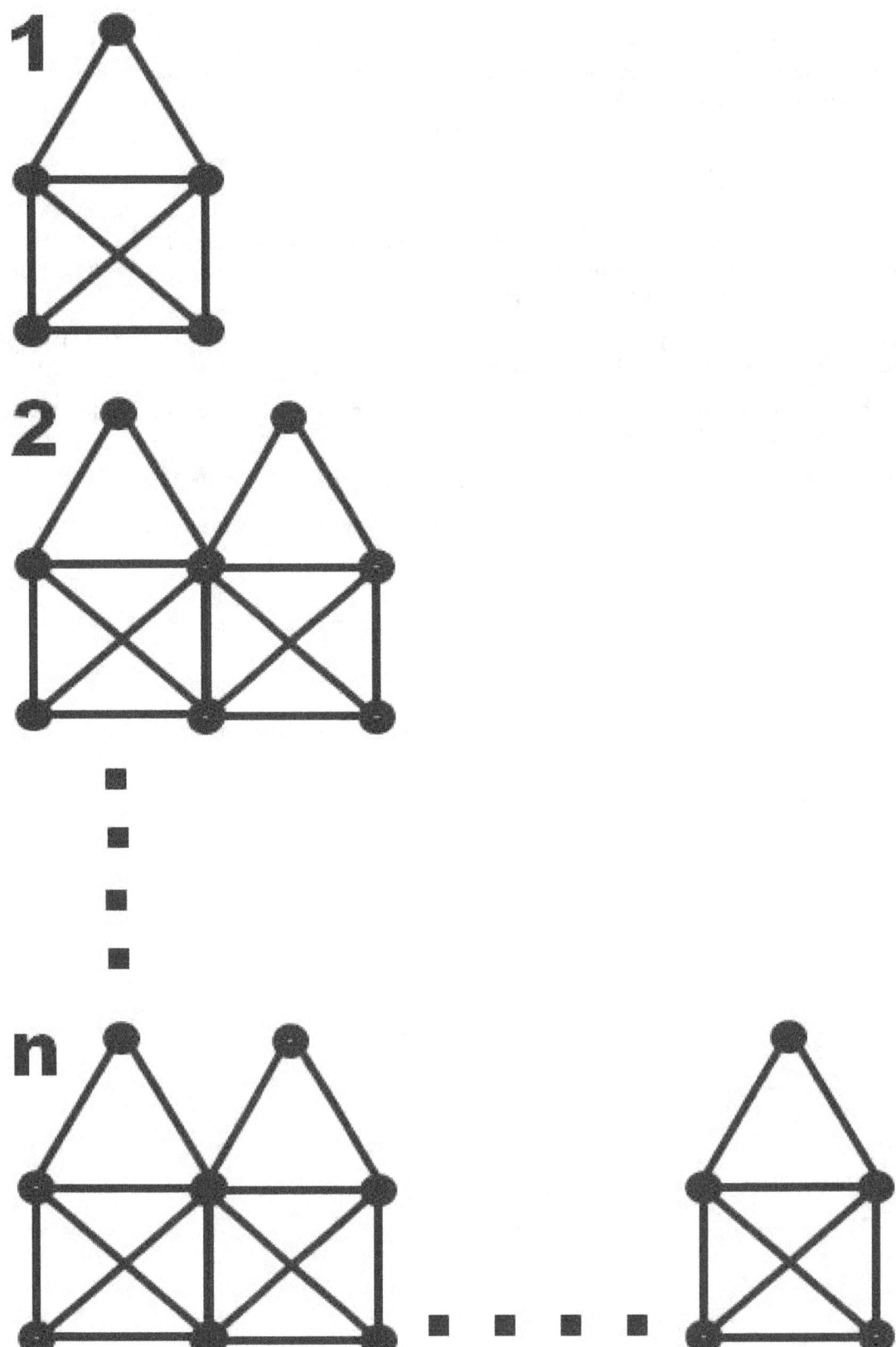

Set of up to n structures.

Chapter 13

Higher-order logic

In mathematics and logic, a **higher-order logic** is a form of predicate logic that is distinguished from first-order logic by additional quantifiers and a stronger semantics. Higher-order logics with their standard semantics are more expressive, but their model-theoretic properties are less well-behaved than those of first-order logic.

First-order logic quantifies only variables that range over individuals; **second-order logic**, in addition, also quantifies over sets; **third-order logic** also quantifies over sets of sets, and so on. For example, the second-order sentence $\forall P((0 \in P \wedge \forall i(i \in P \to i+1 \in P)) \to \forall n(n \in P))$ expresses the principle of mathematical induction. **Higher-order logic** is the union of first-, second-, third-, ..., nth-order logic; *i.e.,* higher-order logic admits quantification over sets that are nested arbitrarily deeply.

13.1 Higher-order simple predicate logic

The term "higher-order logic", abbreviated as **HOL**, is commonly used to mean **higher order simple predicate logic**. Here "simple" indicates that the underlying type theory is simple, not polymorphic or dependent.[1]

There are two possible semantics for HOL.

In the **standard** or **full semantics**, quantifiers over higher-type objects range over *all* possible objects of that type. For example, a quantifier over sets of individuals ranges over the entire powerset of the set of individuals. Thus, in standard semantics, once the set of individuals is specified, this is enough to specify all the quantifiers. HOL with standard semantics is more expressive than first-order logic. For example, HOL admits categorical axiomatizations of the natural numbers, and of the real numbers, which are impossible with first-order logic. However, by a result of Gödel, HOL with standard semantics does not admit an effective, sound, and complete proof calculus.

The model-theoretic properties of HOL with standard semantics are also more complex than those of first-order logic. For example, the Löwenheim number of second-order logic is already larger than the first measurable cardinal, if such a cardinal exists.[2] The Löwenheim number of first-order logic, in contrast, is $\aleph_0$, the smallest infinite cardinal.

In **Henkin semantics**, a separate domain is included in each interpretation for each higher-order type. Thus, for example, quantifiers over sets of individuals may range over only a subset of the powerset of the set of individuals. HOL with these semantics is equivalent to many-sorted first-order logic, rather than being stronger than first-order logic. In particular, HOL with Henkin semantics has all the model-theoretic properties of first-order logic, and has a complete, sound, effective proof system inherited from first-order logic.

13.2 Examples and properties

Higher order logics include the offshoots of Church's Simple Theory of Types[3] and the various forms of Intuitionistic type theory. Gérard Huet has shown that unifiability is undecidable in a type theoretic flavor of third-order logic,[4][5][6] that is, there can be no algorithm to decide whether an arbitrary equation between third-order (let alone arbitrary higher-order) terms has a solution.

Up to a certain notion of isomorphism, the powerset operation is definable in second-order logic. Using this observation, Hintikka established in 1955 that second-order logic can simulate higher-order logics in the sense that for every formula of a higher order-logic one can find an equisatisfiable formula for it in second-order logic.[7]

The term "higher-order logic" is assumed in some context to refer to *classical* higher-order logic. However, modal higher-order logic has been studied as well. According to several logicians, Gödel's ontological proof is best studied (from a technical perspective) in such a context.[8]

13.3   See also

- First-order logic

- Second-order logic

- Higher-order grammar

- Intuitionistic Type Theory

- Many-sorted logic

- Typed lambda calculus

- Modal Logic

13.4   References

[1] Jacobs, 1999, chapter 5

[2] Menachem Magidor and Jouko Väänänen. "On Löwenheim-Skolem-Tarski numbers for extensions of first order logic", Report No. 15 (2009/2010) of the Mittag-Leffler Institute.

[3] Alonzo Church, *A formulation of the simple theory of types*, The Journal of Symbolic Logic 5(2):56–68 (1940)

[4] Huet, Gérard P. (1973). "The Undecidability of Unification in Third Order Logic" (PDF). *Information and Control* **22**: 257–267. doi:10.1016/s0019-9958(73)90301-x.

[5] Huet, Gérard (Sep 1976). *Resolution d'Equations dans des Langages d'Ordre 1,2,...ω* (Ph.D.) (in French). Universite de Paris VII.

[6] Huet, Gérard (2002). "Higher Order Unification 30 years later". In Carreño, V.; Muñoz, C.; Tahar, S. *Proceedings, 15th International Conference TPHOL* (PDF). LNCS **2410**. Springer. pp. 3–12.

[7] SEP entry on HOL

[8] Fitting, Melvin (2002). *Types, Tableaus, and Gödel's God*. Springer Science & Business Media. p. 139. ISBN 978-1-4020-0604-3. Godel's argument is modal and at least second-order, since in his definition of God there is an explicit quantification over properties. [...] [AG96] showed that one could view a part of the argument not as second-order, but as third-order.

- Andrews, Peter B. (2002). *An Introduction to Mathematical Logic and Type Theory: To Truth Through Proof*, 2nd ed, Kluwer Academic Publishers, ISBN 1-4020-0763-9

- Stewart Shapiro, 1991, "Foundations Without Foundationalism: A Case for Second-Order Logic". Oxford University Press., ISBN 0-19-825029-0

- Stewart Shapiro, 2001, "Classical Logic II: Higher Order Logic," in Lou Goble, ed., *The Blackwell Guide to Philosophical Logic*. Blackwell, ISBN 0-631-20693-0

- Lambek, J. and Scott, P. J., 1986. *Introduction to Higher Order Categorical Logic*, Cambridge University Press, ISBN 0-521-35653-9

- Jacobs, Bart (1999). *Categorical Logic and Type Theory*. Studies in Logic and the Foundations of Mathematics 141. North Holland, Elsevier. ISBN 0-444-50170-3.

- Benzmuller, Christoph; Miller, Dale (2014). "Automation of Higher-Order Logic". In Gabbay, Dov M.; Siekmann, Jörg H.; Woods, John. *Handbook of the History of Logic, Volume 9: Computational Logic*. Elsevier. ISBN 978-0-08-093067-1.

13.5   External links

- Andrews, Peter B, Church's Type Theory in Stanford Encyclopedia of Philosophy.

- Miller, Dale, 1991, "Logic: Higher-order," *Encyclopedia of Artificial Intelligence*, 2nd ed.

- Herbert B. Enderton, Second-order and Higher-order Logic in Stanford Encyclopedia of Philosophy, published Dec 20, 2007; substantive revision Mar 4, 2009.

Chapter 14

Infinitary logic

An **infinitary logic** is a logic that allows infinitely long statements and/or infinitely long proofs. Some infinitary logics may have different properties from those of standard first-order logic. In particular, infinitary logics may fail to be compact or complete. Notions of compactness and completeness that are equivalent in finitary logic sometimes are not so in infinitary logics. Therefore for infinitary logics, notions of strong compactness and strong completeness are defined. This article addresses Hilbert-type infinitary logics, as these have been extensively studied and constitute the most straightforward extensions of finitary logic. These are not, however, the only infinitary logics that have been formulated or studied.

Considering whether a certain infinitary logic named Ω-logic is complete promises to throw light on the continuum hypothesis.

14.1   A word on notation and the axiom of choice

As a language with infinitely long formulae is being presented, it is not possible to write expressions down as they should be written. To get around this problem a number of notational conveniences, which, strictly speaking, are not part of the formal language, are used. $\cdots$ is used to point out an expression that is infinitely long. Where it is unclear, the length of the sequence is noted afterwards. Where this notation becomes ambiguous or confusing, suffixes such as $\vee_{\gamma < \delta} A_\gamma$ are used to indicate an infinite disjunction over a set of formulae of cardinality δ. The same notation may be applied to quantifiers for example $\forall_{\gamma < \delta} V_\gamma :$. This is meant to represent an infinite sequence of quantifiers for each V_γ where $\gamma < \delta$.

All usage of suffixes and $\cdots$ are not part of formal infinitary languages. The axiom of choice is assumed (as is often done when discussing infinitary logic) as this is necessary to have sensible distributivity laws.

14.2   Definition of Hilbert-type infinitary logics

A first-order infinitary logic $L\alpha_{,\beta}$, α regular, $\beta = 0$ or $\omega \leq \beta \leq \alpha$, has the same set of symbols as a finitary logic and may use all the rules for formation of formulae of a finitary logic together with some additional ones:

- Given a set of variables $V = \{V_\gamma | \gamma < \delta < \beta\}$ and a formula A_0 then $\forall V_0 : \forall V_1 \cdots (A_0)$ and $\exists V_0 : \exists V_1 \cdots (A_0)$ are formulae (In each case the sequence of quantifiers has length δ).

- Given a set of formulae $A = \{A_\gamma | \gamma < \delta < \alpha\}$ then $(A_0 \vee A_1 \vee \cdots)$ and $(A_0 \wedge A_1 \wedge \cdots)$ are formulae (In each case the sequence has length δ).

The concepts of bound variables apply in the same manner to infinite sentences. Note that the number of brackets in these formulae is always finite. Just as in finitary logic, a formula all of whose variables are bound is referred to as a *sentence*.

A theory T in infinitary logic $L_{\alpha,\beta}$ is a set of statements in the logic. A proof in infinitary logic from a theory T is a sequence of statements of length γ which obeys the following conditions: Each statement is either a logical axiom, an element of T, or is deduced from previous statements using a rule of inference. As before, all rules of inference in finitary logic can be used, together with an additional one:

- Given a set of statements $A = \{A_\gamma | \gamma < \delta < \alpha\}$ which have occurred previously in the proof then the statement $\wedge_{\gamma<\delta} A_\gamma$ can be inferred.

The logical axiom schemata specific to infinitary logic are presented below. Global schemata variables: δ and γ such that $0 < \delta < \alpha$.

- $((\wedge_{\epsilon<\delta}(A_\delta \implies A_\epsilon)) \implies (A_\delta \implies \wedge_{\epsilon<\delta} A_\epsilon))$

- For each $\gamma < \delta$, $((\wedge_{\epsilon<\delta} A_\epsilon) \implies A_\gamma)$

- Chang's distributivity laws (for each γ): $(\vee_{\mu<\gamma}(\wedge_{\delta<\gamma} A_{\mu,\delta}))$, where $\forall \mu \forall \delta \exists \epsilon < \gamma : A_{\mu,\delta} = A_\epsilon$ or $A_{\mu,\delta} = \neg A_\epsilon$, and $\forall g \in \gamma^\gamma \exists \epsilon < \gamma : \{A_\epsilon, \neg A_\epsilon\} \subseteq \{A_{\mu,g(\mu)} : \mu < \gamma\}$

- For $\gamma < \alpha$, $((\wedge_{\mu<\gamma}(\vee_{\delta<\gamma} A_{\mu,\delta})) \implies (\vee_{\epsilon<\gamma^\gamma}(\wedge_{\mu<\gamma} A_{\mu,\gamma_\epsilon(\mu)})))$, where $\{\gamma_\epsilon : \epsilon < \gamma^\gamma\}$ is a well ordering of γ^γ

The last two axiom schemata require the axiom of choice because certain sets must be well orderable. The last axiom schema is strictly speaking unnecessary as Chang's distributivity laws imply it, however it is included as a natural way to allow natural weakenings to the logic.

14.3   Completeness, compactness, and strong completeness

A theory is any set of statements. The truth of statements in models are defined by recursion and will agree with the definition for finitary logic where both are defined. Given a theory T a statement is said to be valid for the theory T if it is true in all models of T.

A logic $L_{\alpha,\beta}$ is complete if for every sentence S valid in every model there exists a proof of S. It is strongly complete if for any theory T for every sentence S valid in T there is a proof of S from T. An infinitary logic can be complete without being strongly complete.

A cardinal $\kappa \neq \omega$ is weakly compact when for every theory T in $L_{\kappa,\kappa}$ containing at most κ many formulas, if every $S \subseteq$ T of cardinality less than κ has a model, then T has a model. A cardinal $\kappa \neq \omega$ is strongly compact when for every theory T in $L_{\kappa,\kappa}$, without restriction on size, if every $S \subseteq$ T of cardinality less than κ has a model, then T has a model.

14.4   Concepts expressible in infinitary logic

In the language of set theory the following statement expresses foundation:

$$\forall_{\gamma<\omega} V_\gamma := \neg \wedge_{\gamma<\omega} V_{\gamma+} \in V_\gamma.$$

Unlike the axiom of foundation, this statement admits no non-standard interpretations. The concept of well foundedness can only be expressed in a logic which allows infinitely many quantifiers in an individual statement. As a consequence many theories, including Peano arithmetic, which cannot be properly axiomatised in finitary logic, can be in a suitable infinitary logic. Other examples include the theories of non-archimedean fields and torsion-free groups. These three theories can be defined without the use of infinite quantification; only infinite junctions are needed.

14.5   Complete infinitary logics

Two infinitary logics stand out in their completeness. These are $L_{\omega,\omega}$ and $L_{\omega_1,\omega}$. The former is standard finitary first-order logic and the latter is an infinitary logic that only allows statements of countable size.

$L_{\omega,\omega}$ is also strongly complete, compact and strongly compact.

$L_{\omega_1,\omega}$ fails to be compact, but it is complete (under the axioms given above). Moreover, it satisfies a variant of the Craig interpolation property.

If $L_{\alpha,\alpha}$ is strongly complete (under the axioms given above) then α is strongly compact (because proofs in these logics cannot use α or more of the given axioms).

14.6   References

- Karp, Carol R. (1964), *Languages with expressions of infinite length*, Amsterdam: North-Holland Publishing Co., MR 0176910

- Barwise, Kenneth Jon(1969), "Infinitary logic and admissible sets",*J. Symbolic Logic***34**(2): 226–252,doi:10.2307 JSTOR 2271099, MR 0406760

Chapter 15

Gödel's completeness theorem

Gödel's completeness theorem is a fundamental theorem in mathematical logic that establishes a correspondence between semantic truth and syntactic provability in first-order logic. It makes a close link between model theory that deals with what is true in different models, and proof theory that studies what can be formally proven in particular formal systems.

It was first proved by Kurt Gödel in 1929. It was then simplified in 1947, when Leon Henkin observed in his Ph.D. thesis that the hard part of the proof can be presented as the Model Existence Theorem (published in 1949). Henkin's proof was simplified by Gisbert Hasenjaeger in 1953.

15.1 Statement of the theorem

15.1.1 Preliminaries

There are numerous deductive systems for first-order logic, including systems of natural deduction and Hilbert-style systems. Common to all deductive systems is the notion of a **formal deduction**. This is a sequence (or, in some cases, a finite tree) of formulas with a specially-designated **conclusion**. The definition of a deduction is such that it is finite and that it is possible to verify algorithmically (by a computer, for example, or by hand) that a given sequence (or tree) of formulas is indeed a deduction.

A first-order formula is called **logically valid** if it is true in every structure for the language of the formula (i.e. for any assignment of values to the variables of the formula). To formally state, and then prove, the completeness theorem, it is necessary to also define a deductive system. A deductive system is called **complete** if every logically valid formula is the conclusion of some formal deduction, and the completeness theorem for a particular deductive system is the theorem that it is complete in this sense. Thus, in a sense, there is a different completeness theorem for each deductive system. A converse to completeness is **soundness,** the fact that only logically valid formulas are provable in the deductive system.

If some specific deductive system of first-order logic is sound and complete, then it is "perfect" (a formula is provable if and only if it is a semantic consequence of the axioms), thus equivalent to any other deductive system with the same quality (any proof in one system can be converted into the other).

15.1.2 Gödel's original formulation

The completeness theorem says that if a formula is logically valid then there is a finite deduction (a formal proof) of the formula.

Gödel's completeness theorem says that a deductive system of first-order predicate calculus is "complete" in the sense that no additional inference rules are required to prove all the logically valid formulas. A converse to completeness is **soundness,** the fact that only logically valid formulas are provable in the deductive system. Together with soundness (whose verification is easy), this theorem implies that a formula is logically valid if and only if it is the conclusion of a formal deduction.

15.1.3 Model existence theorem

The simplest version of this theorem that suffices in practice for most needs, and has connections with the Löwenheim–Skolem theorem, says:

Every consistent, countable first-order theory has a finite or countable model

A more general version can be expressed as :

Every consistent first-order theory with a well-orderable language has a model.

Here, a consistent theory is defined as one in which, for no formula F, both F and ¬F can be proven. See Consistency, the syntactic definition; the semantic definition would be tautological in this context.

This theorem by Henkin is the most directly obtained version of the completeness theorem in its simplest proof.

Given Henkin's theorem, the proof of the completeness theorem is as follows: If $\models A$ is valid, then $\neg A$ does not have models. By the contrapositive of Henkin's, then $\neg A$ is an inconsistent formula. But, by the definition of consistency, if $\neg A$ is inconsistent then it's possible to build a proof of $\vdash A$

15.1.4 More general form

It says that for any first-order theory T with a well-orderable language, and any sentence s in the language of the theory, there is a formal proof of s in T if and only if s is satisfied by every model of T (s is a semantic consequence of T).

This more general theorem is used implicitly, for example, when a sentence is shown to be provable from the axioms of group theory by considering an arbitrary group and showing that the sentence is satisfied by that group. It is deduced from the model existence theorem as follows: if there is no formal proof of a formula then adding its negation to the axioms gives a consistent theory, which has thus a model, so that the formula is not a semantic consequence of the initial theory.

Gödel's original formulation is deduced by taking the particular case of a theory without any axiom.

15.1.5 As a theorem of arithmetic

The Model Existence Theorem and its proof can be formalized in the framework of Peano arithmetic. Precisely, we can systematically define a model of any consistent effective first-order theory T in Peano arithmetic by interpreting each symbol of T by an arithmetical formula whose free variables are the arguments of the symbol. However, the definition expressed by this formula is not recursive.

15.2 Consequences

An important consequence of the completeness theorem is that it is possible to recursively enumerate the semantic consequences of any effective first-order theory, by enumerating all the possible formal deductions from the axioms of the theory, and use this to produce an enumeration of their conclusions.

This comes in contrast with the direct meaning of the notion of semantic consequence, that quantifies over all structures in a particular language, which is clearly not a recursive definition.

Also, it makes the concept of "provability," and thus of "theorem," a clear concept that only depends on the chosen system of axioms of the theory, and not on the choice of a proof system.

15.3   Relationship to the incompleteness theorem

Gödel's incompleteness theorem, another celebrated result, shows that there are inherent limitations in what can be achieved with formal proofs in mathematics. The name for the incompleteness theorem refers to another meaning of *complete* (see model theory – Using the compactness and completeness theorems).

It shows that in any consistent effective theory T containing Peano arithmetic (PA), the formula CT expressing the consistency of T cannot be proven within T.

Applying the completeness theorem to this result, gives the existence of a model of T where the formula CT is false. Such a model (precisely, the set of "natural numbers" it contains) is necessarily non-standard, as it contains the code number of a proof of a contradiction of T. But T is consistent when viewed from the outside. Thus this code number of a proof of contradiction of T must be a non-standard number.

In fact, the model of *any* theory containing PA obtained by the systematic construction of the arithmetical model existence theorem, is *always* non-standard with a non-equivalent provability predicate and a non-equivalent way to interpret its own construction, so that this construction is non-recursive (as recursive definitions would be unambiguous).

Also, there is no recursive non-standard model of PA.

15.4   Relationship to the compactness theorem

The completeness theorem and the compactness theorem are two cornerstones of first-order logic. While neither of these theorems can be proven in a completely effective manner, each one can be effectively obtained from the other.

The compactness theorem says that if a formula φ is a logical consequence of a (possibly infinite) set of formulas Γ then it is a logical consequence of a finite subset of Γ. This is an immediate consequence of the completeness theorem, because only a finite number of axioms from Γ can be mentioned in a formal deduction of φ, and the soundness of the deduction system then implies φ is a logical consequence of this finite set. This proof of the compactness theorem is originally due to Gödel.

Conversely, for many deductive systems, it is possible to prove the completeness theorem as an effective consequence of the compactness theorem.

The ineffectiveness of the completeness theorem can be measured along the lines of reverse mathematics. When considered over a countable language, the completeness and compactness theorems are equivalent to each other and equivalent to a weak form of choice known as weak König's lemma, with the equivalence provable in RCA_0 (a second-order variant of Peano arithmetic restricted to induction over Σ^0_1 formulas). Weak König's lemma is provable in ZF, the system of Zermelo–Fraenkel set theory without axiom of choice, and thus the completeness and compactness theorems for countable languages are provable in ZF. However the situation is different when the language is of arbitrary large cardinality since then, though the completeness and compactness theorems remain provably equivalent to each other in ZF, they are also provably equivalent to a weak form of the axiom of choice known as the ultrafilter lemma. In particular, no theory extending ZF can prove either the completeness or compactness theorems over arbitrary (possibly uncountable) languages without also proving the ultrafilter lemma on a set of same cardinality, knowing that on countable sets, the ultrafilter lemma becomes equivalent to weak König's lemma.

15.5   Completeness in other logics

The completeness theorem is a central property of first-order logic that does not hold for all logics. Second-order logic, for example, does not have a completeness theorem for its standard semantics (but does have the completeness property for Henkin semantics), and the same is true of all higher-order logics. It is possible to produce sound deductive systems for higher-order logics, but no such system can be complete. The set of logically-valid formulas in second-order logic is not enumerable.

Lindström's theorem states that first-order logic is the strongest (subject to certain constraints) logic satisfying both compactness and completeness.

A completeness theorem can be proved for modal logic or intuitionistic logic with respect to Kripke semantics.

15.6   Proofs

Gödel's original proof of the theorem proceeded by reducing the problem to a special case for formulas in a certain syntactic form, and then handling this form with an *ad hoc* argument.

In modern logic texts, Gödel's completeness theorem is usually proved with Henkin's proof, rather than with Gödel's original proof. Henkin's proof directly constructs a term model for any consistent first-order theory. James Margetson (2004) developed a computerized formal proof using the Isabelle theorem prover. Other proofs are also known.

15.7   See also

- Gödel's incompleteness theorems

- Original proof of Gödel's completeness theorem

15.8   Further reading

- Gödel, K (1929). "Über die Vollständigkeit des Logikkalküls". Doctoral dissertation. University Of Vienna. The first proof of the completeness theorem.

- Gödel, K (1930). "Die Vollständigkeit der Axiome des logischen Funktionenkalküls". *Monatshefte für Mathematik* (in German) **37** (1): 349–360. doi:10.1007/BF01696781. JFM 56.0046.04. The same material as the dissertation, except with briefer proofs, more succinct explanations, and omitting the lengthy introduction.

15.9   External links

- Stanford Encyclopedia of Philosophy: "Kurt Gödel"—by Juliette Kennedy.

- MacTutor biography: Kurt Gödel.

- Detlovs, Vilnis, and Podnieks, Karlis, "Introduction to mathematical logic."

Chapter 16

Computable model theory

Computable model theory is a branch of model theory which deals with questions of computability as they apply to model-theoretical structures.

16.1 History

It was developed almost simultaneously by mathematicians in the West, primarily located in the United States and Australia, and Soviet Russia during the middle of the 20th century. Because of the Cold War there was little communication between these two groups and so a number of important results were discovered independently.

16.2 Introduction

Computable model theory introduces the ideas of computable and decidable models and theories and one of the basic problems is discovering whether or not computable or decidable models fulfilling certain model-theoretic conditions can be shown to exist.

16.3 See also

- Vaught conjecture

16.4 References

- Harizanov, V. S. (1998), "Pure Computable Model Theory", in Ershov, Iurii Leonidovich, *Handbook of Recursive Mathematics, Volume 1: Recursive Model Theory*, Studies in Logic and the Foundations of Mathematics **138**, North Holland, pp. 3–114, ISBN 978-0-444-50003-8, MR 1673621.

Chapter 17

Binary operation

Not to be confused with Bitwise operation.

In mathematics, a **binary operation** on a set is a calculation that combines two elements of the set (called operands) to produce another element of the set (more formally, an operation whose arity is two, and whose two domains and one codomain are (subsets of) the same set). Examples include the familiar elementary arithmetic operations of addition, subtraction, multiplication and division. Other examples are readily found in different areas of mathematics, such as vector addition, matrix multiplication and conjugation in groups.

17.1 Terminology

More precisely, a binary operation on a set S is a map which sends elements of the Cartesian product $S \times S$ to S:[1][2][3]

$$f: S \times S \to S.$$

Because the result of performing the operation on a pair of elements of S is again an element of S, the operation is called a **closed** binary operation on S (or sometimes expressed as having the property of closure).[4] If f is not a function, but is instead a partial function, it is called a **partial binary operation**. For instance, division of real numbers is a partial binary operation, because one can't divide by zero: $a/0$ is not defined for any real a. Note however that both in algebra and model theory the binary operations considered are defined on all of $S \times S$.

Sometimes, especially in computer science, the term is used for any binary function.

Binary operations are the keystone of algebraic structures studied in abstract algebra: they are essential in the definitions of groups, monoids, semigroups, rings, and more. Most generally, a *magma* is a set together with some binary operation defined on it.

17.2 Properties and examples

Typical examples of binary operations are the addition (+) and multiplication (×) of numbers and matrices as well as composition of functions on a single set. For instance,

- On the set of real numbers **R**, $f(a, b) = a + b$ is a binary operation since the sum of two real numbers is a real number.

- On the set of natural numbers **N**, $f(a, b) = a + b$ is a binary operation since the sum of two natural numbers is a natural number. This is a different binary operation than the previous one since the sets are different.

- On the set M(2,2) of 2×2 matrices with real entries, $f(A, B) = A + B$ is a binary operation since the sum of two such matrices is another 2×2 matrix.

- On the set M(2,2) of 2×2 matrices with real entries, $f(A, B) = AB$ is a binary operation since the product of two such matrices is another 2×2 matrix.

- For a given set C, let S be the set of all functions $h: C \rightarrow C$. On S, $f(g, h) = g \circ h = g(h(c))$, the composition of the two functions g and h, is a binary operation since the composition of the two functions is another function on the set C (that is, a member of S).

Many binary operations of interest in both algebra and formal logic are commutative, satisfying $f(a, b) = f(b, a)$ for all elements a and b in S, or associative, satisfying $f(f(a, b), c) = f(a, f(b, c))$ for all a, b and c in S. Many also have identity elements and inverse elements.

The first three examples above are commutative and all of the above examples are associative.

On the set of real numbers **R**, subtraction, that is, $f(a, b) = a - b$, is a binary operation which is not commutative since, in general, $a - b \neq b - a$. It is also not associative, since, in general, $a - (b - c) \neq (a - b) - c$; for instance, $1 - (2 - 3) = 2$ but $(1 - 2) - 3 = -4$.

On the set of natural numbers **N**, the binary operation exponentiation, $f(a,b) = a^b$, is not commutative since, in general, $a^b \neq b^a$ and is also not associative since $f(f(a, b), c) \neq f(a, f(b, c))$. For instance, with $a = 2$, $b = 3$ and $c = 2$, $f(2^3,2) = f(8,2) = 64$, but $f(2,3^2) = f(2,9) = 512$. By changing the set **N** to the set of integers **Z**, this binary operation becomes a partial binary operation since it is now undefined when $a = 0$ and b is any negative integer. For either set, this operation has a *right identity* (which is 1) since $f(a, 1) = a$ for all a in the set, which is not an *identity* (two sided identity) since $f(1, b) \neq b$ in general.

Division (/), a partial binary operation on the set of real or rational numbers, is not commutative or associative as well. Tetration ($\uparrow\uparrow$), as a binary operation on the natural numbers, is not commutative nor associative and has no identity element.

17.3   Notation

Binary operations are often written using infix notation such as $a * b$, $a + b$, $a \cdot b$ or (by juxtaposition with no symbol) ab rather than by functional notation of the form $f(a, b)$. Powers are usually also written without operator, but with the second argument as superscript.

Binary operations sometimes use prefix or (probably more often) postfix notation, both of which dispense with parentheses. They are also called, respectively, Polish notation and reverse Polish notation.

17.4   Pair and tuple

A binary operation, ab, depends on the ordered pair (a, b) and so $(ab)c$ (where the parentheses here mean first operate on the ordered pair (a, b) and then operate on the result of that using the ordered pair $((ab), c)$) depends in general on the ordered pair $((a, b), c)$. Thus, for the general, non-associative case, binary operations can be represented with binary trees.

However:

- If the operation is associative, $(ab)c = a(bc)$, then the value of $(ab)c$ depends only on the tuple (a, b, c).

- If the operation is commutative, $ab = ba$, then the value of $(ab)c$ depends only on $\{ \{a, b\}, c\}$, where braces indicate multisets.

- If the operation is both associative and commutative then the value of $(ab)c$ depends only on the multiset $\{a, b, c\}$.

- If the operation is associative, commutative and idempotent, $aa = a$, then the value of $(ab)c$ depends only on the set $\{a, b, c\}$.

17.5 Binary operations as ternary relations

A binary operation f on a set S may be viewed as a *ternary* relation on S, that is, the set of triples $(a, b, f(a,b))$ in $S \times S \times S$ for all a and b in S.

17.6 External binary operations

An **external binary operation** is a binary function from $K \times S$ to S. This differs from a binary operation in the strict sense in that K need not be S; its elements come from *outside*.

An example of an external binary operation is scalar multiplication in linear algebra. Here K is a field and S is a vector space over that field.

An external binary operation may alternatively be viewed as an action; K is acting on S.

Note that the dot product of two vectors is not a binary operation, external or otherwise, as it maps from $S \times S$ to K, where K is a field and S is a vector space over K.

17.7 See also

- Binary operator
- Iterated binary operation
- Operator (programming)
- Ternary operation
- Unary operation

17.8 Notes

[1] Rotman 1973, pg. 1

[2] Hardy & Walker 2002, pg. 176, Definition 67

[3] Fraleigh 1976, pg. 10

[4] Hall 1959, pg. 1

17.9 References

- Fraleigh, John B. (1976), *A First Course in Abstract Algebra* (2nd ed.), Reading: Addison-Wesley, ISBN 0-201-01984-1

- Hall, Jr., Marshall (1959), *The Theory of Groups*, New York: Macmillan

- Hardy, Darel W.; Walker, Carol L. (2003), *Applied Algebra: Codes, Ciphers and Discrete Algorithms*, Upper Saddle River, NJ: Prentice-Hall, ISBN 0-13-067464-8

- Rotman, Joseph J. (1973), *The Theory of Groups: An Introduction* (2nd ed.), Boston: Allyn and Bacon

17.10   External links

- Weisstein, Eric W., "Binary Operation", *MathWorld*.

Chapter 18

Unary operation

In mathematics, a **unary operation** is an operation with only one operand, i.e. a single input. An example is the function $f : A \rightarrow A$, where A is a set. The function f is a unary operation on A.

Common notations are prefix notation (e.g. $+$, $-$, not), postfix notation (e.g. factorial n!), functional notation (e.g. $\sin x$ or $\sin(x)$), and superscripts (e.g. transpose A^{T}). Other notations exist as well. For example, in the case of the square root, a horizontal bar extending the square root sign over the argument can indicate the extent of the argument.

18.1 Unary negative and positive

As unary operations have only one operand they are evaluated before other operations containing them in common mathematics (because certain programming languages do not abide by such rules). Here is an example using negation:

$$3 - -2$$

Here the first '$-$' represents the binary subtraction operation, while the second '$-$' represents the unary negation of the 2 (or '-2' could be taken to mean the integer -2). Therefore, the expression is equal to:

$$3 - (-2) = 5$$

Technically there is also a unary positive but it is not needed since we assume a value to be positive:

$$(+2) = 2$$

Unary positive does not change the sign of a negative operation:

$$(+(-2)) = (-2)$$

In this case a unary negative is needed to change the sign:

$$(-(-2)) = (+2)$$

18.2 Examples from programming languages

18.2.1   C family of languages

In the C family of languages, the following operators are unary:

- Increment: ++x, x++

- Decrement: −−x, x−−

- Address: &x

- Indirection: *x

- Positive: +x

- Negative: −x

- One's complement: ~x

- Logical negation: !x

- Sizeof: sizeof x, sizeof(type-name)

- Cast: (*type-name*) *cast-expression*

18.2.2   Unix Shell (Bash)

In the Unix/Linux shell (bash/sh), **'$'** is a unary operator when used for parameter expansion, replacing the name of a variable by its (sometimes modified) value. For example:

- Simple expansion: $x

- Complex expansion: ${#x}

18.2.3   Other languages

Windows PowerShell

- Increment: ++$x, $x++

- Decrement: −−$x, $x−−

- Positive: +$x

- Negative: −$x

- Logical negation: -not $x

- Invoke in current scope: .$x

- Invoke in new scope: &$x

- Cast: [*type-name*] *cast-expression*

18.3   See also

- Binary operation

- Ternary operation

- Arity

- Operation (mathematics)

- Operator (programming)

18.4   References

- Matt Insall, "Unary Operation", *MathWorld*.

Chapter 19

Arity

In logic, mathematics, and computer science, the **arity** ◀⁾ⁱ/'ærɨti/ of a function or operation is the number of arguments or operands the function or operation accepts. The arity of a relation (or predicate) is the dimension of the domain in the corresponding Cartesian product. (A function of arity n thus has arity $n+1$ considered as a relation.) The term springs from words like unary, binary, ternary, etc. Unary functions or predicates may be also called "monadic"; similarly, binary functions may be called "dyadic".

In mathematics arity may also be named *rank*,[1][2] but this word can have many other meanings in mathematics. In logic and philosophy, arity is also called *adicity* and *degree*.[3][4] In linguistics, arity is usually named *valency*.[5]

In computer programming, there is often a syntactical distinction between operators and functions; syntactical operators usually have arity 0, 1, or 2. Functions vary widely in the number of arguments, though large numbers can become unwieldy. Some programming languages also offer support for variadic functions, i.e., functions syntactically accepting a variable number of arguments.

19.1 Examples

The term "arity" is rarely employed in everyday usage. For example, rather than saying "the arity of the addition operation is 2" or "addition is an operation of arity 2" one usually says "addition is a binary operation". In general, the naming of functions or operators with a given arity follows a convention similar to the one used for n-based numeral systems such as binary and hexadecimal. One combines a Latin prefix with the -ary ending; for example:

- A nullary function takes no arguments.

- A unary function takes one argument.

- A binary function takes two arguments.

- A ternary function takes three arguments.

- An n-ary function takes n arguments.

19.1.1 Nullary

Sometimes it is useful to consider a constant to be an operation of arity 0, and hence call it *nullary*.

Also, in non-functional programming, a function without arguments can be meaningful and not necessarily constant (due to side effects). Often, such functions have in fact some *hidden input* which might be global variables, including the whole state of the system (time, free memory, ...). The latter are important examples which usually also exist in "purely" functional programming languages.

19.1.2 Unary

Examples of unary operators in mathematics and in programming include the unary minus and plus, the increment and decrement operators in C-style languages (not in logical languages), and the factorial, reciprocal, floor, ceiling, fractional part, sign, absolute value, complex conjugate, and norm functions in mathematics. The two's complement, address reference and the logical NOT operators are examples of unary operators in math and programming. According to Quine, a more suitable term is "singulary".[6]

All functions in lambda calculus and in some functional programming languages (especially those descended from ML) are technically unary, but see n-ary below.

19.1.3 Binary

Most operators encountered in programming are of the binary form. For both programming and mathematics these can be the multiplication operator, the addition operator, the division operator. Logical predicates such as *OR*, *XOR*, *AND*, *IMP* are typically used as binary operators with two distinct operands.

19.1.4 Ternary

From C, C++, C#, Java, Perl and variants comes the ternary operator ?:, which is a so-called conditional operator, taking three parameters. Forth also contains a ternary operator, */, which multiplies the first two (one-cell) numbers, dividing by the third, with the intermediate result being a double cell number. This is used when the intermediate result would overflow a single cell. Python has a ternary conditional expression, x if C else y. The dc calculator has several ternary operators, such as |, which will pop three values from the stack and efficiently compute x^y mod z with arbitrary precision. Additionally, many assembly language instructions are ternary or higher, such as MOV %AX, (%BX,%CX), which will load (MOV) into register AX the contents of a calculated memory location that is the sum (parenthesis) of the registers BX and CX.

19.1.5 *n*-ary

From a mathematical point of view, a function of n arguments can always be considered as a function of one single argument which is an element of some product space. However, it may be convenient for notation to consider n-ary functions, as for example multilinear maps (which are not linear maps on the product space, if $n \neq 1$).

The same is true for programming languages, where functions taking several arguments could always be defined as functions taking a single argument of some composite type such as a tuple, or in languages with higher-order functions, by currying.

19.1.6 Variable arity

In computer science, a function accepting a variable number of arguments is called *variadic*. In logic and philosophy, predicates or relations accepting a variable number of arguments are called *multigrade*, anadic, or variably polyadic.[7]

19.2 Other names

There are Latinate names for specific arities, primarily based on Latin distributive numbers meaning "in group of *n*", though some are based on cardinal numbers or ordinal numbers. Only *binary* and *ternary* are both commonly used and derived from distributive numbers.

- *Nullary* means 0-ary (from *nūllus*, zero not being well-understood in antiquity).

- *Unary* means 1-ary (from cardinal *unus*, rather than *singulary* from distributive *singulī*).

- *Binary* means 2-ary.

- *Ternary* means 3-ary.

- *Quaternary* means 4-ary.

- *Quinary* means 5-ary.

- *Senary* means 6-ary.

- *Septenary* means 7-ary.

- *Octonary* means 8-ary (alternatively *octary*).

- *Novenary* means 9-ary (alternatively *nonary*, from ordinal).

- *Denary* means 10-ary (alternatively *decenary*)

- *Polyadic*, *multary* and *multiary* mean 2 or more operands (or parameters).

- *n-ary* means *n* operands (or parameters), but is often used as a synonym of "polyadic".

An alternative nomenclature is derived in a similar fashion from the corresponding Greek roots; for example, *niladic* (or *medadic*), *monadic*, *dyadic*, *triadic*, *polyadic*, and so on. Thence derive the alternative terms *adicity* and *adinity* for the Latin-derived *arity*.

These words are often used to describe anything related to that number (e.g., undenary chess is a chess variant with an 11×11 board, or the Millenary Petition of 1603).

19.3   See also

- Logic of relatives

- Binary relation

- Triadic relation

- Theory of relations

- Signature (logic)

- Parameter

- Parameter

- Cardinality

- Valency

- *n*-ary code

- *n*-ary group

19.4 References

[1] Michiel Hazewinkel (2001). *Encyclopaedia of Mathematics, Supplement III*. Springer. p. 3. ISBN 978-1-4020-0198-7.

[2] Eric Schechter (1997). *Handbook of Analysis and Its Foundations*. Academic Press. p. 356. ISBN 978-0-12-622760-4.

[3] Michael Detlefsen; David Charles McCarty; John B. Bacon (1999). *Logic from A to Z*. Routeledge. p. 7. ISBN 978-0-415-21375-2.

[4] Nino B. Cocchiarella; Max A. Freund (2008). *Modal Logic: An Introduction to its Syntax and Semantics*. Oxford University Press. p. 121. ISBN 978-0-19-536658-7.

[5] David Crystal (2008). *Dictionary of Linguistics and Phonetics* (6th ed.). John Wiley & Sons. p. 507. ISBN 978-1-405-15296-9.

[6] Quine, W. V. O. (1940), *Mathematical logic*, Cambridge, MA: Harvard University Press, p. 13

[7] Oliver, Alex (2004). "Multigrade Predicates". *Mind* **113**: 609–681. doi:10.1093/mind/113.452.609.

19.5 External links

A monograph available free online:

- Burris, Stanley N., and H.P. Sankappanavar, H. P., 1981. *A Course in Universal Algebra*. Springer-Verlag. ISBN 3-540-90578-2. Especially pp. 22–24.

Chapter 20

Ring (mathematics)

This article is about an algebraic structure. For geometric rings, see Annulus (mathematics). For the set theory concept, see Ring of sets.

In mathematics, and more specifically in algebra, a **ring** is an algebraic structure with operations that generalize the arithmetic operations of addition and multiplication. Through this generalization, theorems from arithmetic are extended to non-numerical objects like polynomials, series, matrices and functions.

Rings were first formalized as a common generalization of Dedekind domains that occur in number theory, and of polynomial rings and rings of invariants that occur in algebraic geometry and invariant theory. They are also used in other branches of mathematics such as geometry and mathematical analysis. The formal definition of rings dates from the 1920s.

Briefly, a ring is an abelian group with a second binary operation that is associative, is distributive over the abelian group operation and has an identity element. The abelian group operation is called *addition* and the second binary operation is called *multiplication* by extension from the integers. A familiar example of a ring is the integers. The integers form a commutative ring, since the order in which a pair of elements are multiplied does not change the result. The set of polynomials also forms a commutative ring with the usual operations of addition and multiplication of functions. An example of a ring that is not commutative is the ring of $n \times n$ real square matrices with $n \geq 2$. Finally, a field is a commutative ring in which one can divide by any nonzero element: an example is the field of real numbers.

Whether a ring is commutative or not has profound implication on its behaviour as an abstract object, and the study of such rings is a topic in ring theory. The development of the commutative ring theory, commonly known as commutative algebra, has been greatly influenced by problems and ideas occurring naturally in algebraic number theory and algebraic geometry; important commutative rings include fields, polynomial rings, the coordinate ring of an affine algebraic variety, and the ring of integers of a number field. On the other hand, the noncommutative theory takes examples from representation theory (group rings), functional analysis (operator algebras) and the theory of differential operators (rings of differential operators), and the topology (cohomology ring of a topological space).

20.1 Definition and illustration

The most familiar example of a ring is the set of all integers, **Z**, consisting of the numbers

$$\dots, -5, -4, -3, -2, -1, 0, 1, 2, 3, 4, 5, \dots$$

The familiar properties for addition and multiplication of integers serve as a model for the axioms for rings.

Capitel IX.

Die Zahlringe des Körpers.

§ 31.

Der Zahlring. Das Ringideal und seine wichtigsten Eigenschaften.

Sind ϑ, η, ... irgend welche ganze algebraische Zahlen, deren Rationalitätsbereich der Körper k vom $m$ten Grade ist, so wird das System aller ganzen Functionen von ϑ, η, ..., deren Coefficienten ganze rationale Zahlen sind, ein **Zahlring, Ring** oder **Integritätsbereich***) genannt. Die Addition, Subtraction und Multiplication zweier Zahlen eines Ringes liefert wiederum eine Zahl des Ringes. Der Begriff des Ringes ist mithin gegenüber den drei Rechnungsoperationen der Addition, Subtraction und Multiplication invariant. Der grösste Zahlring des Körpers k ist der durch ω_1, ..., ω_m bestimmte Ring, wo ω_1, ..., ω_m die Zahlen einer Körperbasis bedeuten. Derselbe umfasst alle ganzen Zahlen des Körpers. Jeder Zahlring r enthält m ganze Zahlen ϱ_1, ..., ϱ_m von der Art, dass jede andere Zahl ϱ des Ringes in der Gestalt

$$\varrho = a_1 \varrho_1 + \cdots + a_m \varrho_m$$

dargestellt werden kann, wo a_1, ..., a_m ganze rationale Zahlen sind. Die Zahlen ϱ_1, ..., ϱ_m heissen eine **Basis des Ringes.** Bezeichnen wir die zu ϱ_1, ..., ϱ_m conjugirten Zahlen bez. mit ϱ_1', ..., ϱ_m', ..., $\varrho_1^{(m-1)}$, ..., $\varrho_m^{(m-1)}$, so ist das Quadrat der Determinante

$$\begin{vmatrix} \varrho_1, & \cdots, & \varrho_m \\ \varrho_1', & \cdots, & \varrho_m' \\ \cdots & \cdots & \cdots \\ \varrho_1^{(m-1)}, & \cdots, & \varrho_m^{(m-1)} \end{vmatrix}$$

eine rationale Zahl und heisst die **Discriminante** d_r **des Ringes** r.

Ein **Ringideal** oder ein **Ideal des Ringes** r wird ein solches unendliches System von ganzen algebraischen Zahlen α_1, α_2, ... des Ringes r genannt, welches die Eigenschaft besitzt, dass eine jede lineare Combination $\lambda_1 \alpha_1 + \lambda_2 \alpha_2 + \cdots$ derselben wiederum dem System angehört,

---

*) Nach *Dedekind* „eine Ordnung".

Chapter IX of David Hilbert's Die Theorie der algebraischen Zahlkörper. *The chapter title is* Die Zahlringe des Körpers, *literally "the number rings of the field". The word "ring" is the contraction of "Zahlring".*

20.1.1   Definition

A **ring** is a set R equipped with binary operations[1] $+$ and $\cdot$ satisfying the following three sets of axioms, called the **ring axioms**[2][3][4]

1. R is an abelian group under addition, meaning that

   - $(a + b) + c = a + (b + c)$ for all a, b, c in R ($+$ is associative).
   - $a + b = b + a$ for all a, b in R ($+$ is commutative).
   - There is an element 0 in R such that $a + 0 = a$ for all a in R (0 is the additive identity).
   - For each a in R there exists $-a$ in R such that $a + (-a) = 0$ ($-a$ is the additive inverse of a).

2. R is a monoid under multiplication, meaning that:

   - $(a \cdot b) \cdot c = a \cdot (b \cdot c)$ for all a, b, c in R ($\cdot$ is associative).
   - There is an element 1 in R such that $a \cdot 1 = a$ and $1 \cdot a = a$ for all a in R (1 is the multiplicative identity).[5]

3. Multiplication is distributive with respect to addition:

   - $a \cdot (b + c) = (a \cdot b) + (a \cdot c)$ for all a, b, c in R (left distributivity).
   - $(b + c) \cdot a = (b \cdot a) + (c \cdot a)$ for all a, b, c in R (right distributivity).

20.1.2   Notes on the definition

As explained in § History below, many authors follow an alternative convention in which a ring is not defined to have a multiplicative identity. This article adopts the convention that, unless otherwise stated, a ring is assumed to have such an identity. A structure satisfying all the axioms *except* possibly the existence of a multiplicative identity 1 is called a **rng** (or sometimes **pseudo-ring**). For example, the set of even integers with the usual $+$ and $\cdot$ is a rng, but not a ring.

The operations $+$ and $\cdot$ are called *addition* and *multiplication*, respectively. The multiplication symbol $\cdot$ is often omitted, so the juxtaposition of ring elements is interpreted as multiplication. For example, xy means $x \cdot y$.

Although ring addition is commutative, ring multiplication is not required to be commutative: ab need not necessarily equal ba. Rings that also satisfy commutativity for multiplication (such as the ring of integers) are called **commutative rings**. Books on commutative algebra or algebraic geometry often adopt the convention that "ring" means "commutative ring", to simplify terminology.

The additive group of a ring is the ring equipped just with the structure of addition. Although the definition assumes that the additive group is abelian, this can be inferred from the other ring axioms.[6]

20.1.3   Basic properties

Some basic properties of a ring follow immediately from the axioms:

- The additive identity, the additive inverse of each element, and the multiplicative identity are unique.

- For any element x in a ring R, one has $x0 = 0 = 0x$ and $(-1)x = -x$.

- If $0 = 1$ in a ring R (or more generally, 0 is a unit element), then R has only one element, and is called the zero ring.

- The binomial formula holds for any commuting pair of elements (i.e., any x and y such that $xy = yx$).

20.1.4 Example: Integers modulo 4

See also: Modular arithmetic

Equip the set $\mathbf{Z}_4 = \{\overline{0}, \overline{1}, \overline{2}, \overline{3}\}$ with the following operations:

- The sum $\overline{x} + \overline{y}$ in $\mathbf{Z}_4$ is the remainder when the integer $x + y$ is divided by 4. For example, $\overline{2} + \overline{3} = \overline{1}$ and $\overline{3} + \overline{3} = \overline{2}$
  .

- The product $\overline{x} \cdot \overline{y}$ in $\mathbf{Z}_4$ is the remainder when the integer xy is divided by 4. For example, $\overline{2} \cdot \overline{3} = \overline{2}$ and $\overline{3} \cdot \overline{3} = \overline{1}$.

Then $\mathbf{Z}_4$ is a ring: each axiom follows from the corresponding axiom for $\mathbf{Z}$. If x is an integer, the remainder of x when divided by 4 is an element of $\mathbf{Z}_4$, and this element is often denoted by "x mod 4" or $\overline{x}$, which is consistent with the notation for 0,1,2,3. The additive inverse of any $\overline{x}$ in $\mathbf{Z}_4$ is $\overline{-x}$. For example, $-\overline{3} = \overline{-3} = \overline{1}$.

20.1.5 Example: 2-by-2 matrices

Main article: Matrix ring

The set of 2-by-2 matrices with real number entries is written

$$\mathcal{M}_2(\mathbb{R}) = \left\{ \begin{pmatrix} a & b \\ c & d \end{pmatrix} \middle| a, b, c, d \in \mathbb{R} \right\}.$$

With the operations of matrix addition and matrix multiplication, this set satisfies the above ring axioms. The element $\begin{pmatrix} 1 & 0 \\ 0 & 1 \end{pmatrix}$ is the multiplicative identity of the ring. If $A = \begin{pmatrix} 0 & 1 \\ 1 & 0 \end{pmatrix}$ and $B = \begin{pmatrix} 0 & 1 \\ 0 & 0 \end{pmatrix}$, then $AB = \begin{pmatrix} 0 & 0 \\ 0 & 1 \end{pmatrix}$ while $BA = \begin{pmatrix} 1 & 0 \\ 0 & 0 \end{pmatrix}$; this example shows that the ring is noncommutative.

More generally, for any ring R, commutative or not, and any nonnegative integer n, one may form the ring of n-by-n matrices with entries in R: see matrix ring.

20.2 History

See also: Ring theory § History

20.2.1 Dedekind

The study of rings originated from the theory of polynomial rings and the theory of algebraic integers.[7] In 1871 Richard Dedekind defined the concept of the ring of integers of a number field.[8] In this context, he introduced the terms "ideal" (inspired by Ernst Kummer's notion of ideal number) and "module" and studied their properties. But Dedekind did not use the term "ring" and did not define the concept of a ring in a general setting.

20.2.2 Hilbert

The term "Zahlring" (number ring) was coined by David Hilbert in 1892 and published in 1897.[9] In 19th century German, the word "Ring" could mean "association", which is still used today in English in a limited sense (e.g., spy ring),[10] so

Richard Dedekind, one of the founders of ring theory.

if that were the etymology then it would be similar to the way "group" entered mathematics by being a non-technical word for "collection of related things". According to Harvey Cohn, Hilbert used the term for a ring that had the property of "circling directly back" to an element of itself.[11] Specifically, in a ring of algebraic integers, all high powers of an algebraic integer can be written as an integral combination of a fixed set of lower powers, and thus the powers "cycle back". For instance, if $a^3 - 4a + 1 = 0$ then $a^3 = 4a - 1$, $a^4 = 4a^2 - a$, $a^5 = -a^2 + 16a - 4$, $a^6 = 16a^2 - 8a + 1$, $a^7 = -8a^2 + 65a - 16$, and so on; in general, a^n is going to be an integral linear combination of 1, a, and a^2.

20.2.3 Fraenkel and Noether

The first axiomatic definition of a ring was given by Adolf Fraenkel in 1914,[12][13] but his axioms were stricter than those in the modern definition. For instance, he required every non-zero-divisor to have a multiplicative inverse.[14] In 1921, Emmy Noether gave the modern axiomatic definition of (commutative) ring and developed the foundations of commutative ring theory in her monumental paper *Idealtheorie in Ringbereichen*.[15]

20.2.4 Multiplicative identity: mandatory or optional?

Fraenkel required a ring to have a multiplicative identity 1,[16] whereas Noether did not.[15]

Most or all books on algebra[17][18] up to around 1960 followed Noether's convention of not requiring a 1. Starting in the 1960s, it became increasingly common to see books including the existence of 1 in the definition of ring, especially in advanced books by notable authors such as Artin,[19] Atiyah and MacDonald,[20] Bourbaki,[21] Eisenbud,[22] and Lang.[23] But even today, there remain many books that do not require a 1.

Faced with this terminological ambiguity, some authors have tried to impose their views, while others have tried to adopt more precise terms.

In the first category, we find for instance Gardner and Wiegandt, who argue that if one requires all rings to have a 1, then some consequences include the lack of existence of infinite direct sums of rings, and the fact that proper direct summands of rings are not subrings. They conclude that "in many, maybe most, branches of ring theory the requirement of the existence of a unity element is not sensible, and therefore unacceptable."[24]

In the second category, we find authors who use the following terms:[25][26]

- rings with multiplicative identity: *unital ring, unitary ring, ring with unity, ring with identity*, or *ring with 1*
- rings not requiring multiplicative identity: *rng* or *pseudo-ring*.

20.3 Basic examples

Commutative rings:

- The prototype example is the ring of integers with the two operations of addition and multiplication.
- The rational, real and complex numbers are commutative rings of a type called fields.
- The Gaussian integers form a ring, as do the Eisenstein integers. So does their generalization Kummer ring. cf. quadratic integers.
- The set of all algebraic integers forms a ring. This follows for example from the fact that it is the integral closure of the ring of rational integers in the field of complex numbers. The rings in the previous example are subrings of this ring.
- The polynomial ring $R[X]$ of polynomials over a ring R is also a ring.
- The set of formal power series $R[[X_1, ..., Xn]]$ over a commutative ring R is a ring.
- If S is a set, then the power set of S becomes a ring if we define addition to be the symmetric difference of sets and multiplication to be intersection. This corresponds to a ring of sets and is an example of a Boolean ring.

- The set of all continuous real-valued functions defined on the real line forms a commutative ring. The operations are pointwise addition and multiplication of functions.

- Let X be a set and R a ring. Then the set of all functions from X to R forms a ring, which is commutative if R is commutative. The ring of continuous functions in the previous example is a subring of this ring if X is the real line and R is the field of real numbers.

Noncommutative rings:

- For any ring R and any natural number n, the set of all square n-by-n matrices with entries from R, forms a ring with matrix addition and matrix multiplication as operations. For $n = 1$, this matrix ring is isomorphic to R itself. For $n > 1$ (and R not the zero ring), this matrix ring is noncommutative.

- If G is an abelian group, then the endomorphisms of G form a ring, the endomorphism ring End(G) of G. The operations in this ring are addition and composition of endomorphisms. More generally, if V is a left module over a ring R, then the set of all R-linear maps forms a ring, also called the endomorphism ring and denoted by End$R(V)$.

- If G is a group and R is a ring, the group ring of G over R is a free module over R having G as basis. Multiplication is defined by the rules that the elements of G commute with the elements of R and multiply together as they do in the group G.

- Many rings that appear in analysis are noncommutative. For example, most Banach algebras are noncommutative.

Non-rings:

- The set of natural numbers **N** with the usual operations is not a ring, since (**N**, +) is not even a group (the elements are not all invertible with respect to addition). For instance, there is no natural number which can be added to 3 to get 0 as a result. There is a natural way to make it a ring by adding negative numbers to the set, thus obtaining the ring of integers. The natural numbers (including 0) form an algebraic structure known as a semiring (which has all of the properties of a ring except the additive inverse property).

- Let R be the set of all continuous functions on the real line that vanish outside a bounded interval depending on the function, with addition as usual but with multiplication defined as convolution:

$$(f * g)(x) = \int_{-\infty}^{\infty} f(y)g(x-y)dy.$$

  Then R is a rng, but not a ring: the Dirac delta function has the property of a multiplicative identity, but it is not a function and hence is not an element of R.

20.4   Basic concepts

20.4.1   Elements in a ring

A left zero divisor of a ring R is an element a in the ring such that there exists a nonzero element b of R such that $ab = 0$.[27] A right zero divisor is defined similarly.

A nilpotent element is an element a such that $a^n = 0$ for some $n > 0$. One example of a nilpotent element is a nilpotent matrix. A nilpotent element in a nonzero ring is necessarily a zero divisor.

An idempotent e is an element such that $e^2 = e$. One example of an idempotent element is a projection in linear algebra.

A unit is an element a having a multiplicative inverse; in this case the inverse is unique, and is denoted by a^{-1} . The set of units of a ring is a group under ring multiplication; this group is denoted by $R^\times$ or R^* or $U(R)$. For example, if R is the ring of all square matrices of size n over a field, then $R^\times$ consists of the set of all invertible matrices of size n, and is called the general linear group.

20.4.2 Subring

Main article: Subring

A subset S of R is said to be a subring if it can be regarded as a ring with the addition and the multiplication restricted from R to S. Equivalently, S is a subring if it is not empty, and for any x, y in S, xy, $x + y$ and $-x$ are in S. If all rings have been assumed, by convention, to have a multiplicative identity, then to be a subring one would also require S to share the same identity element as R.[28] So if all rings have been assumed to have a multiplicative identity, then a proper ideal is not a subring.

For example, the ring $\mathbf{Z}$ of integers is a subring of the field of real numbers and also a subring of the ring of polynomials $\mathbf{Z}[X]$ (in both cases, $\mathbf{Z}$ contains 1, which is the multiplicative identity of the larger rings). On the other hand, the subset of even integers $2\mathbf{Z}$ does not contain the identity element 1 and thus does not qualify as a subring.

An intersection of subrings is a subring. The smallest subring containing a given subset E of R is called a subring generated by E. Such a subring exists since it is the intersection of all subrings containing E.

For a ring R, the smallest subring containing 1 is called the *characteristic subring* of R. It can be obtained by adding copies of 1 and -1 together many times in any mixture. It is possible that $n \cdot 1 = 1 + 1 + \ldots + 1$ (n times) can be zero. If n is the smallest positive integer such that this occurs, then n is called the *characteristic* of R. In some rings, $n \cdot 1$ is never zero for any positive integer n, and those rings are said to have *characteristic zero*.

Given a ring R, let $Z(R)$ denote the set of all elements x in R such that x commutes with every element in R: $xy = yx$ for any y in R. Then $Z(R)$ is a subring of R; called the center of R. More generally, given a subset X of R, let S be the set of all elements in R that commute with every element in X. Then S is a subring of R, called the centralizer (or commutant) of X. The center is the centralizer of the entire ring R. Elements or subsets of the center are said to be *central* in R; they generate a subring of the center.

20.4.3 Ideal

Main article: Ideal (ring theory)

The definition of an ideal in a ring is analogous to that of normal subgroup in a group. But, in actuality, it plays a role of an idealized generalization of an element in a ring; hence, the name "ideal". Like elements of rings, the study of ideals is central to structural understanding of a ring.

Let R be a ring. A nonempty subset I of R is then said to be a **left ideal** in R if, for any x, y in I and r in R, $x + y$ and rx are in I. If RI denotes the span of I over R; i.e., the set of finite sums

$$r_1 x_1 + \cdots + r_n x_n, \quad r_i \in R, \quad x_i \in I,$$

then I is a left ideal if $RI \subseteq I$. Similarly, I is said to be **right ideal** if $IR \subseteq I$. A subset I is said to be a **two-sided ideal** or simply **ideal** if it is both a left ideal and right ideal. A one-sided or two-sided ideal is then an additive subgroup of R. If E is a subset of R, then RE is a left ideal, called the left ideal generated by E; it is the smallest left ideal containing E. Similarly, one can consider the right ideal or the two-sided ideal generated by a subset of R.

If x is in R, then Rx and xR are left ideals and right ideals, respectively; they are called the principal left ideals and right ideals generated by x. The principal ideal RxR is written as (x). For example, the set of all positive and negative multiples of 2 along with 0 form an ideal of the integers, and this ideal is generated by the integer 2. In fact, every ideal of the ring of integers is principal.

Like a group, a ring is said to be a simple if it is nonzero and it has no proper nonzero two-sided ideals. A commutative simple ring is precisely a field.

Rings are often studied with special conditions set upon their ideals. For example, a ring in which there is no strictly increasing infinite chain of left ideals is called a left Noetherian ring. A ring in which there is no strictly decreasing infinite chain of left ideals is called a left Artinian ring. It is a somewhat surprising fact that a left Artinian ring is left Noetherian (the Hopkins–Levitzki theorem). The integers, however, form a Noetherian ring which is not Artinian.

For commutative rings, the ideals generalize the classical notion of divisibility and decomposition of an integer into prime numbers in algebra. A proper ideal P of R is called a prime ideal if for any elements $x, y \in R$ we have that $xy \in P$ implies either $x \in P$ or $y \in P$. Equivalently, P is prime if for any ideals I, J we have that $IJ \subseteq P$ implies either $I \subseteq P$ or $J \subseteq P$. This latter formulation illustrates the idea of ideals as generalizations of elements.

20.4.4   Homomorphism

Main article: Ring homomorphism

A **homomorphism** from a ring $(R, +, \cdot)$ to a ring $(S, \ddagger, *)$ is a function f from R to S that preserves the ring operations; namely, such that, for all a, b in R the following identities hold:

- $f(a + b) = f(a) \ddagger f(b)$

- $f(a \cdot b) = f(a) * f(b)$

- $f(1R) = 1S$

If one is working with not necessarily unital rings, then the third condition is dropped.

A ring homomorphism is said to be an **isomorphism** if there exists an inverse homomorphism to f (i.e., a ring homomorphism which is an inverse function). Any bijective ring homomorphism is a ring isomorphism. Two rings R, S are said to be isomorphic if there is an isomorphism between them and in that case one writes $R \simeq S$. A ring homomorphism between the same ring is called an endomorphism and an isomorphism between the same ring an automorphism.

Examples:

- The function that maps each integer x to its remainder modulo 4 (a number in $\{0, 1, 2, 3\}$) is a homomorphism from the ring $\mathbf{Z}$ to the quotient ring $\mathbf{Z}/4\mathbf{Z}$ ("quotient ring" is defined below).

- If u is a unit element in a ring R, then $R \to R, x \mapsto uxu^{-1}$ is a ring homomorphism, called an inner automorphism of R.

- Let R be a commutative ring of prime characteristic p. Then $x \mapsto x^p$ is a ring endmorphism of R called the Frobenius homomorphism.

- The Galois group of a field extension L/K is the set of all automorphisms of L whose restrictions to K are the identity.

- For any ring R, there are a unique ring homomorphism $\mathbf{Z} \to R$ and a unique ring homomorphism $R \to 0$.

- An epimorphism (i.e., right-cancelable morphism) of rings need not be surjective. For example, the unique map $\mathbb{Z} \to \mathbb{Q}$ is an epimorphism.

- An algebra homomorphism from a k-algebra to the endomorphism algebra of a vector space over k is called a representation of the algebra.

Given a ring homomorphism $f : R \to S$, the set of all elements mapped to 0 by f is called the kernel of f. The kernel is a two-sided ideal of R. The image of f, on the other hand, is not always an ideal, but it is always a subring of S.

To give a ring homomorphism from a commutative ring R to a ring A with image contained in the center of A is the same as to give a structure of an algebra over R to A (in particular gives a structure of A-module).

20.4.5 Quotient ring

Main article: Quotient ring

The **quotient ring** of a ring, is analogous to the notion of a quotient group of a group. More formally, given a ring $(R, +, \cdot)$ and a two-sided ideal I of $(R, +, \cdot)$, the **quotient ring** (or **factor ring**) R/I is the set of cosets of I (with respect to the additive group of $(R, +, \cdot)$; i.e. cosets with respect to $(R, +)$) together with the operations:

$$(a + I) + (b + I) = (a + b) + I \text{ and}$$
$$(a + I)(b + I) = (ab) + I.$$

for every a, b in R.

Like the case of a quotient group, there is a canonical map $p : R \to R/I$ given by $x \mapsto x + I$. It is surjective and satisfies the universal property: if $f : R \to S$ is a ring homomorphism such that $f(I) = 0$, then there is a unique $\overline{f} : R/I \to S$ such that $f = \overline{f} \circ p$. In particular, taking I to be the kernel, one sees that the quotient ring $R/\ker f$ is isomorphic to the image of f; the fact known as the first isomorphism theorem. The last fact implies that actually *any* surjective ring homomorphism satisfies the universal property since the image of such a map is a quotient ring.

20.5 Ring action: a module over a ring

In group theory, one can consider the action of a group on a set. To give a group action, say, G acting on a set S, is to give a group homomorphism from G to the automorphism group of S (that is, the symmetric group of S.)

In much the same way, one can consider a **ring action**; that is, a ring homomorphism f from a ring R to the endomorphism ring of an abelian group M. One usually writes rm or $r \cdot m$ for $f(r)m$ and calls M a left module over R. If R is a field, this amounts to giving a structure of a vector space on M.

In particular, a ring R is a left module over R itself through $l : R \to \text{End}(R)$, $l(r)x = rx$ (called the left regular representation of R). Some ring-theoretic concepts can be stated in a module-theoretic language: for example, a subset of a ring R is a left ideal of R if and only if it is an R-submodule with respect to the left R-module structure of R. A left ideal is principal if and only if it is a cyclic submodule.

A **Z**-module is the same thing as an abelian group; this allows one to use the module theory to study abelian groups. For example, in general, if M is a left module over a ring R that is cyclic; i.e., $M = Rx$ for some x, then M is isomorphic to the quotient of R by the kernel of $R \to M, r \mapsto rx$. In particular, if R is **Z**, then any cyclic group (which is cyclic as **Z**-module) is of the form **Z**/n**Z**, recovering the usual classification of cyclic groups.

See § Domains for an example of an application to linear algebra.

Any ring homomorphism induces the structure of a module: if $f : R \to S$ is a ring homomorphism, then S is a left module over R by the formula: $r \cdot s = f(r)s$. A module that is also a ring is called an algebra over the base ring (provided the base ring is central).

Example: Geometrically, a module can be viewed as an algebraic counterpart of a vector bundle. Let E be a vector bundle over a compact space, and $\Gamma(E)$ the space of its sections. Then $\Gamma(E)$ is a module over the ring R of continuous functions on the base space. Swan's theorem states that, via Γ, the category of vector bundles is equivalent to the category of finitely generated projective R-modules ("projective" corresponds to local trivialization.)

In application, one often cooks up a ring by summing up modules. Continuing the above geometric example, let L be a line bundle on an algebraic variety ($\Gamma(L)$ is a module over the coordinate ring of the variety). Then the direct sum of modules

$$\oplus_{n \geq 0} \Gamma(L^{\otimes n})$$

has the structure of a commutative ring; it is called the section ring of L. A particularly important case is when L is the canonical line bundle and then R is the canonical ring of the base variety.

20.6   Constructions

20.6.1   Direct product

Main article: Direct product of rings

Let R and S be rings. Then the product $R \times S$ can be equipped with the following natural ring structure:

- $(r_1, s_1) + (r_2, s_2) = (r_1 + r_2, s_1 + s_2)$

- $(r_1, s_1) \cdot (r_2, s_2) = (r_1 \cdot r_2, s_1 \cdot s_2)$

for every r_1, r_2 in R and s_1, s_2 in S. The ring $R \times S$ with the above operations of addition and multiplication and the multiplicative identity $(1, 1)$ is called the **direct product** of R with S. The same construction also works for an arbitrary family of rings: if R_i are rings indexed by a set I, then $\prod_{i \in I} R_i$ is a ring with componentwise addition and multiplication.

Let R be a commutative ring and $\mathfrak{a}_1, \cdots, \mathfrak{a}_n$ be ideals such that $\mathfrak{a}_i + \mathfrak{a}_j = (1)$ whenever $i \neq j$. Then the Chinese remainder theorem says there is a canonical ring isomorphism:

$$R/(\cap \mathfrak{a}_i) \simeq \prod R/\mathfrak{a}_i, \quad x \mapsto (x \bmod \mathfrak{a}_1, \ldots, x \bmod \mathfrak{a}_n)$$

A "finite" direct product may also be viewed as a direct sum of ideals.[29] Namely, let $R_i, 1 \leq i \leq n$ be rings, $R_i \to R = \prod R_i$ the inclusions with the images $\mathfrak{a}_i$ (in particular $\mathfrak{a}_i$ are rings though not subrings). Then $\mathfrak{a}_i$ are ideals of R and

$$R = \mathfrak{a}_1 \oplus \cdots \oplus \mathfrak{a}_n, \quad \mathfrak{a}_i \mathfrak{a}_j = 0, i \neq j, \quad \mathfrak{a}_i^2 \subseteq \mathfrak{a}_i$$

as a direct sum of abelian groups (because for abelian groups finite products are the same as direct sums). Clearly the direct sum of such ideals also defines a product of rings that is isomorphic to R. Equivalently, the above can be done through central idempotents. Assume R has the above decomposition. Then we can write

$$1 = e_1 + \cdots + e_n, \quad e_i \in \mathfrak{a}_i.$$

By the conditions on $\mathfrak{a}_i$, one has that e_i are central idempotents and $e_i e_j = 0, i \neq j$ (orthogonal). Again, one can reverse the construction. Namely, if one is given a partition of 1 in orthogonal central idempotents, then let $\mathfrak{a}_i = R e_i$, which are two-sided ideals. If each e_i is not a sum of orthogonal central idempotents,[30] then their direct sum is isomorphic to R.

An important application of an infinite direct product is the construction of a projective limit of rings (see below). Another application is a restricted product of a family of rings (cf. adele ring).

20.6.2   Polynomial ring

Main article: Polynomial ring

Given a symbol t (called a variable) and a commutative ring R, the set of polynomials

$$R[t] = \left\{ a_n t^n + a_{n-1} t^{n-1} + \cdots + a_1 t + a_0 \mid n \geq 0, a_j \in R \right\}$$

forms a commutative ring with the usual addition and multiplication, containing R as a subring. It is called the polynomial ring over R. More generally, the set $R[t_1, \ldots, t_n]$ of all polynomials in variables $t_1, \ldots, t_n$ forms a commutative ring, containing $R[t_i]$ as subrings.

If R is an integral domain, then $R[t]$ is also an integral domain; its field of fractions is the field of rational functions. If R is a noetherian ring, then $R[t]$ is a noetherian ring. If R is a unique factorization domain, then $R[t]$ is a unique factorization domain. Finally, R is a field if and only if $R[t]$ is a principal ideal domain.

Let $R \subseteq S$ be commutative rings. Given an element x of S, one can consider the ring homomorphism

$$R[t] \to S, \quad f \mapsto f(x)$$

(i.e., the substitution). If $S=R[t]$ and $x=t$, then $f(t)=f$. Because of this, the polynomial f is often also denoted by $f(t)$. The image of the map $f \mapsto f(x)$ is denoted by $R[x]$; it is the same thing as the subring of S generated by R and x.

Example: $k[t^2, t^3]$ denotes the image of the homomorphism

$$k[x,y] \to k[t], \ f \mapsto f(t^2, t^3).$$

In other words, it is the subalgebra of $k[t]$ generated by t^2 and t^3.

Example: let f be a polynomial in one variable; i.e., an element in a polynomial ring R. Then $f(x + h)$ is an element in $R[h]$ and $f(x + h) - f(x)$ is divisible by h in that ring. The result of substituting zero to h in $(f(x + h) - f(x))/h$ is $f'(x)$, the derivative of f at x.

The substitution is a special case of the universal property of a polynomial ring. The property states: given a ring homomorphism $\phi : R \to S$ and an element x in S there exists a unique ring homomorphism $\overline{\phi} : R[t] \to S$ such that $\overline{\phi}(t) = x$ and $\overline{\phi}$ restricts to ϕ .[31] For example, choosing a basis, a symmetric algebra satisfies the universal property and so is a polynomial ring.

To give an example, let S be the ring of all functions from R to itself; the addition and the multiplication are those of functions. Let x be the identity function. Each r in R defines a constant function, giving rise to the homomorphism $R \to S$. The universal property says that this map extends uniquely to

$$R[t] \to S, \quad f \mapsto \overline{f}$$

(t maps to x) where $\overline{f}$ is the polynomial function defined by f. The resulting map is injective if and only if R is infinite.

Given a non-constant monic polynomial f in $R[t]$, there exists a ring S containing R such that f is a product of linear factors in $S[t]$.[32]

Let k be an algebraically closed field. The Hilbert's Nullstellensatz (theorem of zeros) states that there is a natural one-to-one correspondence between the set of all prime ideals in $k[t_1, \ldots, t_n]$ and the set of closed subvarieties of k^n . In particular, many local problems in algebraic geometry may be attacked through the study of the generators of an ideal in a polynomial ring. (cf. Gröbner basis.)

There are some other related constructions. A formal power series ring $R[\![t]\!]$ consists of formal power series

$$\sum_0^\infty a_i t^i, \quad a_i \in R$$

together with multiplication and addition that mimic those for convergent series. It contains $R[t]$ as a subring. Note a formal power series ring does not have the universal property of a polynomial ring; a series may not converge after a substitution. The important advantage of a formal power series ring over a polynomial ring is that it is local (in fact, complete).

20.6.3 Matrix ring and endomorphism ring

Main articles: Matrix ring and Endomorphism ring

Let R be a ring (not necessarily commutative). The set of all square matrices of size n with entries in R forms a ring with the entry-wise addition and the usual matrix multiplication. It is called the matrix ring and is denoted by $\mathrm{M}n(R)$. Given a right R-module U, the set of all R-linear maps from U to itself forms a ring with addition that is of function and multiplication that is of composition of functions; it is called the endomorphism ring of U and is denoted by $\mathrm{End}_R(U)$.

As in linear algebra, a matrix ring may be canonically interpreted as an endomorphism ring: $\mathrm{End}_R(R^n) \simeq \mathrm{M}_n(R)$. This is a special case of the following fact: If $f : \oplus_1^n U \to \oplus_1^n U$ is an R-linear map, then f may be written as a matrix with entries f_{ij} in $S = \mathrm{End}_R(U)$, resulting in the ring isomorphism:

$$\mathrm{End}_R(\oplus_1^n U) \to \mathrm{M}_n(S), \quad f \mapsto (f_{ij}).$$

Any ring homomorphism $R \to S$ induces $\mathrm{M}n(R) \to \mathrm{M}n(S)$; in fact, any ring homomorphism between matrix rings arises in this way.[33]

Schur's lemma says that if U is a simple right R-module, then $\mathrm{End}_R(U)$ is a division ring.[34] If $U = \bigoplus_{i=1}^{r} U_i^{\oplus m_i}$ is a direct sum of mi-copies of simple R-modules U_i, then

$$\mathrm{End}_R(U) \simeq \bigoplus_1^r \mathrm{M}_{m_i}(\mathrm{End}_R(U_i))$$

The Artin–Wedderburn theorem states any semisimple ring (cf. below) is of this form.

A ring R and the matrix ring $\mathrm{M}n(R)$ over it are Morita equivalent: the category of right modules of R is equivalent to the category of right modules over $\mathrm{M}n(R)$.[33] In particular, two-sided ideals in R correspond in one-to-one to two-sided ideals in $\mathrm{M}n(R)$.

Examples:

- The automorphisms of the projective line over a ring are given by homographies from the 2 x 2 matrix ring.

20.6.4 Limits and colimits of rings

Let Ri be a sequence of rings such that Ri is a subring of Ri_{+1} for all i. Then the union (or filtered colimit) of Ri is the ring $\varinjlim R_i$ defined as follows: it is the disjoint union of all Ri's modulo the equivalence relation $x \sim y$ if and only if $x = y$ in Ri for sufficiently large i.

Examples of colimits:

- A polynomial ring in infinitely many variables: $R[t_1, t_2, \cdots] = \varinjlim R[t_1, t_2, \cdots, t_m]$.

- The algebraic closure of finite fields of the same characteristic $\overline{\mathbf{F}}_p = \varinjlim \mathbf{F}_{p^m}$.

- The field of formal Laurent series over a field k: $k((t)) = \varinjlim t^{-m} k[\![t]\!]$ (it is the field of fractions of the formal power series ring $k[\![t]\!]$.)

- The function field of an algebraic variety over a field k is $\varinjlim k[U]$ where the limit runs over all the coordinate rings $k[U]$ of nonempty open subsets U (more succinctly it is the stalk of the structure sheaf at the generic point.)

Any commutative ring is the colimit of finitely generated subrings.

A projective limit (or a filtered limit) of rings is defined as follows. Suppose we're given a family of rings R_i, i running over positive integers, say, and ring homomorphisms $R_j \to R_i, j \geq i$ such that $R_i \to R_i$ are all the identities and $R_k \to R_j \to R_i$ is $R_k \to R_i$ whenever $k \geq j \geq i$. Then $\varprojlim R_i$ is the subring of $\prod R_i$ consisting of (x_n) such that x_j maps to x_i under $R_j \to R_i, j \geq i$.

For an example of a projective limit, see #completion.

20.6.5 Localization

The localization generalizes the construction of the field of fractions of an integral domain to an arbitrary ring and modules. Given a (not necessarily commutative) ring R and a subset S of R, there exists a ring $R[S^{-1}]$ together with the ring homomorphism $R \to R[S^{-1}]$ that "inverts" S; that is, the homomorphism maps elements in S to unit elements in $R[S^{-1}]$, and, moreover, any ring homomorphism from R that "inverts" S uniquely factors through $R[S^{-1}]$.[35] The ring $R[S^{-1}]$ is called the **localization** of R with respect to S. For example, if R is a commutative ring and f an element in R, then the localization $R[f^{-1}]$ consists of elements of the form r/f^n, $r \in R$, $n \geq 0$ (to be precise, $R[f^{-1}] = R[t]/(tf - 1)$.)[36]

The localization is frequently applied to a commutative ring R with respect to the complement of a prime ideal (or a union of prime ideals) in R. In that case $S = R - \mathfrak{p}$, one often writes $R_\mathfrak{p}$ for $R[S^{-1}]$. $R_\mathfrak{p}$ is then a local ring with the maximal ideal $\mathfrak{p}R_\mathfrak{p}$. This is the reason for the terminology "localization". The field of fractions of an integral domain R is the localization of R at the prime ideal zero. If $\mathfrak{p}$ is a prime ideal of a commutative ring R, then the field of fractions of $R/\mathfrak{p}$ is the same as the residue field of the local ring $R_\mathfrak{p}$ and is denoted by $k(\mathfrak{p})$.

If M is a left R-module, then the localization of M with respect to S is given by a change of rings $M[S^{-1}] = R[S^{-1}] \otimes_R M$.

The most important properties of localization are the following: when R is a commutative ring and S a multiplicatively closed subset

- $\mathfrak{p} \mapsto \mathfrak{p}[S^{-1}]$ is a bijection between the set of all prime ideals in R disjoint from S and the set of all prime ideals in $R[S^{-1}]$.[37]

- $R[S^{-1}] = \varinjlim R[f^{-1}]$, f running over elements in S with partial ordering given by divisibility.[38]

- The localization is exact:

  $0 \to M'[S^{-1}] \to M[S^{-1}] \to M''[S^{-1}] \to 0$ is exact over $R[S^{-1}]$ whenever $0 \to M' \to M \to M'' \to 0$ is exact over R.

- Conversely, if $0 \to M'_\mathfrak{m} \to M_\mathfrak{m} \to M''_\mathfrak{m} \to 0$ is exact for any maximal ideal $\mathfrak{m}$, then $0 \to M' \to M \to M'' \to 0$ is exact.

- A remark: localization is no help in proving a global existence. One instance of this is that if two modules are isomorphic at all prime ideals, it does not follow that they are isomorphic. (One way to explain this is that the localization allows one to view a module as a sheaf over prime ideals and a sheaf is inherently a local notion.)

In category theory, a localization of a category amounts to making some morphisms isomorphisms. An element in a commutative ring R may be thought of as an endomorphism of any R-module. Thus, categorically, a localization of R with respect to a subset S of R is a functor from the category of R-modules to itself that sends elements of S viewed as endomorphisms to automorphisms and is universal with respect to this property. (Of course, R then maps to $R[S^{-1}]$ and R-modules map to $R[S^{-1}]$ -modules.)

20.6.6 Completion

Let R be a commutative ring, and let I be an ideal of R. The **completion** of R at I is the projective limit $\hat{R} = \varprojlim R/I^n$; it is a commutative ring. The canonical homomorphisms from R to the quotients R/I^n induce a homomorphism $R \to \hat{R}$. The latter homomorphism is injective if R is a noetherian integral domain and I is a proper ideal, or if R is a noetherian local ring with maximal ideal I, by Krull's intersection theorem.[39] The construction is especially useful when I is a maximal ideal.

The basic example is the completion $\mathbf{Z}p$ of $\mathbf{Z}$ at the principal ideal (p) generated by a prime number p; it is called the ring of p-adic integers. The completion can in this case be constructed also from the p-adic absolute value on $\mathbf{Q}$. The p-adic absolute value on $\mathbf{Q}$ is a map $x \mapsto |x|$ from $\mathbf{Q}$ to $\mathbf{R}$ given by $|n|_p = p^{-v_p(n)}$ where $v_p(n)$ denotes the exponent of p in the prime factorization of a nonzero integer n into prime numbers (we also put $|0|_p = 0$ and $|m/n|_p = |m|_p/|n|_p$). It defines a distance function on $\mathbf{Q}$ and the completion of $\mathbf{Q}$ as a metric space is denoted by $\mathbf{Q}p$. It is again a field since the field operations extend to the completion. The subring of $\mathbf{Q}p$ consisting of elements x with $|x|_p \leq 1$ is isomorphic to $\mathbf{Z}p$.

Similarly, the formal power series ring $R[\![t]\!]$ is the completion of $R[t]$ at (t) .

See also: Hensel's lemma.

A complete ring has much simpler structure than a commutative ring. This owns to the Cohen structure theorem, which says, roughly, that a complete local ring tends to look like a formal power series ring or a quotient of it. On the other hand, the interaction between the integral closure and completion has been among the most important aspects that distinguish modern commutative ring theory from the classical one developed by the likes of Noether. Pathological examples found by Nagata led to the reexamination of the roles of Noetherian rings and motivated, among other things, the definition of excellent ring.

20.6.7   Rings with generators and relations

The most general way to construct a ring is by specifying generators and relations. Let F be a free ring (i.e., free algebra over the integers) with the set X of symbols; i.e., F consists of polynomials with integral coefficients in noncommuting variables that are elements of X. A free ring satisfies the universal property: any function from the set X to a ring R factors through F so that $F \rightarrow R$ is the unique ring homomorphism. Just as in the group case, every ring can be represented as a quotient of a free ring.[40]

Now, we can impose relations among symbols in X by taking a quotient. Explicitly, if E is a subset of F, then the quotient ring of F by the ideal generated by E is called the ring with generators X and relations E. If we used a ring, say, A as a base ring instead of $\mathbf{Z}$, then the resulting ring will be over A. For example, if $E = \{xy - yx \mid x, y \in X\}$, then the resulting ring will be the usual polynomial ring with coefficients in A in variables that are elements of X (It is also the same thing as the symmetric algebra over A with symbols X.)

In the category-theoretic terms, the formation $S \mapsto$ set the by generated ring free the$S$ is the left adjoint functor of the forgetful functor from the category of rings to **Set** (and it is often called the free ring functor.)

Let A, B be algebras over a commutative ring R. Then the tensor product of R-modules $A \otimes_R B$ is a R-module. We can turn it to a ring by extending linearly $(x \otimes u)(y \otimes v) = xy \otimes uv$. See also: tensor product of algebras, change of rings.

20.7   Special kinds of rings

20.7.1   Domains

A nonzero ring with no nonzero zero-divisors is called a domain. A commutative domain is called an integral domain. The most important integral domains are principal ideals domains, PID for short, and fields. A principal ideal domain is an integral domain in which every ideal is principal. An important class of integral domains that contain a PID is a unique factorization domain (UFD), an integral domain in which every nonunit element is a product of prime elements (an element is prime if it generates a prime ideal.) The fundamental question in algebraic number theory is on the extent to which the ring of (generalized) integers in a number field, where an "ideal" admits prime factorization, fails to be a PID.

Among theorems concerning a PID, the most important one is the structure theorem for finitely generated modules over a principal ideal domain. The theorem may be illustrated by the following application to linear algebra.[41] Let V be a finite-dimensional vector space over a field k and $f : V \rightarrow V$ a linear map with minimal polynomial q. Then, since $k[t]$ is a unique factorization domain, q factors into powers of distinct irreducible polynomials (i.e., prime elements):

$$q = p_1^{e_1}...p_s^{e_s}.$$

Letting $t \cdot v = f(v)$, we make V a $k[t]$-module. The structure theorem then says V is a direct sum of cyclic modules, each of which is isomorphic to the module of the form $k[t]/(p_i^{k_j})$. Now, if $p_i(t) = t - \lambda_i$, then such a cyclic module (for p_i) has a basis in which the restriction of f is represented by a Jordan matrix. Thus, if, say, k is algebraically closed, then all p_i 's are of the form $t - \lambda_i$ and the above decomposition corresponds to the Jordan canonical form of f.

In algebraic geometry, UFD's arise because of smoothness. More precisely, a point in a variety (over a perfect field) is smooth if the local ring at the point is a regular local ring. A regular local ring is a UFD.[42]

The following is a chain of class inclusions that describes the relationship between rings, domains and fields:

- **Commutative rings ⊃ integral domains ⊃ integrally closed domains ⊃ unique factorization domains ⊃ principal ideal domains ⊃ Euclidean domains ⊃ fields**

20.7.2 Division ring

A division ring is a ring such that every non-zero element is a unit. A commutative division ring is a field. A prominent example of a division ring that is not a field is the ring of quaternions. Any centralizer in a division ring is also a division ring. In particular, the center of a division ring is a field. It turned out that every *finite* domain (in particular finite division ring) is a field; in particular commutative (the Wedderburn's little theorem).

Every module over a division ring is a free module (has a basis); consequently, much of linear algebra can be carried out over a division ring instead of a field.

The study of conjugacy classes figures prominently in the classical theory of division rings. Cartan famously asked the following question: given a division ring D and a proper sub-division-ring S that is not contained in the center, does each inner automorphism of D restrict to an automorphism of S? The answer is negative: this is the Cartan–Brauer–Hua theorem.

A cyclic algebra, introduced by L. E. Dickson, is a generalization of a quaternion algebra.

20.7.3 Semisimple rings

A ring is called a semisimple ring if it is semisimple as a left module (or right module) over itself; i.e., a direct sum of simple modules. A ring is called a semiprimitive ring if its Jacobson radical is zero. (The Jacobson radical is the intersection of all maximal left ideals.) A ring is semisimple if and only if it is artinian and is semiprimitive.

An algebra over a field k is artinian if and only if it has finite dimension. Thus, a semisimple algebra over a field is necessarily finite-dimensional, while a simple algebra may have infinite-dimension; e.g., the ring of differential operators.

Any module over a semisimple ring is semisimple. (Proof: any free module over a semisimple ring is clearly semisimple and any module is a quotient of a free module.)

Examples of semisimple rings:

- A matrix ring over a division ring is semisimple (actually simple).

- The group ring $k[G]$ of a finite group G over a field k is semisimple if the characteristic of k does not divide the order of G. (Maschke's theorem)

- The Weyl algebra (over a field) is a simple ring; it is not semisimple since it has infinite dimension and thus not artinian.

- Clifford algebras are semisimple.

Semisimplicity is closely related to separability. An algebra A over a field k is said to be separable if the base extension $A \otimes_k F$ is semisimple for any field extension F/k . If A happens to be a field, then this is equivalent to the usual definition in field theory (cf. separable extension.)

20.7.4   Central simple algebra and Brauer group

Main article: Central simple algebra

For a field k, a k-algebra is central if its center is k and is simple if it is a simple ring. Since the center of a simple k-algebra is a field, any simple k-algebra is a central simple algebra over its center. In this section, a central simple algebra is assumed to have finite dimension. Also, we mostly fix the base field; thus, an algebra refers to a k-algebra. The matrix ring of size n over a ring R will be denoted by R_n .

The Skolem–Noether theorem states any automorphism of a central simple algebra is inner.

Two central simple algebras A and B are said to be *similar* if there are integers n and m such that $A \otimes_k k_n \approx B \otimes_k k_m$.[43] Since $k_n \otimes_k k_m \simeq k_{nm}$, the similarity is an equivalence relation. The similarity classes $[A]$ with the multiplication $[A][B] = [A \otimes_k B]$ form an abelian group called the Brauer group of k and is denoted by $\mathrm{Br}(k)$. By the Artin–Wedderburn theorem, a central simple algebra is the matrix ring of a division ring; thus, each similarity class is represented by a unique division ring.

For example, $\mathrm{Br}(k)$ is trivial if k is a finite field or an algebraically closed field (more generally quasi-algebraically closed field; cf. Tsen's theorem). $\mathrm{Br}(\mathbb{R})$ has order 2 (a special case of the theorem of Frobenius). Finally, if k is a nonarchimedean local field (e.g., $\mathbb{Q}_p$), then $\mathrm{Br}(k) = \mathbb{Q}/\mathbb{Z}$ through the invariant map.

Now, if F is a field extension of k, then the base extension $- \otimes_k F$ induces $\mathrm{Br}(k) \to \mathrm{Br}(F)$. Its kernel is denoted by $\mathrm{Br}(F/k)$. It consists of $[A]$ such that $A \otimes_k F$ is a matrix ring over F (i.e., A is split by F.) If the extension is finite and Galois, then $\mathrm{Br}(F/k)$ is canonically isomorphic to $H^2(\mathrm{Gal}(F/k), k^*)$.[44]

Azumaya algebras generalize the notion of central simple algebras to a commutative local ring.

20.7.5   Valuation ring

Main article: valuation ring

If K is a field, a valuation v is a group homomorphism from the multiplicative group K^* to a totally ordered abelian group G such that, for any f, g in K with $f + g$ nonzero, $v(f + g) \geq \min\{v(f), v(g)\}$. The valuation ring of v is the subring of K consisting of zero and all nonzero f such that $v(f) \geq 0$.

Examples:

- The field of formal Laurent series $k((t))$ over a field k comes with the valuation v such that $v(f)$ is the least degree of a nonzero term in f; the valuation ring of v is the formal power series ring $k[\![t]\!]$.

- More generally, given a field k and a totally ordered abelian group G, let $k((G))$ be the set of all functions from G to k whose supports (the sets of points at which the functions are nonzero) are well ordered. It is a field with the multiplication given by convolution:

$$(f * g)(t) = \sum_{s \in G} f(s)g(t - s)$$

  It also comes with the valuation v such that $v(f)$ is the least element in the support of f. The subring consisting of elements with finite support is called the group ring of G (which makes sense even if G is not commutative). If G is the ring of integers, then we recover the previous example (by identifying f with the series whose n-th coefficient is $f(n)$.)

See also: Novikov ring and uniserial ring.

20.8 Rings with extra structure

A ring may be viewed as an abelian group (by using the addition operation), with extra structure: namely, ring multiplication. In the same way, there are other mathematical objects which may be considered as rings with extra structure. For example:

- An associative algebra is a ring that is also a vector space over a field K such that the scalar multiplication distributes over the ring multiplication. For instance, the set of n-by-n matrices over the real field $\mathbf{R}$ has dimension n^2 as a real vector space.

- A ring R is a topological ring if its set of elements R is given a topology which makes the addition map ($+$: $R \times R \to R$) and the multiplication map ($\cdot : R \times R \to R$) to be both continuous as maps between topological spaces (where $X \times X$ inherits the product topology or any other product in the category). For example, n-by-n matrices over the real numbers could be given either the Euclidean topology, or the Zariski topology, and in either case one would obtain a topological ring.

- A λ-ring is a commutative ring R together with operations $\lambda^n \colon R \to R$ that are like n-th exterior powers:

$$\lambda^n(x + y) = \sum_0^n \lambda^i(x)\lambda^{n-i}(y)$$

For example, $\mathbf{Z}$ is a λ-ring with $\lambda^n(x) = \binom{x}{n}$, the binomial coefficients. The notion plays a central rule in the algebraic approach to the Riemann–Roch theorem.

20.9 Some examples of the ubiquity of rings

Many different kinds of mathematical objects can be fruitfully analyzed in terms of some associated ring.

20.9.1 Cohomology ring of a topological space

To any topological space X one can associate its integral cohomology ring

$$H^*(X, \mathbb{Z}) = \bigoplus_{i=0}^{\infty} H^i(X, \mathbb{Z}),$$

a graded ring. There are also homology groups $H_i(X, \mathbb{Z})$ of a space, and indeed these were defined first, as a useful tool for distinguishing between certain pairs of topological spaces, like the spheres and tori, for which the methods of point-set topology are not well-suited. Cohomology groups were later defined in terms of homology groups in a way which is roughly analogous to the dual of a vector space. To know each individual integral homology group is essentially the same as knowing each individual integral cohomology group, because of the universal coefficient theorem. However, the advantage of the cohomology groups is that there is a natural product, which is analogous to the observation that one can multiply pointwise a k-multilinear form and an l-multilinear form to get a $(k + l)$-multilinear form.

The ring structure in cohomology provides the foundation for characteristic classes of fiber bundles, intersection theory on manifolds and algebraic varieties, Schubert calculus and much more.

20.9.2 Burnside ring of a group

To any group is associated its Burnside ring which uses a ring to describe the various ways the group can act on a finite set. The Burnside ring's additive group is the free abelian group whose basis are the transitive actions of the group and whose addition is the disjoint union of the action. Expressing an action in terms of the basis is decomposing an action into its transitive constituents. The multiplication is easily expressed in terms of the representation ring: the multiplication in the Burnside ring is formed by writing the tensor product of two permutation modules as a permutation module. The ring structure allows a formal way of subtracting one action from another. Since the Burnside ring is contained as a finite index subring of the representation ring, one can pass easily from one to the other by extending the coefficients from integers to the rational numbers.

20.9.3   Representation ring of a group ring

To any group ring or Hopf algebra is associated its representation ring or "Green ring". The representation ring's additive group is the free abelian group whose basis are the indecomposable modules and whose addition corresponds to the direct sum. Expressing a module in terms of the basis is finding an indecomposable decomposition of the module. The multiplication is the tensor product. When the algebra is semisimple, the representation ring is just the character ring from character theory, which is more or less the Grothendieck group given a ring structure.

20.9.4   Function field of an irreducible algebraic variety

To any irreducible algebraic variety is associated its function field. The points of an algebraic variety correspond to valuation rings contained in the function field and containing the coordinate ring. The study of algebraic geometry makes heavy use of commutative algebra to study geometric concepts in terms of ring-theoretic properties. Birational geometry studies maps between the subrings of the function field.

20.9.5   Face ring of a simplicial complex

Every simplicial complex has an associated face ring, also called its Stanley–Reisner ring. This ring reflects many of the combinatorial properties of the simplicial complex, so it is of particular interest in algebraic combinatorics. In particular, the algebraic geometry of the Stanley–Reisner ring was used to characterize the numbers of faces in each dimension of simplicial polytopes.

20.10   Category theoretical description

Main article: Category of rings

Every ring can be thought of as a monoid in **Ab**, the category of abelian groups (thought of as a monoidal category under the tensor product of $\mathbb{Z}$-modules). The monoid action of a ring R on an abelian group is simply an R-module. Essentially, an R-module is a generalization of the notion of a vector space – where rather than a vector space over a field, one has a "vector space over a ring".

Let $(A, +)$ be an abelian group and let $\text{End}(A)$ be its endomorphism ring (see above). Note that, essentially, $\text{End}(A)$ is the set of all morphisms of A, where if f is in $\text{End}(A)$, and g is in $\text{End}(A)$, the following rules may be used to compute $f + g$ and $f \cdot g$:

- $(f + g)(x) = f(x) + g(x)$

- $(f \cdot g)(x) = f(g(x))$

where $+$ as in $f(x) + g(x)$ is addition in A, and function composition is denoted from right to left. Therefore, associated to any abelian group, is a ring. Conversely, given any ring, $(R, +, \cdot)$, $(R, +)$ is an abelian group. Furthermore, for every r in R, right (or left) multiplication by r gives rise to a morphism of $(R, +)$, by right (or left) distributivity. Let $A = (R, +)$. Consider those endomorphisms of A, that "factor through" right (or left) multiplication of R. In other words, let $\text{End}_R(A)$

be the set of all morphisms m of A, having the property that $m(r \cdot x) = r \cdot m(x)$. It was seen that every r in R gives rise to a morphism of A: right multiplication by r. It is in fact true that this association of any element of R, to a morphism of A, as a function from R to $\text{End}R(A)$, is an isomorphism of rings. In this sense, therefore, any ring can be viewed as the endomorphism ring of some abelian X-group (by X-group, it is meant a group with X being its set of operators).[45] In essence, the most general form of a ring, is the endomorphism group of some abelian X-group.

Any ring can be seen as a preadditive category with a single object. It is therefore natural to consider arbitrary preadditive categories to be generalizations of rings. And indeed, many definitions and theorems originally given for rings can be translated to this more general context. Additive functors between preadditive categories generalize the concept of ring homomorphism, and ideals in additive categories can be defined as sets of morphisms closed under addition and under composition with arbitrary morphisms.

20.11 Generalization

Algebraists have defined structures more general than rings by weakening or dropping some of ring axioms.

20.11.1 Rng

A rng is the same as a ring, except that the existence of a multiplicative identity is not assumed.[46]

20.11.2 Nonassociative ring

A nonassociative ring is an algebraic structure that satisfies all of the ring axioms but the associativity and the existence of a multiplicative identity. A notable example is a Lie algebra. There exists some structure theory for such algebras that generalizes the analogous results for Lie algebras and associative algebras.

20.11.3 Semiring

A semiring is obtained by weakening the assumption that $(R,+)$ is an abelian group to the assumption that $(R,+)$ is a commutative monoid, and adding the axiom that $0 \cdot a = a \cdot 0 = 0$ for all a in R (since it no longer follows from the other axioms).

Example: a tropical semiring.

20.12 Other ring-like objects

20.12.1 Ring object in a category

Let C be a category with finite products. Let pt denote a terminal object of C (an empty product). A **ring object** in C is an object R equipped with morphisms $R \times R \xrightarrow{a} R$ (addition), $R \times R \xrightarrow{m} R$ (multiplication), pt $\xrightarrow{0} R$ (additive identity), $R \xrightarrow{i} R$ (additive inverse), and pt $\xrightarrow{1} R$ (multiplicative identity) satisfying the usual ring axioms. Equivalently, a ring object is an object R equipped with a factorization of its functor of points $h_R = \text{Hom}(-, R) : C^{\text{op}} \to \textbf{Sets}$ through the category of rings: $C^{\text{op}} \to \textbf{Rings} \xrightarrow{\text{forgetful}} \textbf{Sets}$.

20.12.2 Ring scheme

In algebraic geometry, a **ring scheme** over a base scheme S is a ring object in the category of S-schemes. One example is the ring scheme Wn over Spec $\textbf{Z}$, which for any commutative ring A returns the ring W$n(A)$ of p-isotypic Witt vectors of length n over A.[47]

20.12.3   Ring spectrum

In algebraic topology, a ring spectrum is a spectrum X together with a multiplication $\mu\colon X \wedge X \to X$ and a unit map $S \to X$ from the sphere spectrum S, such that the ring axiom diagrams commute up to homotopy. In practice, it is common to define a ring spectrum as a monoid object in a good category of spectra such as the category of symmetric spectra.

20.13   See also

- Algebra over a commutative ring

- Algebraic structure

- Categorical ring

- Category of rings

- Glossary of ring theory

- Nonassociative ring

- Ring theory

- Semiring

- Spectrum of a ring

- Simplicial commutative ring

Special types of rings:

- Boolean ring

- Commutative ring

- Dedekind ring

- Differential ring

- Division ring (skew field)

- Exponential ring

- Field

- Integral domain

- Lie ring

- Local ring

- Noetherian and artinian rings

- Ordered ring

- Principal ideal domain (PID)

- Reduced ring

- Regular ring

- Ring of periods

- Ring theory

- SBI ring

- Unique factorization domain (UFD)

- Valuation ring and discrete valuation ring

- Zero ring

20.14   Notes

^ **a:** Some authors only require that a ring be a semigroup under multiplication; that is, do not require that there be a multiplicative identity (1). See the section Notes on the definition for more details.

^ **b:** Elements which do have multiplicative inverses are called units, see Lang 2002, §II.1, p. 84.

^ **c:** The closure axiom is already implied by the condition that +/• be a binary operation. Some authors therefore omit this axiom. Lang 2002

^ **d:** The transition from the integers to the rationals by adding fractions is generalized by the quotient field.

^ **e:** Many authors include commutativity of rings in the set of *ring axioms* (see above) and therefore refer to "commutative rings" as just "rings".

20.15   Citations

[1] Implicit in the assumption that "+" is a binary operation is that 1) $a + b$ is **defined** for all ordered pairs (a,b) of elements a and b of R; 2) "+" is **well-defined**, that is, if $a + b = c_1$ and $a + b = c_2$, then $c_1 = c_2$; and 3) R is **closed** under "+", meaning that for any a and b in R, the value of $a + b$ is defined to be an element of R. The same applies to multiplication. Closure would be an axiom, however, only if, instead of binary operations on R, we had functions "+" and "·" *a priori* taking values in some larger set S.

[2] Nicolas Bourbaki (1970). "§I.8". *Algebra*. Springer-Verlag.

[3] Saunders MacLane; Garrett Birkhoff (1967). *Algebra*. AMS Chelsea. p. 85.

[4] Serge Lang (2002). *Algebra* (Third ed.). Springer-Verlag. p. 83.

[5] The existence of 1 is not assumed by some authors. In this article, and more generally in Wikipedia, we adopt the most common convention of the existence of a multiplicative identity, and use the term rng if this existence is not required. See next subsection

[6] I. M. Isaacs, Algebra: A Graduate Course, AMS, 1994, p. 160.

[7] The development of Ring Theory

[8] Kleiner 1998, p. 27.

[9] Hilbert 1897.

[10]

[11] Cohn, Harvey (1980), *Advanced Number Theory*, New York: Dover Publications, p. 49, ISBN 978-0-486-64023-5

[12] Fraenkel, pp. 143–145

[13] Jacobson (2009), p. 86, footnote 1.

[14] Fraenkel, p. 144, axiom $R_{8)}$.

[15] Noether, p. 29.

[16] Fraenkel, p. 144, axiom $R_{7)}$.

[17] Van der Waerden, 1930.

[18] Zariski and Samuel, 1958.

[19] Artin, p. 346.

[20] Atiyah and MacDonald, p. 1.

[21] Bourbaki, p. 96.

[22] Eisenbud, p. 11.

[23] Lang, p. 83.

[24] Gardner and Wiegandt 2003.

[25] Wilder 1965, p. 176.

[26] Rotman 1998, p. 7.

[27] This is the definition of Bourbaki. Some other authors such as Lang require a zero divisor to be nonzero.

[28] In the unital case, like addition and multiplication, the multiplicative identity must be restricted from the original ring. The definition is also equivalent to requiring the set-theoretic inclusion is a ring homomorphism.

[29] Cohn 2003, Theorem 4.5.1

[30] such a central idempotent is called centrally primitive.

[31] Jacobson 1974, Theorem 2.10

[32] Bourbaki Algèbre commutative, Ch 5. §1, Lemma 2

[33] Cohn 2003, 4.4

[34] Lang 2002, Ch. XVII. Proposition 1.1.

[35] Cohn 1995, Proposition 1.3.1.

[36] Eisenbud 2004, Exercise 2.2

[37] Milne 2012, Proposition 6.4

[38] Milne 2012, The end of Chapter 7

[39] Atiyah and Macdonald, Theorem 10.17 and its corollaries.

[40] Cohn 1995, pg. 242.

[41] Lang 2002, Ch XIV, §2

[42] Weibel, Ch 1, Theorem 3.8

[43] Milne CFT, Ch IV, §2

[44] Serre, J-P ., Applications algébriques de la cohomologie des groupes, I, II, Séminaire Henri Cartan, 1950/51

[45] Jacobson (2009), p. 162, Theorem 3.2.

[46] Jacobson 2009.

[47] Serre, p. 44.

20.16 References

20.16.1 General references

- Artin, Michael (1991). *Algebra*. Prentice-Hall.

- Atiyah, Michael; Macdonald, Ian G. (1969). *Introduction to commutative algebra*. Addison–Wesley.

- Bourbaki, N. (1998). *Algebra I, Chapters 1-3*. Springer.

- Cohn, Paul Moritz (2003), *Basic algebra: groups, rings, and fields*, Springer, ISBN 978-1-85233-587-8.

- Eisenbud, David (1995). *Commutative algebra with a view toward algebraic geometry*. Springer.

- Herstein, I. N. (1994) [reprint of the 1968 original]. *Noncommutative rings*. Carus Mathematical Monographs **15**. With an afterword by Lance W. Small. Mathematical Association of America. ISBN 0-88385-015-X.

- Gardner, J.W.; Wiegandt, R. (2003). *Radical Theory of Rings*. Chapman & Hall/CRC Pure and Applied Mathematics. ISBN 0824750330.

- Jacobson, Nathan (2009). *Basic algebra* **1** (2nd ed.). Dover. ISBN 978-0-486-47189-1.

- Jacobson, Nathan (1964). "Structure of rings". *American Mathematical Society Colloquium Publications* (Revised ed.) **37**.

- Jacobson, Nathan (1943). "The Theory of Rings". *American Mathematical Society Mathematical Surveys* **I**.

- Kaplansky, Irving (1974), *Commutative rings* (Revised ed.), University of Chicago Press, ISBN 0-226-42454-5, MR 0345945.

- Lam, Tsit Yuen (2001). *A first course in noncommutative rings*. Graduate Texts in Mathematics **131** (2nd ed.). Springer. ISBN 0-387-95183-0.

- Lam, Tsit Yuen (2003). *Exercises in classical ring theory*. Problem Books in Mathematics (2nd ed.). Springer. ISBN 0-387-00500-5.

- Lam, Tsit Yuen (1999). *Lectures on modules and rings*. Graduate Texts in Mathematics **189**. Springer. ISBN 0-387-98428-3.

- Lang, Serge (2002), *Algebra*, Graduate Texts in Mathematics **211** (Revised third ed.), New York: Springer-Verlag, ISBN 978-0-387-95385-4, Zbl 0984.00001, MR 1878556.

- Matsumura, Hideyuki (1989). *Commutative Ring Theory*. Cambridge Studies in Advanced Mathematics (2nd ed.). Cambridge University Press. ISBN 978-0-521-36764-6.

- Milne, J. "A primer of commutative algebra".

- Rotman, Joseph (1998), *Galois Theory* (2nd ed.), Springer, ISBN 0-387-98541-7.

- van der Waerden, Bartel Leendert (1930), *Moderne Algebra. Teil I*, Die Grundlehren der mathematischen Wissenschaften **33**, Springer, ISBN 978-3-540-56799-8, MR 0009016MR 0037277MR 0069787MR 0122834MR 0177027MR 0263581.

- Wilder, Raymond Louis (1965). *Introduction to Foundations of Mathematics*. Wiley.

- Zariski, Oscar; Samuel, Pierre (1958). *Commutative Algebra* **1**. Van Nostrand.

20.16.2 Special references

- Balcerzyk, Stanisław; Józefiak, Tadeusz (1989), *Commutative Noetherian and Krull rings*, Mathematics and its Applications, Chichester: Ellis Horwood Ltd., ISBN 978-0-13-155615-7.

- Balcerzyk, Stanisław; Józefiak, Tadeusz (1989), *Dimension, multiplicity and homological methods*, Mathematics and its Applications, Chichester: Ellis Horwood Ltd., ISBN 978-0-13-155623-2.

- Ballieu, R. (1947). "Anneaux finis; systèmes hypercomplexes de rang trois sur un corps commutatif". *Ann. Soc. Sci. Bruxelles* **I** (61): 222–227.

- Berrick, A. J.; Keating, M. E. (2000). *An Introduction to Rings and Modules with K-Theory in View*. Cambridge University Press.

- Cohn, Paul Moritz (1995), *Skew Fields: Theory of General Division Rings*, Encyclopedia of Mathematics and its Applications **57**, Cambridge University Press, ISBN 9780521432177.

- Eisenbud, David (1995), *Commutative algebra. With a view toward algebraic geometry.*, Graduate Texts in Mathematics **150**, Springer, ISBN 978-0-387-94268-1, MR 1322960.

- Gilmer, R.; Mott, J. (1973). "Associative Rings of Order". *Proc. Japan Acad.* **49**: 795–799. doi:10.3792/pja/1195519146.

- Harris, J. W.; Stocker, H. (1998). *Handbook of Mathematics and Computational Science*. Springer.

- Jacobson, Nathan (1945), "Structure theory of algebraic algebras of bounded degree", *Annals of Mathematics* (Annals of Mathematics) **46** (4): 695–707, doi:10.2307/1969205, ISSN 0003-486X, JSTOR 1969205.

- Knuth, D. E. (1998). *The Art of Computer Programming*. Vol. 2: Seminumerical Algorithms (3rd ed.). Addison–Wesley.

- Korn, G. A.; Korn, T. M. (2000). *Mathematical Handbook for Scientists and Engineers*. Dover.

- Milne, J. "Class field theory".

- Nagata, Masayoshi (1962) [1975 reprint], *Local rings*, Interscience Tracts in Pure and Applied Mathematics **13**, Interscience Publishers, ISBN 978-0-88275-228-0, MR 0155856.

- Pierce, Richard S. (1982). *Associative algebras*. Graduate Texts in Mathematics **88**. Springer. ISBN 0-387-90693-2.

- Serre, Jean-Pierre (1979), *Local fields*, Graduate Texts in Mathematics **67**, Springer.

- Springer, Tonny A. (1977), *Invariant theory*, Lecture Notes in Mathematics **585**, Springer.

- Weibel, Charles. "The K-book: An introduction to algebraic K-theory".

- Zariski, Oscar; Samuel, Pierre (1975). *Commutative algebra*. Graduate Texts in Mathematics. 28-29. Springer. ISBN 0-387-90089-6.

20.16.3 Primary sources

- Fraenkel, A. (1914). "Über die Teiler der Null und die Zerlegung von Ringen". *J. reine angew. Math.* **145**: 139–176.

- Hilbert, David (1897). "Die Theorie der algebraischen Zahlkörper". *Jahresbericht der Deutschen Mathematiker Vereinigung* **4**.

- Noether, Emmy (1921). "Idealtheorie in Ringbereichen". *Math. Annalen* **83**: 24–66. doi:10.1007/bf01464225.

20.16.4 Historical references

- History of ring theory at the MacTutor Archive

- Birkhoff, G. and Mac Lane, S. A Survey of Modern Algebra, 5th ed. New York: Macmillian, 1996.

- Bronshtein, I. N. and Semendyayev, K. A. Handbook of Mathematics, 4th ed. New York: Springer-Verlag, 2004. ISBN 3-540-43491-7.

- Faith, Carl, *Rings and things and a fine array of twentieth century associative algebra.* Mathematical Surveys and Monographs, 65. American Mathematical Society, Providence, RI, 1999. xxxiv+422 pp. ISBN 0-8218-0993-8.

- Itô, K. (Ed.). "Rings." §368 in Encyclopedic Dictionary of Mathematics, 2nd ed., Vol. 2. Cambridge, MA: MIT Press, 1986.

- Kleiner, I., "The Genesis of the Abstract Ring Concept", Amer. Math. Monthly 103, 417–424, 1996.

- Kleiner, I., "From numbers to rings: the early history of ring theory", *Elem. Math.* **53** (1998), 18–35.

- Renteln, P. and Dundes, A. "Foolproof: A Sampling of Mathematical Folk Humor." Notices Amer. Math. Soc. 52, 24–34, 2005.

- Singmaster, D. and Bloom, D. M. "Problem E1648." Amer. Math. Monthly 71, 918–920, 1964.

- Van der Waerden, B. L. A History of Algebra. New York: Springer-Verlag, 1985.

Chapter 21

Monoid

Not to be confused with monad.
For the object in category theory, see Monoid (category theory).

In abstract algebra, a branch of mathematics, a **monoid** is an algebraic structure with a single associative binary operation and an identity element. Monoids are studied in semigroup theory as they are semigroups with identity. Monoids occur in several branches of mathematics; for instance, they can be regarded as categories with a single object. Thus, they capture the idea of function composition within a set. Monoids are also commonly used in computer science, both in its foundational aspects and in practical programming. The set of strings built from a given set of characters is a free monoid. The transition monoid and syntactic monoid are used in describing finite state machines, whereas trace monoids and history monoids provide a foundation for process calculi and concurrent computing. Some of the more important results in the study of monoids are the Krohn–Rhodes theorem and the star height problem. The history of monoids, as well as a discussion of additional general properties, are found in the article on semigroups.

21.1 Definition

Suppose that S is a set and $\bullet$ is some binary operation $S \times S \to S$, then S with $\bullet$ is a **monoid** if it satisfies the following two axioms:

Associativity For all a, b and c in S, the equation $(a \bullet b) \bullet c = a \bullet (b \bullet c)$ holds.

Identity element There exists an element e in S such that for every element a in S, the equations $e \bullet a = a \bullet e = a$ hold.

In other words, a monoid is a semigroup with an identity element. It can also be thought of as a magma with associativity and identity. The identity element of a monoid is unique.[1] A monoid in which each element has an inverse is a group.

Depending on the context, the symbol for the binary operation may be omitted, so that the operation is denoted by juxtaposition; for example, the monoid axioms may be written $(ab)c = a(bc)$ and $ea = ae = a$. This notation does not imply that it is numbers being multiplied.

21.2 Monoid structures

21.2.1 Submonoids

A **submonoid** of a monoid $(M, \bullet)$ is a subset N of M that is closed under the monoid operation and contains the identity element e of M.[2][3] Symbolically, N is a submonoid of M if $N \subseteq M$, $x \bullet y \in N$ whenever $x, y \in N$, and $e \in N$. N is thus a monoid under the binary operation inherited from M.

21.2.2 Generators

A subset *S* of *M* is said to be a **generator** of *M* if *M* is the smallest set containing *S* that is closed under the monoid operation, or equivalently *M* is the result of applying the finitary closure operator to *S*. If there is a generator of *M* that has finite cardinality, then *M* is said to be **finitely generated**. Not every set *S* will generate a monoid, as the generated structure may lack an identity element.

21.2.3 Commutative monoid

A monoid whose operation is commutative is called a **commutative monoid** (or, less commonly, an **abelian monoid**). Commutative monoids are often written additively. Any commutative monoid is endowed with its **algebraic** preordering $\leq$, defined by $x \leq y$ if there exists z such that $x + z = y$.[4] An **order-unit** of a commutative monoid *M* is an element *u* of *M* such that for any element *x* of *M*, there exists a positive integer *n* such that $x \leq nu$. This is often used in case *M* is the positive cone of a partially ordered abelian group *G*, in which case we say that *u* is an order-unit of *G*.

21.2.4 Partially commutative monoid

A monoid for which the operation is commutative for some, but not all elements is a trace monoid; trace monoids commonly occur in the theory of concurrent computation.

21.3 Examples

- Out of the 16 possible binary Boolean operators, each of the four that has a two sided identity is also commutative and associative and thus makes the set {False, True} a commutative monoid. Under the standard definitions, AND and XNOR have the identity True while XOR and OR have the identity False. The monoids from AND and OR are also idempotent while those from XOR and XNOR are not.

- The natural numbers, **N**, form a commutative monoid under addition (identity element zero), or multiplication (identity element one). A submonoid of **N** under addition is called a numerical monoid.

- The positive integers, $\mathbf{N} \setminus \{0\}$, form a commutative monoid under multiplication (identity element one).

- Given a set A, all subsets of A form a commutative monoid under intersection operation (identity element is A itself).

- Given a set A, all subsets of A form a commutative monoid under union operation (identity element is the empty set).

- Generalizing the previous example, every bounded semilattice is an idempotent commutative monoid.

    - In particular, any bounded lattice can be endowed with both a meet- and a join- monoid structure. The identity elements are the lattice's top and its bottom, respectively. Being lattices, Heyting algebras and Boolean algebras are endowed with these monoid structures.

- Every singleton set {*x*} closed under a binary operation • forms the trivial (one-element) monoid, which is also the trivial group.

- Every group is a monoid and every abelian group a commutative monoid.

- Any semigroup *S* may be turned into a monoid simply by adjoining an element *e* not in *S* and defining $e \bullet s = s = s \bullet e$ for all $s \in S$. This conversion of any semigroup to the monoid is done by the free functor between the category of semigroups and the category of monoids.[5]

- Thus, an idempotent monoid (sometimes known as *find-first*) may be formed by adjoining an identity element e to the left zero semigroup over a set S. The opposite monoid (sometimes called *find-last*) is formed from the right zero semigroup over S.

  - Adjoin an identity e to the left-zero semigroup with two elements $\{lt; gt\}$. Then the resulting idempotent monoid $\{lt; e; gt\}$ models the lexicographical order of a sequence given the orders of its elements, with e representing equality.

- The elements of any unital ring, with addition or multiplication as the operation.

  - The integers, rational numbers, real numbers or complex numbers, with addition or multiplication as operation.[6]

  - The set of all n by n matrices over a given ring, with matrix addition or matrix multiplication as the operation.

- The set of all finite strings over some fixed alphabet Σ forms a monoid with string concatenation as the operation. The empty string serves as the identity element. This monoid is denoted Σ^* and is called the **free monoid** over Σ.

- Given any monoid M, the *opposite monoid* M^{op} has the same carrier set and identity element as M, and its operation is defined by $x \bullet^{op} y = y \bullet x$. Any commutative monoid is the opposite monoid of itself.

- Given two sets M and N endowed with monoid structure (or, in general, any finite number of monoids, M_1, ..., Mk), their cartesian product $M \times N$ is also a monoid (respectively, $M_1 \times ... \times Mk$). The associative operation and the identity element are defined pairwise.[7]

- Fix a monoid M. The set of all functions from a given set to M is also a monoid. The identity element is a constant function mapping any value to the identity of M; the associative operation is defined pointwise.

- Fix a monoid M with the operation $\bullet$ and identity element e, and consider its power set $P(M)$ consisting of all subsets of M. A binary operation for such subsets can be defined by $S \bullet T = \{ s \bullet t : s \in S, t \in T \}$. This turns $P(M)$ into a monoid with identity element $\{e\}$. In the same way the power set of a group G is a monoid under the product of group subsets.

- Let S be a set. The set of all functions $S \to S$ forms a monoid under function composition. The identity is just the identity function. It is also called the **full transformation monoid** of S. If S is finite with n elements, the monoid of functions on S is finite with n^n elements.

- Generalizing the previous example, let C be a category and X an object in C. The set of all endomorphisms of X, denoted $EndC(X)$, forms a monoid under composition of morphisms. For more on the relationship between category theory and monoids see below.

- The set of homeomorphism classes of compact surfaces with the connected sum. Its unit element is the class of the ordinary 2-sphere. Furthermore, if a denotes the class of the torus, and b denotes the class of the projective plane, then every element c of the monoid has a unique expression the form $c = na + mb$ where n is the integer ≥ 0 and $m = 0, 1,$ or 2. We have $3b = a + b$.

- Let $\langle f \rangle$ be a cyclic monoid of order n, that is, $\langle f \rangle = \{f^0, f^1, \ldots, f^{n-1}\}$. Then $f^n = f^k$ for some $0 \leq k < n$. In fact, each such k gives a distinct monoid of order n, and every cyclic monoid is isomorphic to one of these.

Moreover, f can be considered as a function on the points $\{0, 1, 2, \ldots, n-1\}$ given by

$$\begin{bmatrix} 0 & 1 & 2 & \ldots & n-2 & n-1 \\ 1 & 2 & 3 & \ldots & n-1 & k \end{bmatrix}$$

or, equivalently

$$f(i) := \begin{cases} i+1, & \text{if } 0 \leq i < n-1 \\ k, & \text{if } i = n-1. \end{cases}$$

Multiplication of elements in $\langle f \rangle$ is then given by function composition.

Note also that when $k = 0$ then the function f is a permutation of $\{0, 1, 2, \ldots, n-1\}$ and gives the unique cyclic group of order n.

21.4 Properties

In a monoid, one can define positive integer powers of an element x : $x^1 = x$, and $x^n = x \bullet \ldots \bullet x$ (n times) for $n > 1$. The rule of powers $x^{n+p} = x^n \bullet x^p$ is obvious.

From the definition of a monoid, one can show that the identity element e is unique. Then, for any x, one can set $x^0 = e$ and the rule of powers is still true with nonnegative exponents.

It is possible to define invertible elements: an element x is called invertible if there exists an element y such that $x \bullet y = e$ and $y \bullet x = e$. The element y is called the inverse of x. If y and z are inverses of x, then by associativity $y = (zx)y = z(xy) = z$. Thus inverses, if they exist, are unique.[8]

If y is the inverse of x, one can define negative powers of x by setting $x^{-1} = y$ and $x^{-n} = y \bullet \ldots \bullet y$ (n times) for $n > 1$. And the rule of exponents is still verified for all n, p rational integers. This is why the inverse of x is usually written x^{-1}. The set of all invertible elements in a monoid M, together with the operation $\bullet$, forms a group. In that sense, every monoid contains a group (possibly only the trivial group consisting of only the identity).

However, not every monoid sits inside a group. For instance, it is perfectly possible to have a monoid in which two elements a and b exist such that $a \bullet b = a$ holds even though b is not the identity element. Such a monoid cannot be embedded in a group, because in the group we could multiply both sides with the inverse of a and would get that $b = e$, which isn't true. A monoid $(M, \bullet)$ has the cancellation property (or is cancellative) if for all a, b and c in M, $a \bullet b = a \bullet c$ always implies $b = c$ and $b \bullet a = c \bullet a$ always implies $b = c$. A commutative monoid with the cancellation property can always be embedded in a group via the Grothendieck construction. That is how the additive group of the integers (a group with operation +) is constructed from the additive monoid of natural numbers (a commutative monoid with operation + and cancellation property). However, a non-commutative cancellative monoid need not be embeddable in a group.

If a monoid has the cancellation property and is *finite*, then it is in fact a group. Proof: Fix an element x in the monoid. Since the monoid is finite, $x^n = x^m$ for some $m > n > 0$. But then, by cancellation we have that $x^{m-n} = e$ where e is the identity. Therefore $x \bullet x^{m-n-1} = e$, so x has an inverse.

The right- and left-cancellative elements of a monoid each in turn form a submonoid (i.e. obviously include the identity and not so obviously are closed under the operation). This means that the cancellative elements of any commutative monoid can be extended to a group.

It turns out that requiring the cancellative property in a monoid is not required to perform the Grothendieck construction – commutativity is sufficient. However, if the original monoid has an absorbing element then its Grothendieck group is the trivial group. Hence the homomorphism is, in general, not injective.

An **inverse monoid** is a monoid where for every a in M, there exists a unique a^{-1} in M such that $a = a \bullet a^{-1} \bullet a$ and $a^{-1} = a^{-1} \bullet a \bullet a^{-1}$. If an inverse monoid is cancellative, then it is a group.

In the opposite direction, a **zerosumfree monoid** is an additively written monoid in which $a + b = 0$ implies that $a = 0$ and $b = 0$:[9] equivalently, that no element other than zero has an additive inverse.

21.5 Acts and operator monoids

Main article: monoid act

Let M be a monoid, with the binary operation denoted by $\bullet$ and the identity element denoted by e. Then a (left) **M-act** (or left act over M) is a set X together with an operation $\cdot : M \times X \to X$ which is compatible with the monoid structure as follows:

- for all x in X: $e \cdot x = x$;

- for all a, b in M and x in X: $a \cdot (b \cdot x) = (a \bullet b) \cdot x$.

This is the analogue in monoid theory of a (left) group action. Right M-acts are defined in a similar way. A monoid with an act is also known as an **operator monoid**. Important examples include transition systems of semiautomata. A transformation semigroup can be made into an operator monoid by adjoining the identity transformation.

21.6 Monoid homomorphisms

×	0	1	2	3	4	5	6	7	8	9	⋯
0	0	0	0	0	0	0	0	0	0	0	⋯
1	0	1	2	3	4	5	6	7	8	9	⋯
2	0	2	4	6	8	10	12	14	16	18	⋯
3	0	3	6	9	12	15	18	21	24	27	⋯
4	0	4	8	12	16	20	24	28	32	36	
5	0	5	10	15	20	25	30	35	40	45	
6	0	6	12	18	24	30	36	42	48	54	⋯
7	0	7	14	21	28	35	42	49	56	63	⋯
8	0	8	16	24	32	40	48	56	64	72	⋯
9	0	9	18	27	36	45	54	63	72	81	⋯
⋮	⋮	⋮	⋮	⋮	⋮		⋮	⋮	⋮	⋱	

+	0	1	2	3	4	⋯
0	0	1	2	3	4	⋯
1	1	2	3	4	5	⋯
2	2	3	4	5	6	⋯
3	3	4	5	6	7	⋯
4	4	5	6	7	8	⋯
⋮	⋮	⋮	⋮	⋮	⋱	

Example monoid homomorphism x $\mapsto 2^x$ *from (N, +, 0) to (N, ×, 1). It is injective, but not surjective.*

A homomorphism between two monoids $(M, *)$ and $(N, \bullet)$ is a function $f : M \to N$ such that

- $f(x * y) = f(x) \bullet f(y)$ for all x, y in M

- $f(eM) = eN$,

where *eM* and *eN* are the identities on M and N respectively. Monoid homomorphisms are sometimes simply called **monoid morphisms**.

Not every semigroup homomorphism is a monoid homomorphism, since it may not map the identity to the identity of the target monoid, even though the element it maps the identity to will be an identity of the image of the mapping. In contrast, a semigroup homomorphisms between groups is always a group homomorphism, as it necessarily preserves the identity. Since for monoids this isn't always true, it is necessary to state this as a separate requirement.

A bijective monoid homomorphism is called a monoid isomorphism. Two monoids are said to be isomorphic if there is a monoid isomorphism between them.

21.7 Equational presentation

Main article: Presentation of a monoid

Monoids may be given a **presentation**, much in the same way that groups can be specified by means of a group presentation. One does this by specifying a set of generators Σ, and a set of relations on the free monoid Σ^*. One does this by extending (finite) binary relations on Σ^* to monoid congruences, and then constructing the quotient monoid, as above.

Given a binary relation $R \subset \Sigma^* \times \Sigma^*$, one defines its symmetric closure as $R \cup R^{-1}$. This can be extended to a symmetric relation $E \subset \Sigma^* \times \Sigma^*$ by defining $x \sim_E y$ if and only if $x = sut$ and $y = svt$ for some strings u, v, s, $t \in \Sigma^*$ with $(u,v) \in R \cup R^{-1}$. Finally, one takes the reflexive and transitive closure of E, which is then a monoid congruence.

In the typical situation, the relation R is simply given as a set of equations, so that $R = \{u_1 = v_1, \cdots, u_n = v_n\}$. Thus, for example,

$$\langle p, q \mid pq = 1 \rangle$$

is the equational presentation for the bicyclic monoid, and

$$\langle a, b \mid aba = baa, bba = bab \rangle$$

is the plactic monoid of degree 2 (it has infinite order). Elements of this plactic monoid may be written as $a^i b^j (ba)^k$ for integers i, j, k, as the relations show that ba commutes with both a and b.

21.8 Relation to category theory

Monoids can be viewed as a special class of categories. Indeed, the axioms required of a monoid operation are exactly those required of morphism composition when restricted to the set of all morphisms whose source and target is a given object.[10] That is,

> *A monoid is, essentially, the same thing as a category with a single object.*

More precisely, given a monoid $(M, \bullet)$, one can construct a small category with only one object and whose morphisms are the elements of M. The composition of morphisms is given by the monoid operation $\bullet$.

Likewise, monoid homomorphisms are just functors between single object categories.[10] So this construction gives an equivalence between the category of (small) monoids **Mon** and a full subcategory of the category of (small) categories **Cat**. Similarly, the category of groups is equivalent to another full subcategory of **Cat**.

In this sense, category theory can be thought of as an extension of the concept of a monoid. Many definitions and theorems about monoids can be generalised to small categories with more than one object. For example, a quotient of a category with one object is just a quotient monoid.

Monoids, just like other algebraic structures, also form their own category, **Mon**, whose objects are monoids and whose morphisms are monoid homomorphisms.[10]

There is also a notion of monoid object which is an abstract definition of what is a monoid in a category. A monoid object in **Set** is just a monoid.

21.9   Monoids in computer science

In computer science, many abstract data types can be endowed with a monoid structure. In a common pattern, a sequence of elements of a monoid is "folded" or "accumulated" to produce a final value. For instance, many iterative algorithms need to update some kind of "running total" at each iteration; this pattern may be elegantly expressed by a monoid operation. Alternatively, the associativity of monoid operations ensures that the operation can be parallelized by employing a prefix sum or similar algorithm, in order to utilize multiple cores or processors efficiently.

Given a sequence of values of type M with identity element ε and associative operation $*$, the *fold* operation is defined as follows:

$$\text{fold} : M^* \to M = l \mapsto \begin{cases} \varepsilon & \text{if } l = \text{nil} \\ m * \text{fold } l' & \text{if } l = \text{cons } m \, l' \end{cases}$$

In addition, any data structure can be 'folded' in a similar way, given a serialization of its elements. For instance, the result of "folding" a binary tree might differ depending on pre-order vs. post-order tree traversal.

21.10   Complete monoids

A **complete monoid** is a commutative monoid equipped with an infinitary sum operation Σ_I for any index set I such that:[11][12][13][14]

$$\sum_{i \in \emptyset} m_i = 0; \quad \sum_{i \in \{j\}} m_i = m_j; \quad \sum_{i \in \{j,k\}} m_i = m_j + m_k \quad \text{for } j \neq k$$

and

$$\sum_{j \in J} \sum_{i \in I_j} m_i = \sum_{i \in I}(m_i) \text{ if } \bigcup_{j \in J} I_j = I \text{ and } I_j \cap I_{j'} = \emptyset \text{ for } j \neq j'$$

A **continuous monoid** is an ordered commutative monoid in which every directed set has a least upper bound compatible with the monoid operation:

$$a + \sup S = \sup(a + S) .$$

These two concepts are closely related: a continuous monoid is a complete monoid in which the infinitary sum may be defined as

$$\sum_I a_i = \sup \sum_E a_i$$

where the supremum on the right runs over all finite subsets E of I and each sum on the right is a finite sum in the monoid.[14]

21.11   See also

- Green's relations

- Monad (functional programming)

- Semiring and Kleene algebra

- Star height problem

- Vedic square

21.12 Notes

[1] If both e_1 and e_2 satisfy the above equations, then $e_1 = e_1 \cdot e_2 = e_2$.

[2] Jacobson (2009)

[3] Some authors omit the requirement that a submonoid must contain the identity element from its definition, requiring only that it have *an* identity element, which can be distinct from that of *M*.

[4] Gondran, Michel; Minoux, Michel (2008). *Graphs, Dioids and Semirings: New Models and Algorithms*. Operations Research/Computer Science Interfaces Series **41**. Dordrecht: Springer-Verlag. p. 13. ISBN 978-0-387-75450-5. Zbl 1201.16038.

[5] Rhodes, John; Steinberg, Benjamin (2009), *The q-theory of Finite Semigroups: A New Approach*, Springer Monographs in Mathematics **71**, Springer, p. 22, ISBN 9780387097817.

[6] Jacobson (2009), p. 29, examples 1, 2, 4 & 5.

[7] Jacobson (2009), p. 35

[8] Jacobson, I.5. p. 22

[9] Wehrung, Friedrich (1996). "Tensor products of structures with interpolation". *Pacific Journal of Mathematics* **176** (1): 267–285. Zbl 0865.06010.

[10] Awodey, Steve (2006). *Category Theory*. Oxford Logic Guides **49**. Oxford University Press. p. 10. ISBN 0-19-856861-4. Zbl 1100.18001.

[11] Droste, M., & Kuich, W. (2009). Semirings and Formal Power Series. *Handbook of Weighted Automata*, 3–28. doi:10.1007/978-3-642-01492-5_1, pp. 7-10

[12] Hebisch, Udo (1992). "Eine algebraische Theorie unendlicher Summen mit Anwendungen auf Halbgruppen und Halbringe". *Bayreuther Mathematische Schriften* (in German) **40**: 21–152. Zbl 0747.08005.

[13] Kuich, Werner (1990). "ω-continuous semirings, algebraic systems and pushdown automata". In Paterson, Michael S. *Automata, Languages and Programming: 17th International Colloquium, Warwick University, England, July 16-20, 1990, Proceedings*. Lecture Notes in Computer Science **443**. Springer-Verlag. pp. 103–110. ISBN 3-540-52826-1.

[14] Kuich, Werner (2011). "Algebraic systems and pushdown automata". In Kuich, Werner. *Algebraic foundations in computer science. Essays dedicated to Symeon Bozapalidis on the occasion of his retirement*. Lecture Notes in Computer Science **7020**. Berlin: Springer-Verlag. pp. 228–256. ISBN 978-3-642-24896-2. Zbl 1251.68135.

21.13 References

- Howie, John M. (1995), *Fundamentals of Semigroup Theory*, London Mathematical Society Monographs. New Series **12**, Oxford: Clarendon Press, ISBN 0-19-851194-9, Zbl 0835.20077

- Jacobson, Nathan (1951), *Lectures in Abstract Algebra* **I**, D. Van Nostrand Company, ISBN 0-387-90122-1

- Jacobson, Nathan (2009), *Basic algebra* **1** (2nd ed.), Dover, ISBN 978-0-486-47189-1

- Kilp, Mati; Knauer, Ulrich; Mikhalev, Alexander V. (2000), *Monoids, acts and categories. With applications to wreath products and graphs. A handbook for students and researchers*, de Gruyter Expositions in Mathematics **29**, Berlin: Walter de Gruyter, ISBN 3-11-015248-7, Zbl 0945.20036

- Lothaire, M. (1997), *Combinatorics on words*, Encyclopedia of Mathematics and Its Applications **17**, Perrin, D.; Reutenauer, C.; Berstel, J.; Pin, J. E.; Pirillo, G.; Foata, D.; Sakarovitch, J.; Simon, I.; Schützenberger, M. P.; Choffrut, C.; Cori, R.; Lyndon, Roger; Rota, Gian-Carlo. Foreword by Roger Lyndon (2nd ed.), Cambridge University Press, doi:10.1017/CBO9780511566097, ISBN 0-521-59924-5, MR 1475463, Zbl 0874.20040

21.14   External links

- Hazewinkel, Michiel, ed. (2001), "Monoid", *Encyclopedia of Mathematics*, Springer, ISBN 978-1-55608-010-4

- Weisstein, Eric W., "Monoid", *MathWorld*.

- Monoid at PlanetMath.org.

Chapter 22

Semigroup

In mathematics, a **semigroup** is an algebraic structure consisting of a set together with an associative binary operation. The binary operation of a semigroup is most often denoted multiplicatively: $x \cdot y$, or simply xy, denotes the result of applying the semigroup operation to the ordered pair (x, y). Associativity is formally expressed as that $(x \cdot y) \cdot z = x \cdot (y \cdot z)$ for all x, y and z in the semigroup.

The name "semigroup" originates in the fact that a semigroup generalizes a group by preserving only associativity and closure under the binary operation from the axioms defining a group.[note 1] From the opposite point of view (of adding rather than removing axioms), a semigroup is an associative magma. As in the case of groups or magmas, the semigroup operation need not be commutative, so $x \cdot y$ is not necessarily equal to $y \cdot x$; a typical example of associative but non-commutative operation is matrix multiplication. If the semigroup operation is commutative, then the semigroup is called a *commutative semigroup* or (less often than in the analogous case of groups) it may be called an *abelian semigroup*.

A monoid is an algebraic structure intermediate between groups and semigroups, and is a semigroup having an identity element, thus obeying all but one of the axioms of a group; existence of inverses is not required of a monoid. A natural example is strings with concatenation as the binary operation, and the empty string as the identity element. Restricting to non-empty strings gives an example of a semigroup that is not a monoid. Positive integers with addition form a commutative semigroup that is not a monoid. Whereas the non-negative integers do form a monoid. A semigroup without an identity element can be easily turned into a monoid by just adding an identity element. Consequently, monoids are studied in the theory of semigroups rather than in group theory. Semigroups should not be confused with quasigroups, which are a generalization of groups in a different direction; the operation in a quasigroup need not be associative but quasigroups preserve from groups a notion of division. Division in semigroups (or in monoids) is not possible in general.

The formal study of semigroups began in the early 20th century. Early results include a Cayley theorem for semigroups realizing any semigroup as transformation semigroup, in which arbitrary functions replace the role of bijections from group theory. Other fundamental techniques of studying semigroups like Green's relations do not imitate anything in group theory though. A deep result in the classification of finite semigroups is Krohn–Rhodes theory. The theory of finite semigroups has been of particular importance in theoretical computer science since the 1950s because of the natural link between finite semigroups and finite automata via the syntactic monoid. In probability theory, semigroups are associated with Markov processes.[1] In other areas of applied mathematics, semigroups are fundamental models for linear time-invariant systems. In partial differential equations, a semigroup is associated to any equation whose spatial evolution is independent of time. There are numerous special classes of semigroups, semigroups with additional properties, which appear in particular applications. Some of these classes are even closer to groups by exhibiting some additional but not all properties of a group. Of these we mention: regular semigroups, orthodox semigroups, semigroups with involution, inverse semigroups and cancellative semigroups. There also interesting classes of semigroups that do not contain any groups except the trivial group; examples of the latter kind are bands and their commutative subclass—semilattices, which are also ordered algebraic structures.

22.1   Definition

A semigroup is a set S together with a binary operation " $\cdot$ " (that is, a function $\cdot : S \times S \to S$) that satisfies the associative property:

> For all $a, b, c \in S$, the equation $(a \cdot b) \cdot c = a \cdot (b \cdot c)$ holds.

More succinctly, a semigroup is an associative magma.

22.2   Examples of semigroups

- Empty semigroup: the empty set forms a semigroup with the empty function as the binary operation.

- Semigroup with one element: there is essentially only one, the singleton $\{a\}$ with operation $a \cdot a = a$.

- Semigroup with two elements: there are five which are essentially different.

- The set of positive integers with addition. (With 0 included, this becomes a monoid.)

- The set of integers with minimum or maximum. (With positive/negative infinity included, this becomes a monoid.)

- Square nonnegative matrices of a given size with matrix multiplication.

- Any ideal of a ring with the multiplication of the ring.

- The set of all finite strings over a fixed alphabet Σ with concatenation of strings as the semigroup operation — the so-called "free semigroup over Σ". With the empty string included, this semigroup becomes the free monoid over Σ.

- A probability distribution F together with all convolution powers of F, with convolution as the operation. This is called a convolution semigroup.

- A monoid is a semigroup with an identity element.

- A group is a monoid in which every element has an inverse element.

- Transformation semigroups and monoids

- The set of continuous functions from a topological space to itself

22.3   Basic concepts

22.3.1   Identity and zero

If it has both a left identity and a right identity, a semigroup (and indeed magma) has at most one identity element, which is then two-sided. A semigroup with identity is called a monoid. A semigroup may have multiple left identities but no right identity,[note 2] or vice versa. A semigroup without identity may be embedded in a monoid formed by adjoining an element $e \notin S$ to S and defining $e \cdot s = s \cdot e = s$ for all $s \in S \cup \{e\}$.[2][3] The notation S^1 denotes a monoid obtained from S by adjoining an identity *if necessary* ($S^1 = S$ for a monoid).[3]

Similarly, every magma has at most one absorbing element, which in semigroup theory is called a **zero**. Analogous to the above construction, for every semigroup S, one can define S^0, a semigroup with 0 that embeds S.

22.3.2 Subsemigroups and ideals

The semigroup operation induces an operation on the collection of its subsets: given subsets A and B of a semigroup S, their product $A \cdot B$, written commonly as AB, is the set $\{\, ab \mid a$ in A and b in $B \,\}$. (This notion is defined identically as it is for groups.) In terms of this operations, a subset A is called

- a **subsemigroup** if AA is a subset of A,

- a **right ideal** if AS is a subset of A, and

- a **left ideal** if SA is a subset of A.

If A is both a left ideal and a right ideal then it is called an **ideal** (or a **two-sided ideal**).

If S is a semigroup, then the intersection of any collection of subsemigroups of S is also a subsemigroup of S. So the subsemigroups of S form a complete lattice.

An example of semigroup with no minimal ideal is the set of positive integers under addition. The minimal ideal of a commutative semigroup, when it exists, is a group.

Green's relations, a set of five equivalence relations that characterise the elements in terms of the principal ideals they generate, are important tools for analysing the ideals of a semigroup and related notions of structure.

The subset with the property that its every element commutes with any other element of the semigroup is called the **center** of the semigroup.[4] The center of a semigroup is actually a subsemigroup.[5]

22.3.3 Homomorphisms and congruences

A **semigroup homomorphism** is a function that preserves semigroup structure. A function $f \colon S \to T$ between two semigroups is a homomorphism if the equation

$$f(ab) = f(a)f(b).$$

holds for all elements a, b in S, i.e. the result is the same when performing the semigroup operation after or before applying the map f.

A semigroup homomorphism between monoids preserves identity if it is a monoid homomorphism. But there are semigroup homomorphisms which are not monoid homomorphisms, e.g. the canonical embedding of a semigroup S without identity into S^1. Conditions characterizing monoid homomorphisms are discussed further. Let $f : S_0 \to S_1$ be a semigroup homomorphism. The image of f is also a semigroup. If S_0 is a monoid with an identity element e_0, then $f(e_0)$ is the identity element in the image of f. If S_1 is also a monoid with an identity element e_1 and e_1 belongs to the image of f, then $f(e_0) = e_1$, i.e. f is a monoid homomorphism. Particularly, if f is surjective, then it is a monoid homomorphism.

Two semigroups S and T are said to be isomorphic if there is a bijection $f : S \leftrightarrow T$ with the property that, for any elements a, b in S, $f(ab) = f(a)f(b)$. Isomorphic semigroups have the same structure.

A **semigroup congruence** $\sim$ is an equivalence relation that is compatible with the semigroup operation. That is, a subset $\sim \,\subseteq S \times S$ that is an equivalence relation and $x \sim y$ and $u \sim v$ implies $xu \sim yv$ for every x, y, u, v in S. Like any equivalence relation, a semigroup congruence $\sim$ induces congruence classes

$$[a]_\sim = \{x \in S \mid x \sim a\}$$

and the semigroup operation induces a binary operation $\circ$ on the congruence classes:

$$[u]_\sim \circ [v]_\sim = [uv]_\sim$$

Because $\sim$ is a congruence, the set of all congruence classes of $\sim$ forms a semigroup with $\circ$, called the **quotient semi-group** or **factor semigroup**, and denoted $S/\sim$. The mapping $x \mapsto [x]_\sim$ is a semigroup homomorphism, called the **quotient map**, **canonical surjection** or **projection**; if S is a monoid then quotient semigroup is a monoid with identity $[1]_\sim$. Conversely, the kernel of any semigroup homomorphism is a semigroup congruence. These results are nothing more than a particularization of the first isomorphism theorem in universal algebra. Congruence classes and factor monoids are the objects of study in string rewriting systems.

A **nuclear congruence** on S is one which is the kernel of an endomorphism of S.[6]

A semigroup S satisfies the **maximal condition on congruences** if any family of congruences on S, ordered by inclusion, has a maximal element. By Zorn's lemma, this is equivalent to saying that the ascending chain condition holds: there is no infinite strictly ascending chain of congruences on S.[7]

Every ideal I of a semigroup induces a subsemigroup, the Rees factor semigroup via the congruence $x \rho y \Leftrightarrow$ either $x = y$ or both x and y are in I.

22.4   Structure of semigroups

For any subset A of S there is a smallest subsemigroup T of S which contains A, and we say that A **generates** T. A single element x of S generates the subsemigroup $\{ x^n \mid n$ is a positive integer $\}$. If this is finite, then x is said to be of **finite order**, otherwise it is of **infinite order**. A semigroup is said to be **periodic** if all of its elements are of finite order. A semigroup generated by a single element is said to be monogenic (or cyclic). If a monogenic semigroup is infinite then it is isomorphic to the semigroup of positive integers with the operation of addition. If it is finite and nonempty, then it must contain at least one idempotent. It follows that every nonempty periodic semigroup has at least one idempotent.

A subsemigroup which is also a group is called a **subgroup**. There is a close relationship between the subgroups of a semigroup and its idempotents. Each subgroup contains exactly one idempotent, namely the identity element of the subgroup. For each idempotent e of the semigroup there is a unique maximal subgroup containing e. Each maximal subgroup arises in this way, so there is a one-to-one correspondence between idempotents and maximal subgroups. Here the term *maximal subgroup* differs from its standard use in group theory.

More can often be said when the order is finite. For example, every nonempty finite semigroup is periodic, and has a minimal ideal and at least one idempotent. The number of finite semigroups of a given size (greater than 1) is (obviously) larger than the number of groups of the same size. For example, of the sixteen possible "multiplication tables" for a set of two elements {a, b}, eight form semigroups[note 3] whereas only four of these are monoids and only two form groups. For more on the structure of finite semigroups, see Krohn–Rhodes theory.

22.5   Special classes of semigroups

Main article: Special classes of semigroups

- A monoid is a semigroup with identity.

- A subsemigroup is a subset of a semigroup that is closed under the semigroup operation.

- A band is a semigroup the operation of which is idempotent.

- A cancellative semigroup is one having the cancellation property:[8] $a \cdot b = a \cdot c$ implies $b = c$ and similarly for $b \cdot a = c \cdot a$.

- A semilattice is a semigroup whose operation is idempotent and commutative.

- 0-simple semigroups.

- Transformation semigroups: any finite semigroup S can be represented by transformations of a (state-) set Q of at most $|S| + 1$ states. Each element x of S then maps Q into itself $x: Q \to Q$ and sequence xy is defined by $q(xy) = (qx)y$ for each q in Q. Sequencing clearly is an associative operation, here equivalent to function composition. This representation is basic for any automaton or finite state machine (FSM).

- The bicyclic semigroup is in fact a monoid, which can be described as the free semigroup on two generators p and q, under the relation $pq = 1$.

- C_0-semigroups.

- Regular semigroups. Every element x has at least one inverse y satisfying $xyx=x$ and $yxy=y$; the elements x and y are sometimes called "mutually inverse".

- Inverse semigroups are regular semigroups where every element has exactly one inverse. Alternatively, a regular semigroup is inverse if and only if any two idempotents commute.

- Affine semigroup: a semigroup that is isomorphic to a finitely-generated subsemigroup of Z^d. These semigroups have applications to commutative algebra.

22.6   Structure theorem for commutative semigroups

There is a structure theorem for commutative semigroups in terms of semilattices.[9] A semilattice (or more precisely a meet-semilattice) $(L, \leq)$ is a partially ordered set where every pair of elements $a, b \in L$ has a greatest lower bound, denoted $a \wedge b$. The operation $\wedge$ makes L into a semigroup satisfying the additional idempotence law $a \wedge a = a$.

Given a homomorphism $f : S \to L$ from an arbitrary semigroup to a semilattice, each inverse image $S_a = f^{-1}\{a\}$ is a (possibly empty) semigroup. Moreover S becomes **graded** by L, in the sense that

$S_a S_b \subseteq S_{a \wedge b}$

If f is onto, the semilattice L is isomorphic to the quotient of S by the equivalence relation $\sim$ such that $x \sim y$ iff $f(x) = f(y)$. This equivalence relation is a semigroup congruence, as defined above.

Whenever we take the quotient of a commutative semigroup by a congruence, we get another commutative semigroup. The structure theorem says that for any commutative semigroup S, there is a finest congruence $\sim$ such that the quotient of S by this equivalence relation is a semilattice. Denoting this semilattice by L, we get a homomorphism f from S onto L. As mentioned, S becomes graded by this semilattice.

Furthermore, the components S_a are all Archimedean semigroups. An Archimedean semigroup is one where given any pair of elements x, y, there exists an element z and $n > 0$ such that $x^n = yz$.

The Archimedean property follows immediately from the ordering in the semilattice L, since with this ordering we have $f(x) \leq f(y)$ if and only if $x^n = yz$ for some z and $n > 0$.

22.7   Group of fractions

The **group of fractions** or **group completion** of a semigroup S is the group $G = G(S)$ generated by the elements of S as generators and all equations $xy = z$ which hold true in S as relations.[10] There is an obvious semigroup homomorphism j : $S \to G(S)$ which sends each element of S to the corresponding generator. This has a universal property for morphisms from S to a group:[11] given any group H and any semigroup homomorphism $k : S \to H$, there exists a unique group homomorphism $f : G \to H$ with $k=fj$. We may think of G as the "most general" group that contains a homomorphic image of S.

An important question is to characterize those semigroups for which this map is an embedding. This need not always be the case: for example, take S to be the semigroup of subsets of some set X with set-theoretic intersection as the binary operation (this is an example of a semilattice). Since $A.A = A$ holds for all elements of S, this must be true for all generators of $G(S)$ as well: which is therefore the trivial group. It is clearly necessary for embeddability that S have the cancellation property. When S is commutative this condition is also sufficient[12] and the Grothendieck group of the semigroup provides a construction of the group of fractions. The problem for non-commutative semigroups can be traced to the first substantial paper on semigroups.[13][14] Anatoly Maltsev gave necessary and conditions for embeddability in 1937.[15]

22.8   Semigroup methods in partial differential equations

Further information: C0-semigroup

Semigroup theory can be used to study some problems in the field of partial differential equations. Roughly speaking, the semigroup approach is to regard a time-dependent partial differential equation as an ordinary differential equation on a function space. For example, consider the following initial/boundary value problem for the heat equation on the spatial interval $(0, 1) \subset \mathbf{R}$ and times $t \geq 0$:

$$\begin{cases} \partial_t u(t,x) = \partial_x^2 u(t,x), & x \in (0,1), t > 0; \\ u(t,x) = 0, & x \in \{0,1\}, t > 0; \\ u(t,x) = u_0(x), & x \in (0,1), t = 0. \end{cases}$$

Let $X = L^2((0, 1)\ \mathbf{R})$ be the L^p space of square-integrable real-valued functions with domain the interval $(0, 1)$ and let A be the second-derivative operator with domain

$$D(A) = \left\{ u \in H^2((0,1); \mathbf{R}) \big| u(0) = u(1) = 0 \right\},$$

where H^2 is a Hardy space. Then the above initial/boundary value problem can be interpreted as an initial value problem for an ordinary differential equation on the space X:

$$\begin{cases} \dot{u}(t) = Au(t); \\ u(0) = u_0. \end{cases}$$

On an heuristic level, the solution to this problem "ought" to be $u(t) = \exp(tA)u_0$. However, for a rigorous treatment, a meaning must be given to the exponential of tA. As a function of t, $\exp(tA)$ is a semigroup of operators from X to itself, taking the initial state u_0 at time $t = 0$ to the state $u(t) = \exp(tA)u_0$ at time t. The operator A is said to be the infinitesimal generator of the semigroup.

22.9   History

The study of semigroups trailed behind that of other algebraic structures with more complex axioms such as groups or rings. A number of sources[16][17] attribute the first use of the term (in French) to J.-A. de Séguier in *Élements de la Théorie des Groupes Abstraits* (Elements of the Theory of Abstract Groups) in 1904. The term is used in English in 1908 in Harold Hinton's *Theory of Groups of Finite Order*.

Anton Suschkewitsch obtained the first non-trivial results about semigroups. His 1928 paper *Über die endlichen Gruppen ohne das Gesetz der eindeutigen Umkehrbarkeit* (*On finite groups without the rule of unique invertibility*) determined the structure of finite simple semigroups and showed that the minimal ideal (or Green's relations J-class) of a finite semigroup is simple.[17] From that point on, the foundations of semigroup theory were further laid by David Rees, James Alexander Green, Evgenii Sergeevich Lyapin, Alfred H. Clifford and Gordon Preston. The latter two published a two-volume monograph on semigroup theory in 1961 and 1967 respectively. In 1970, a new periodical called *Semigroup Forum* (currently edited by Springer Verlag) became one of the few mathematical journals devoted entirely to semigroup theory.

In recent years researchers in the field have become more specialized with dedicated monographs appearing on important classes of semigroups, like inverse semigroups, as well as monographs focusing on applications in algebraic automata theory, particularly for finite automata, and also in functional analysis.

22.10 Generalizations

If the associativity axiom of a semigroup is dropped, the result is a magma, which is nothing more than a set M equipped with a binary operation $M \times M \to M$.

Generalizing in a different direction, an **n-ary semigroup** (also **n-semigroup**, **polyadic semigroup** or **multiary semigroup**) is a generalization of a semigroup to a set G with a n-ary operation instead of a binary operation.[18] The associative law is generalized as follows: ternary associativity is $(abc)de = a(bcd)e = ab(cde)$, i.e. the string $abcde$ with any three adjacent elements bracketed. N-ary associativity is a string of length $n + (n - 1)$ with any n adjacent elements bracketed. A 2-ary semigroup is just a semigroup. Further axioms lead to an n-ary group.

A third generalization is the semigroupoid, in which the requirement that the binary relation be total is lifted. As categories generalize monoids in the same way, a semigroupoid behaves much like a category but lacks identities.

Infinitary generalizations of commutative semigroups have sometimes been considered by various authors. [19]

22.11 See also

- Absorbing element

- Biordered set

- Empty semigroup

- Identity element

- Light's associativity test

- Semigroup ring

- Weak inverse

- Quantum dynamical semigroup

22.12 Notes

[1] The closure axiom is implied by the definition of a binary operation on a set. Some authors thus omit it and specify three *laws* for a group and only one law (associativity) for semigroup.

[2] For instance, the semigroup of three elements e, f, g with, for any x, $ex = x$, $fx = f$, and $gx = f$ has exactly one left identity but no right identity.

[3] Namely: the trivial semigroup in which (for all x and y) $xy =$ a and its counterpart in which $xy =$ b, the semigroups based on multiplication modulo 2 (choosing a or b as the identity element 1), the groups equivalent to addition modulo 2 (choosing a or b to be the identity element 0), and the semigroups in which the elements are either both left identities or both right identities.

22.13   Citations

[1] (Feller 1971)

[2] Jacobson 2009, p. 30, ex. 5

[3] Lawson 1998, p. 20

[4] Kilp, Mati; Knauer, U.; Mikhalev, Aleksandr V. (2000). *Monoids, Acts, and Categories: With Applications to Wreath Products and Graphs : a Handbook for Students and Researchers*. Walter de Gruyter. p. 25. ISBN 978-3-11-015248-7. Zbl 0945.20036.

[5] Lîapin, E. S. (1968). *Semigroups*. American Mathematical Soc. p. 96. ISBN 978-0-8218-8641-0.

[6] Lothaire 2011, p. 463

[7] Lothaire 2011, p. 465

[8] Clifford & Preston 1967, p. 3

[9] Grillet 2001

[10] Farb, B. (2006), *Problems on mapping class groups and related topics*, Amer. Math. Soc., p. 357, ISBN 0-8218-3838-5

[11] Auslander, M.; Buchsbaum, D. A. (1974). *Groups, rings, modules*. Harper & Row. p. 50. ISBN 0-06-040387-X.

[12] Clifford & Preston 1961, p. 34

[13] (Suschkewitsch 1928)

[14] Preston, G. B. (1990), *Personal reminiscences of the early history of semigroups*, retrieved 2009-05-12

[15] Maltsev, A. (1937), "On the immersion of an algebraic ring into a field", *Math. Annalen* **113**: 686–691, doi:10.1007/BF01571659.

[16] Earliest Known Uses of Some of the Words of Mathematics

[17] An account of Suschkewitsch's paper by Christopher Hollings

[18] Dudek, W.A. (2001), "On some old problems in *n*-ary groups", *Quasigroups and Related Systems* **8**: 15–36

[19] See references in Udo Hebisch and Hanns Joachim Weinert, *Semirings and Semifields*, in particular, Section 10, *Semirings with infinite sums*, in M. Hazewinkel, Handbook of Algebra, Vol. 1, Elsevier, 1996. Notice that in this context the authors use the term *semimodule* in place of *semigroup*.

22.14   References

General references

- Howie, John M. (1995), *Fundamentals of Semigroup Theory*, Clarendon Press, ISBN 0-19-851194-9, Zbl 0835.20077.

- Clifford, A. H.; Preston, G. B. (1961), *The Algebraic Theory of Semigroups* **1**, American Mathematical Society, ISBN 978-0-8218-0271-7, Zbl 0111.03403.

- Clifford, A. H.; Preston, G. B. (1967), *The Algebraic Theory of Semigroups* **2**, American Mathematical Society, ISBN 978-0-8218-0272-4, Zbl 0178.01203.

- Grillet, Pierre A. (1995), *Semigroups: An Introduction to the Structure Theory*, Marcel Dekker, ISBN 978-0-8247-9662-4, Zbl 0830.20079.

- Grillet, Pierre A. (2001), *Commutative Semigroups*, Springer Verlag, ISBN 978-0-7923-7067-3, Zbl 1040.20048.

- Hollings, Christopher (2014), *Mathematics across the Iron Curtain: A History of the Algebraic Theory of Semigroups*, American Mathematical Society, ISBN 978-1-4704-1493-1, Zbl 06329297.

- Petrich, Mario (1973), *Introduction to Semigroups*, Charles E. Merrill, ISBN 0-675-09062-8, Zbl 0321.20037.

Specific references

- Feller, William (1971), *An introduction to probability theory and its applications* **II** (2nd ed.), Wiley, MR 0270403.

- Hille, Einar; Phillips, Ralph S. (1974), *Functional analysis and semi-groups*, American Mathematical Society, ISBN 0821874640, MR 0423094.

- Suschkewitsch, Anton (1928), "Über die endlichen Gruppen ohne das Gesetz der eindeutigen Umkehrbarkeit", *Mathematische Annalen* **99** (1): 30–50, doi:10.1007/BF01459084, ISSN 0025-5831, MR 1512437.

- Kantorovitz, Shmuel (2009), *Topics in Operator Semigroups*, Springer, ISBN 978-0-8176-4932-6, Zbl 1187.47003.

- Jacobson, Nathan (2009), *Basic algebra* **1** (2nd ed.), Dover, ISBN 978-0-486-47189-1

- Lawson, M.V. (1998), *Inverse semigroups: the theory of partial symmetries*, World Scientific, ISBN 978-981-02-3316-7, Zbl 1079.20505

- Lothaire, M. (2011) [2002], *Algebraic combinatorics on words*, Encyclopedia of Mathematics and Its Applications **90**, Cambridge University Press, ISBN 978-0-521-18071-9, Zbl 1221.68183

Chapter 23

Magma (algebra)

For other uses, see Magma (disambiguation).

In abstract algebra, a **magma** (or **groupoid**; not to be confused with groupoids in category theory) is a basic kind of algebraic structure. Specifically, a magma consists of a set, M, equipped with a single binary operation, $M \times M \to M$. The binary operation must be closed by definition but no other properties are imposed.

23.1 History and terminology

The term *groupoid* was introduced in 1926 by Heinrich Brandt describing his Brandt groupoid (translated from the German *Gruppoid*). The term was then appropriated by B. A. Hausmann and Øystein Ore (1937)[1] in the sense (of a set with a binary operation) used in this article. In a couple of reviews of subsequent papers in Zentralblatt, Brandt strongly disagreed with this overloading of terminology. The Brandt groupoid is a groupoid in the sense used in category theory, but not in the sense used by Hausmann and Ore. Nevertheless, influential books in semigroup theory, including Clifford and Preston (1961) and Howie (1995) use groupoid in the sense of Hausmann and Ore. Hollings (2014) writes that the term *groupoid* is "perhaps most often used in modern mathematics" in the sense given to it in category theory.[2]

According to Bergman and Hausknecht (1996): "There is no generally accepted word for a set with a not necessarily associative binary operation. The word *groupoid* is used by many universal algebraists, but workers in category theory and related areas object strongly this usage because they use same word to mean "category in which all morphisms are invertible". The term *magma* was used by Serre [Lie Algebras and Lie Groups, 1965]."[3] It also appears in Bourbaki's *Éléments de mathématique*, Algèbre, chapitres 1 à 3, 1970.[4]

23.2 Definition

A magma is a set M matched with an operation, •, that sends any two elements $a, b \in M$ to another element, $a • b$. The symbol, •, is a general placeholder for a properly defined operation. To qualify as a magma, the set and operation $(M, •)$ must satisfy the following requirement (known as the *magma or closure axiom*):

> For all a, b in M, the result of the operation $a • b$ is also in M.

And in mathematical notation:

$$\forall\, a, b \in M : a • b \in M.$$

If • is instead a partial operation, then S is called a **partial magma**[5] or more often a partial groupoid.[5][6]

23.3 Morphism of magmas

A morphism of magmas is a function, $f : M \rightarrow N$, mapping magma, M, to magma, N, that preserves the binary operation:

$$f (x \bullet_M y) = f(x) \bullet_N f(y)$$

where $\bullet_M$ and $\bullet_N$ denote the binary operation on M and N respectively.

23.4 Notation and combinatorics

The magma operation may be applied repeatedly, and in the general, non-associative case, the order matters, which is notated with parentheses. Also, the operation, $\bullet$, is often omitted and notated by juxtaposition:

$$(a \bullet (b \bullet c)) \bullet d = (a(bc))d$$

A shorthand is often used to reduce the number of parentheses, in which the innermost operations and pairs of parentheses are omitted, being replaced just with juxtaposition, $xy \bullet z = (x \bullet y) \bullet z$. For example, the above is abbreviated to the following expression, still containing parentheses:

$$(a \bullet bc)d.$$

A way to avoid completely the use of parentheses is prefix notation, in which the same expression would be written $\bullet\bullet a \bullet bcd$.

The set of all possible strings consisting of symbols denoting elements of the magma, and sets of balanced parentheses is called the Dyck language. The total number of different ways of writing n applications of the magma operator is given by the Catalan number, Cn. Thus, for example, $C_2 = 2$, which is just the statement that $(ab)c$ and $a(bc)$ are the only two ways of pairing three elements of a magma with two operations. Less trivially, $C_3 = 5$: $((ab)c)d$, $(a(bc))d$, $(ab)(cd)$, $a((bc)d)$, and $a(b(cd))$.

The number of non-isomorphic magmas having 0, 1, 2, 3, 4, ... elements are 1, 1, 10, 3330, 178981952, ... (sequence A001329 in OEIS). The corresponding numbers of non-isomorphic and non-antiisomorphic magmas are 1, 1, 7, 1734, 89521056, ... (sequence A001424 in OEIS).[7]

23.5 Free magma

A **free magma**, MX, on a set, X, is the "most general possible" magma generated by X (i.e., there are no relations or axioms imposed on the generators; see free object). It can be described as the set of non-associative words on X with parentheses retained:[8]

It can also be viewed, in terms familiar in computer science, as the magma of binary trees with leaves labelled by elements of X. The operation is that of joining trees at the root. It therefore has a foundational role in syntax.

A free magma has the universal property such that, if $f : X \rightarrow N$ is a function from X to any magma, N, then there is a unique extension of f to a morphism of magmas, f'

$$f' : MX \rightarrow N.$$

See also: free semigroup, free group, Hall set, Wedderburn–Etherington number

Magma

divisibility   associativity

Quasigroup   Semigroup

identity   identity

Loop   Monoid

associativity   invertibility

Group

23.6 Types of magmas

Magmas are not often studied as such; instead there are several different kinds of magmas, depending on what axioms one might require of the operation. Commonly studied types of magmas include:

Quasigroups Magmas where division is always possible

Loops Quasigroups with identity elements

Semigroups Magmas where the operation is associative

Semilattices Semigroups where the operation is commutative and idempotent

Monoids Semigroups with identity elements

Groups Monoids with inverse elements, or equivalently, associative loops or non-empty associative quasigroups

Abelian groups Groups where the operation is commutative

Note that each of divisibility and invertibility imply the cancellation property.

23.7 Classification by properties

A magma $(S, \bullet)$, with $x, y, u, z \in S$, is called

Medial If it satisfies the identity, $xy \bullet uz \equiv xu \bullet yz$

Left semimedial If it satisfies the identity, $xx \bullet yz \equiv xy \bullet xz$

Right semimedial If it satisfies the identity, $yz \bullet xx \equiv yx \bullet zx$

Semimedial If it is both left and right semimedial

Left distributive If it satisfies the identity, $x \bullet yz \equiv xy \bullet xz$

Right distributive If it satisfies the identity, $yz \bullet x \equiv yx \bullet zx$

Autodistributive If it is both left and right distributive

Commutative If it satisfies the identity, $xy \equiv yx$

Idempotent If it satisfies the identity, $xx \equiv x$

Unipotent If it satisfies the identity, $xx \equiv yy$

Zeropotent If it satisfies the identities, $xx \bullet y \equiv xx \equiv y \bullet xx$[9]

Alternative If it satisfies the identities $xx \bullet y \equiv x \bullet xy$ and $x \bullet yy \equiv xy \bullet y$

Power-associative If the submagma generated by any element is associative

A semigroup, or associative If it satisfies the identity, $x \bullet yz \equiv xy \bullet z$

A left unar If it satisfies the identity, $xy \equiv xz$

A right unar If it satisfies the identity, $yx \equiv zx$

Semigroup with zero multiplication, or null semigroup If it satisfies the identity, $xy \equiv uv$

Unital If it has an identity element

Left-cancellative If, for all x, y, and, z, $xy = xz$ implies $y = z$

Right-cancellative If, for all x, y, and, z, $yx = zx$ implies $y = z$

Cancellative If it is both right-cancellative and left-cancellative

A semigroup with left zeros If it is a semigroup and, for all x, the identity, $x \equiv xy$, holds

A semigroup with right zeros If it is a semigroup and, for all x, the identity, $x \equiv yx$, holds

Trimedial If any triple of (not necessarily distinct) elements generates a medial submagma

Entropic If it is a homomorphic image of a medial cancellation magma.[10]

23.8 Generalizations

See *n*-ary group.

23.9   See also

- Magma category

- Auto magma object

- Universal algebra

- Magma computer algebra system, named after the object of this article.

- Commutative non-associative magmas

- Algebraic structures whose axioms are all identities

- Groupoid algebra

23.10   References

[1] Hausmann, B. A.; Ore, Øystein (October 1937), "Theory of quasi-groups", *American Journal of Mathematics* **59** (4): 983–1004, JSTOR 2371362

[2] Hollings, Christopher (2014), *Mathematics across the Iron Curtain: A History of the Algebraic Theory of Semigroups*, American Mathematical Society, pp. 142–3, ISBN 978-1-4704-1493-1

[3] Bergman, George M.; Hausknecht, Adam O. (1996), *Cogroups and Co-rings in Categories of Associative Rings*, American Mathematical Society, p. 61, ISBN 978-0-8218-0495-7

[4] Bourbaki, N. (1998) [1970], "Alebraic Structures: §1.1 Laws of Composition: Definition 1", *Algebra I: Chapters 1–3*, Springer, p. 1, ISBN 978-3-540-64243-5

[5] Müller-Hoissen, Folkert; Pallo, Jean Marcel; Stasheff, Jim, eds. (2012), *Associahedra, Tamari Lattices and Related Structures: Tamari Memorial Festschrift*, Springer, p. 11, ISBN 978-3-0348-0405-9

[6] Evseev, A. E. (1988), "A survey of partial groupoids", in Silver, Ben, *Nineteen Papers on Algebraic Semigroups*, American Mathematical Society, ISBN 0-8218-3115-1

[7] Weisstein, Eric W., "Groupoid", *MathWorld*.

[8] Rowen, Louis Halle (2008), "Definition 21B.1.", *Graduate Algebra: Noncommutative View*, Graduate Studies in Mathematics, American Mathematical Society, p. 321, ISBN 0-8218-8408-5

[9] Kepka, T.; Němec, P. (1996), "Simple balanced groupoids" (PDF), *Acta Universitatis Palackianae Olomucensis. Facultas Rerum Naturalium. Mathematica* **35** (1): 53–60

[10] Ježek, Jaroslav; Kepka, Tomáš (1981), "Free entropic groupoids" (PDF), *Commentationes Mathematicae Universitatis Carolinae* **22** (2): 223–233, MR 620359.

- M. Hazewinkel (2001), "Magma", in Hazewinkel, Michiel, *Encyclopedia of Mathematics*, Springer, ISBN 978-1-55608-010-4

- M. Hazewinkel (2001), "Groupoid", in Hazewinkel, Michiel, *Encyclopedia of Mathematics*, Springer, ISBN 978-1-55608-010-4

- M. Hazewinkel (2001), "Free magma", in Hazewinkel, Michiel, *Encyclopedia of Mathematics*, Springer, ISBN 978-1-55608-010-4

- Weisstein, Eric W., "Groupoid", *MathWorld*.

23.11   Further reading

- Bruck, Richard Hubert (1971), *A survey of binary systems* (3rd ed.), Springer, ISBN 978-0-387-03497-3

Chapter 24

First-order logic

First-order logic is a formal system used in mathematics, philosophy, linguistics, and computer science. It is also known as **first-order predicate calculus**, the **lower predicate calculus**, **quantification theory**, and predicate logic. First-order logic uses quantified variables over (non-logical) objects. This distinguishes it from propositional logic which does not use quantifiers.

A theory about some topic is usually first-order logic together with a specified domain of discourse over which the quantified variables range, finitely many functions which map from that domain into it, finitely many predicates defined on that domain, and a recursive set of axioms which are believed to hold for those things. Sometimes "theory" is understood in a more formal sense, which is just a set of sentences in first-order logic.

The adjective "first-order" distinguishes first-order logic from higher-order logic in which there are predicates having predicates or functions as arguments, or in which one or both of predicate quantifiers or function quantifiers are permitted.[1] In first-order theories, predicates are often associated with sets. In interpreted higher-order theories, predicates may be interpreted as sets of sets.

There are many deductive systems for first-order logic that are sound (all provable statements are true in all models) and complete (all statements which are true in all models are provable). Although the logical consequence relation is only semidecidable, much progress has been made in automated theorem proving in first-order logic. First-order logic also satisfies several metalogical theorems that make it amenable to analysis in proof theory, such as the Löwenheim–Skolem theorem and the compactness theorem.

First-order logic is the standard for the formalization of mathematics into axioms and is studied in the foundations of mathematics. Mathematical theories, such as number theory and set theory, have been formalized into first-order axiom schemas such as Peano arithmetic and Zermelo–Fraenkel set theory (ZF) respectively.

No first-order theory, however, has the strength to describe uniquely a structure with an infinite domain, such as the natural numbers or the real line. A uniquely describing, i.e. categorical, axiom system for such a structure can be obtained in stronger logics such as second-order logic.

For a history of first-order logic and how it came to dominate formal logic, see José Ferreirós (2001).

24.1 Introduction

While propositional logic deals with simple declarative propositions, first-order logic additionally covers predicates and quantification.

A predicate takes an entity or entities in the domain of discourse as input and outputs either True or False. Consider the two sentences "Socrates is a philosopher" and "Plato is a philosopher". In propositional logic, these sentences are viewed as being unrelated and are denoted, for example, by p and q. However, the predicate "is a philosopher" occurs in both sentences which have a common structure of "a is a philosopher". The variable a is instantiated as "Socrates" in the first sentence and is instantiated as "Plato" in the second sentence. The use of predicates, such as "is a philosopher" in this example, distinguishes first-order logic from propositional logic.

Predicates can be compared. Consider, for example, the first-order formula "if a is a philosopher, then a is a scholar". This formula is a conditional statement with "a is a philosopher" as hypothesis and "a is a scholar" as conclusion. The truth of this formula depends on which object is denoted by a, and on the interpretations of the predicates "is a philosopher" and "is a scholar".

Variables can be quantified over. The variable a in the previous formula can be quantified over, for instance, in the first-order sentence "For every a, if a is a philosopher, then a is a scholar". The universal quantifier "for every" in this sentence expresses the idea that the claim "if a is a philosopher, then a is a scholar" holds for *all* choices of a.

The *negation* of the sentence "For every a, if a is a philosopher, then a is a scholar" is logically equivalent to the sentence "There exists a such that a is a philosopher and a is not a scholar". The existential quantifier "there exists" expresses the idea that the claim "a is a philosopher and a is not a scholar" holds for *some* choice of a.

The predicates "is a philosopher" and "is a scholar" each take a single variable. Predicates can take several variables. In the first-order sentence "Socrates is the teacher of Plato", the predicate "is the teacher of" takes two variables.

To interpret a first-order formula, one specifies what each predicate means and the entities that can instantiate the predicated variables. These entities form the domain of discourse or universe, which is usually required to be a nonempty set. Given that the interpretation with the domain of discourse as consisting of all human beings and the predicate "is a philosopher" understood as "have written the Republic", the sentence "There exists a such that a is a philosopher" is seen as being true, as witnessed by Plato.

24.2   Syntax

There are two key parts of first-order logic. The syntax determines which collections of symbols are legal expressions in first-order logic, while the semantics determine the meanings behind these expressions.

24.2.1   Alphabet

Unlike natural languages, such as English, the language of first-order logic is completely formal, so that it can be mechanically determined whether a given expression is legal. There are two key types of legal expressions: **terms**, which intuitively represent objects, and **formulas**, which intuitively express predicates that can be true or false. The terms and formulas of first-order logic are strings of **symbols** which together form the **alphabet** of the language. As with all formal languages, the nature of the symbols themselves is outside the scope of formal logic; they are often regarded simply as letters and punctuation symbols.

It is common to divide the symbols of the alphabet into **logical symbols**, which always have the same meaning, and **non-logical symbols**, whose meaning varies by interpretation. For example, the logical symbol $\land$ always represents "and"; it is never interpreted as "or". On the other hand, a non-logical predicate symbol such as Phil(x) could be interpreted to mean "x is a philosopher", "x is a man named Philip", or any other unary predicate, depending on the interpretation at hand.

Logical symbols

There are several logical symbols in the alphabet, which vary by author but usually include:

- The quantifier symbols $\forall$ and $\exists$

- The logical connectives: $\land$ for conjunction, $\lor$ for disjunction, $\rightarrow$ for implication, $\leftrightarrow$ for biconditional, $\neg$ for negation. Occasionally other logical connective symbols are included. Some authors use Cpq, instead of $\rightarrow$, and Epq, instead of $\leftrightarrow$, especially in contexts where $\rightarrow$ is used for other purposes. Moreover, the horseshoe $\supset$ may replace $\rightarrow$; the triple-bar $\equiv$ may replace $\leftrightarrow$; a tilde ($\sim$), Np, or Fpq, may replace $\neg$; $\|$, or Apq may replace $\lor$; and $\&$, Kpq, or the middle dot, ·, may replace ∧, especially if these symbols are not available for technical reasons. (*Note*: the aforementioned symbols C*pq*, E*pq*, N*p*, A*pq*, and K*pq* are used in Polish notation.)

- Parentheses, brackets, and other punctuation symbols. The choice of such symbols varies depending on context.

- An infinite set of **variables**, often denoted by lowercase letters at the end of the alphabet x, y, z, ... Subscripts are often used to distinguish variables: x_0, x_1, x_2, ...

- An **equality symbol** (sometimes, **identity symbol**) =; see the section on equality below.

It should be noted that not all of these symbols are required – only one of the quantifiers, negation and conjunction, variables, brackets and equality suffice. There are numerous minor variations that may define additional logical symbols:

- Sometimes the truth constants T, V*pq*, or ⊤, for "true" and F, O*pq*, or ⊥, for "false" are included. Without any such logical operators of valence 0, these two constants can only be expressed using quantifiers.

- Sometimes additional logical connectives are included, such as the Sheffer stroke, D*pq* (NAND), and exclusive or, J*pq*.

Non-logical symbols

The non-logical symbols represent predicates (relations), functions and constants on the domain of discourse. It used to be standard practice to use a fixed, infinite set of non-logical symbols for all purposes. A more recent practice is to use different non-logical symbols according to the application one has in mind. Therefore it has become necessary to name the set of all non-logical symbols used in a particular application. This choice is made via a **signature**.[2]

The traditional approach is to have only one, infinite, set of non-logical symbols (one signature) for all applications. Consequently, under the traditional approach there is only one language of first-order logic.[3] This approach is still common, especially in philosophically oriented books.

1. For every integer $n \geq 0$ there is a collection of *n*-**ary**, or *n*-**place**, **predicate symbols**. Because they represent relations between n elements, they are also called **relation symbols**. For each arity n we have an infinite supply of them:

    $P^n{}_0, P^n{}_1, P^n{}_2, P^n{}_3, \ldots$

2. For every integer $n \geq 0$ there are infinitely many *n*-ary **function symbols**:

    $f^n{}_0, f^n{}_1, f^n{}_2, f^n{}_3, \ldots$

In contemporary mathematical logic, the signature varies by application. Typical signatures in mathematics are $\{1, \times\}$ or just $\{\times\}$ for groups, or $\{0, 1, +, \times, <\}$ for ordered fields. There are no restrictions on the number of non-logical symbols. The signature can be empty, finite, or infinite, even uncountable. Uncountable signatures occur for example in modern proofs of the Löwenheim-Skolem theorem.

In this approach, every non-logical symbol is of one of the following types.

1. A **predicate symbol** (or **relation symbol**) with some **valence** (or **arity**, number of arguments) greater than or equal to 0. These are often denoted by uppercase letters P, Q, R,... .

    - Relations of valence 0 can be identified with propositional variables. For example, P, which can stand for any statement.

    - For example, $P(x)$ is a predicate variable of valence 1. One possible interpretation is "x is a man".

    - $Q(x,y)$ is a predicate variable of valence 2. Possible interpretations include "x is greater than y" and "x is the father of y".

2. A **function symbol**, with some valence greater than or equal to 0. These are often denoted by lowercase letters f, g, h,... .

- Examples: $f(x)$ may be interpreted as for "the father of x". In arithmetic, it may stand for "-x". In set theory, it may stand for "the power set of x". In arithmetic, $g(x,y)$ may stand for "x+y". In set theory, it may stand for "the union of x and y".

- Function symbols of valence 0 are called **constant symbols**, and are often denoted by lowercase letters at the beginning of the alphabet a, b, c,... . The symbol a may stand for Socrates. In arithmetic, it may stand for 0. In set theory, such a constant may stand for the empty set.

The traditional approach can be recovered in the modern approach by simply specifying the "custom" signature to consist of the traditional sequences of non-logical symbols.

24.2.2   Formation rules

The formation rules define the terms and formulas of first order logic. When terms and formulas are represented as strings of symbols, these rules can be used to write a formal grammar for terms and formulas. These rules are generally context-free (each production has a single symbol on the left side), except that the set of symbols may be allowed to be infinite and there may be many start symbols, for example the variables in the case of terms.

Terms

The set of **terms** is inductively defined by the following rules:

1. **Variables.** Any variable is a term.

2. **Functions.** Any expression $f(t_1,...,tn)$ of n arguments (where each argument ti is a term and f is a function symbol of valence n) is a term. In particular, symbols denoting individual constants are 0-ary function symbols, and are thus terms.

Only expressions which can be obtained by finitely many applications of rules 1 and 2 are terms. For example, no expression involving a predicate symbol is a term.

Formulas

The set of **formulas** (also called well-formed formulas [4] or **wff**s) is inductively defined by the following rules:

1. **Predicate symbols.** If P is an n-ary predicate symbol and $t1$, ..., tn are terms then $P(t_1,...,t_n)$ is a formula.

2. **Equality.** If the equality symbol is considered part of logic, and $t1$ and t_2 are terms, then $t_1 = t_2$ is a formula.

3. **Negation.** If φ is a formula, then $\neg \varphi$ is a formula.

4. **Binary connectives.** If φ and ψ are formulas, then $(\varphi \to \psi)$ is a formula. Similar rules apply to other binary logical connectives.

5. **Quantifiers.** If φ is a formula and x is a variable, then $\forall x \varphi$ (for all x, φ holds) and $\exists x \varphi$ (there exists x such that φ) are formulas.

Only expressions which can be obtained by finitely many applications of rules 1–5 are formulas. The formulas obtained from the first two rules are said to be **atomic formulas**.

For example,

$$\forall x \forall y (P(f(x)) \rightarrow \neg (P(x) \rightarrow Q(f(y), x, z)))$$

is a formula, if f is a unary function symbol, P a unary predicate symbol, and Q a ternary predicate symbol. On the other hand, $\forall x \, x \rightarrow$ is not a formula, although it is a string of symbols from the alphabet.

The role of the parentheses in the definition is to ensure that any formula can only be obtained in one way by following the inductive definition (in other words, there is a unique parse tree for each formula). This property is known as **unique readability** of formulas. There are many conventions for where parentheses are used in formulas. For example, some authors use colons or full stops instead of parentheses, or change the places in which parentheses are inserted. Each author's particular definition must be accompanied by a proof of unique readability.

This definition of a formula does not support defining an if-then-else function ite(c, a, b), where "c" is a condition expressed as a formula, that would return "a" if c is true, and "b" if it is false. This is because both predicates and functions can only accept terms as parameters, but the first parameter is a formula. Some languages built on first-order logic, such as SMT-LIB 2.0, add this.[5]

Notational conventions

For convenience, conventions have been developed about the precedence of the logical operators, to avoid the need to write parentheses in some cases. These rules are similar to the order of operations in arithmetic. A common convention is:

- $\neg$ is evaluated first

- $\wedge$ and $\vee$ are evaluated next

- Quantifiers are evaluated next

- $\rightarrow$ is evaluated last.

Moreover, extra punctuation not required by the definition may be inserted to make formulas easier to read. Thus the formula

$$(\neg \forall x P(x) \rightarrow \exists x \neg P(x))$$

might be written as

$$(\neg [\forall x P(x)]) \rightarrow \exists x [\neg P(x)].$$

In some fields, it is common to use infix notation for binary relations and functions, instead of the prefix notation defined above. For example, in arithmetic, one typically writes "2 + 2 = 4" instead of "=(+(2,2),4)". It is common to regard formulas in infix notation as abbreviations for the corresponding formulas in prefix notation.

The definitions above use infix notation for binary connectives such as $\rightarrow$. A less common convention is Polish notation, in which one writes $\rightarrow$, $\wedge$, and so on in front of their arguments rather than between them. This convention allows all punctuation symbols to be discarded. Polish notation is compact and elegant, but rarely used in practice because it is hard for humans to read it. In Polish notation, the formula

$$\forall x \forall y (P(f(x)) \rightarrow \neg (P(x) \rightarrow Q(f(y), x, z)))$$

becomes "∀x∀y→Pfx¬→ PxQfyxz".

24.2.3 Free and bound variables

Main article: Free variables and bound variables

In a formula, a variable may occur **free** or **bound**. Intuitively, a variable is free in a formula if it is not quantified: in $\forall y\, P(x, y)$, variable x is free while y is bound. The free and bound variables of a formula are defined inductively as follows.

1. **Atomic formulas.** If φ is an atomic formula then x is free in φ if and only if x occurs in φ. Moreover, there are no bound variables in any atomic formula.

2. **Negation.** x is free in $\neg\, \varphi$ if and only if x is free in φ. x is bound in $\neg\, \varphi$ if and only if x is bound in φ.

3. **Binary connectives.** x is free in $(\varphi \rightarrow \psi)$ if and only if x is free in either φ or ψ. x is bound in $(\varphi \rightarrow \psi)$ if and only if x is bound in either φ or ψ. The same rule applies to any other binary connective in place of $\rightarrow$.

4. **Quantifiers.** x is free in $\forall\, y\, \varphi$ if and only if x is free in φ and x is a different symbol from y. Also, x is bound in $\forall\, y\, \varphi$ if and only if x is y or x is bound in φ. The same rule holds with $\exists$ in place of $\forall$.

For example, in $\forall x \forall y\, (P(x) \rightarrow Q(x, f(x), z))$, x and y are bound variables, z is a free variable, and w is neither because it does not occur in the formula.

Free and bound variables of a formula need not be disjoint sets: x is both free and bound in $P(x) \rightarrow \forall x\, Q(x)$.

Freeness and boundness can be also specialized to specific occurrences of variables in a formula. For example, in $P(x) \rightarrow \forall x\, Q(x)$, the first occurrence of x is free while the second is bound. In other words, the x in $P(x)$ is free while the x in $\forall x\, Q(x)$ is bound.

A formula in first-order logic with no free variables is called a **first-order sentence**. These are the formulas that will have well-defined truth values under an interpretation. For example, whether a formula such as $\mathrm{Phil}(x)$ is true must depend on what x represents. But the sentence $\exists x\, \mathrm{Phil}(x)$ will be either true or false in a given interpretation.

24.2.4 Examples

Ordered abelian groups

In mathematics the language of ordered abelian groups has one constant symbol 0, one unary function symbol $-$, one binary function symbol $+$, and one binary relation symbol $\leq$. Then:

- The expressions $+(x, y)$ and $+(x, +(y, -(z)))$ are **terms**. These are usually written as $x + y$ and $x + y - z$.

- The expressions $+(x, y) = 0$ and $\leq(+(x, +(y, -(z))), +(x, y))$ are **atomic formulas**.

  These are usually written as $x + y = 0$ and $x + y - z \leq x + y$.

- The expression $(\forall x \forall y \leq(+(x, y), z) \rightarrow \forall x\, \forall y +(x, y) = 0)$ is a **formula**, which is usually written as $\forall x \forall y (x + y \leq z) \rightarrow \forall x \forall y (x + y = 0)$.

Loving relation

English sentences like "everyone loves someone" can be formalized by first-order logic formulas like $\forall x \exists y\, L(x, y)$. This is accomplished by abbreviating the relation "x loves y" by $L(x, y)$. Using just the two quantifiers $\forall$ and $\exists$ and the loving relation symbol L, but no logical connectives and no function symbols (including constants), formulas with 8 different meanings can be built. The following diagrams show models for each of them, assuming that there are exactly five individuals $a,...,e$ who can love (vertical axis) and be loved (horizontal axis). A small red box at row x and column y indicates $L(x,y)$. Only for the formulas 9 and 10 is the model unique, all other formulas may be satisfied by several models.

Each model, represented by a logical matrix, satisfies the formulas in its caption in a "minimal" way, i.e. whitening any red cell in any matrix would make it non-satisfying the corresponding formula. For example, formula 1 is also satisfied by the matrices at 3, 6, and 10, but not by those at 2, 4, 5, and 7. Conversely, the matrix shown at 6 satisfies 1, 2, 5, 6, 7, and 8, but not 3, 4, 9, and 10.

Some formulas imply others, i.e. *all* matrices satisfying the antecedent (LHS) also satisfy the conclusion (RHS) of the implication — e.g. formula 3 implies formula 1, i.e.: each matrix fulfilling formula 3 also fulfills formula 1, but not vice versa (see the Hasse diagram for this ordering relation). In contrast, only some matrices,[6] which satisfy formula 2, happen to satisfy also formula 5, whereas others,[7] also satisfying formula 2, do not; therefore formula 5 is not a logical consequence of formula 2.

The sequence of the quantifiers is important! So it is instructive to distinguish formulas 1: $\forall x \, \exists y \, L(y,x)$, and 3: $\exists x \, \forall y \, L(x,y)$. In both cases everyone is loved; but in the first case everyone (x) is loved by someone (y), in the second case everyone (y) is loved by just exactly one person (x).

24.3 Semantics

An interpretation of a first-order language assigns a denotation to all non-logical constants in that language. It also determines a domain of discourse that specifies the range of the quantifiers. The result is that each term is assigned an object that it represents, and each sentence is assigned a truth value. In this way, an interpretation provides semantic meaning to the terms and formulas of the language. The study of the interpretations of formal languages is called formal semantics. What follows is a description of the standard or Tarskian semantics for first-order logic. (It is also possible to define game semantics for first-order logic, but aside from requiring the axiom of choice, game semantics agree with Tarskian semantics for first-order logic, so game semantics will not be elaborated herein.)

The domain of discourse D is a nonempty set of "objects" of some kind. Intuitively, a first-order formula is a statement about these objects; for example, $\exists x P(x)$ states the existence of an object x such that the predicate P is true where referred to it. The domain of discourse is the set of considered objects. For example, one can take D to be the set of integer numbers.

The interpretation of a function symbol is a function. For example, if the domain of discourse consists of integers, a function symbol f of arity 2 can be interpreted as the function that gives the sum of its arguments. In other words, the symbol f is associated with the function $I(f)$ which, in this interpretation, is addition.

The interpretation of a constant symbol is a function from the one-element set D^0 to D, which can be simply identified with an object in D. For example, an interpretation may assign the value $I(c) = 10$ to the constant symbol c.

The interpretation of an n-ary predicate symbol is a set of n-tuples of elements of the domain of discourse. This means that, given an interpretation, a predicate symbol, and n elements of the domain of discourse, one can tell whether the predicate is true of those elements according to the given interpretation. For example, an interpretation $I(P)$ of a binary predicate symbol P may be the set of pairs of integers such that the first one is less than the second. According to this interpretation, the predicate P would be true if its first argument is less than the second.

24.3.1 First-order structures

Main article: Structure (mathematical logic)

The most common way of specifying an interpretation (especially in mathematics) is to specify a **structure** (also called a **model**; see below). The structure consists of a nonempty set D that forms the domain of discourse and an interpretation I of the non-logical terms of the signature. This interpretation is itself a function:

- Each function symbol f of arity n is assigned a function $I(f)$ from D^n to D. In particular, each constant symbol of the signature is assigned an individual in the domain of discourse.

- Each predicate symbol P of arity n is assigned a relation *I(P)* over D^n or, equivalently, a function from D^n to $\{true, false\}$. Thus each predicate symbol is interpreted by a Boolean-valued function on *D*.

24.3.2   Evaluation of truth values

A formula evaluates to true or false given an interpretation, and a **variable assignment** μ that associates an element of the domain of discourse with each variable. The reason that a variable assignment is required is to give meanings to formulas with free variables, such as $y = x$. The truth value of this formula changes depending on whether x and y denote the same individual.

First, the variable assignment μ can be extended to all terms of the language, with the result that each term maps to a single element of the domain of discourse. The following rules are used to make this assignment:

1. **Variables.** Each variable x evaluates to $\mu(x)$

2. **Functions.** Given terms $t_1, \ldots, t_n$ that have been evaluated to elements $d_1, \ldots, d_n$ of the domain of discourse, and a n-ary function symbol f, the term $f(t_1, \ldots, t_n)$ evaluates to $(I(f))(d_1, \ldots, d_n)$.

Next, each formula is assigned a truth value. The inductive definition used to make this assignment is called the T-schema.

1. **Atomic formulas (1).** A formula $P(t_1, ., t_n)$ is associated the value true or false depending on whether $\langle v_1, .., v_n \rangle \in I(P)$, where $v_1, \ldots, v_n$ are the evaluation of the terms $t_1, \ldots, t_n$ and $I(P)$ is the interpretation of P , which by assumption is a subset of D^n .

2. **Atomic formulas (2).** A formula $t_1 = t_2$ is assigned true if t_1 and t_2 evaluate to the same object of the domain of discourse (see the section on equality below).

3. **Logical connectives.** A formula in the form $\neg\phi$, $\phi \to \psi$, etc. is evaluated according to the truth table for the connective in question, as in propositional logic.

4. **Existential quantifiers.** A formula $\exists x \phi(x)$ is true according to M and μ if there exists an evaluation μ' of the variables that only differs from μ regarding the evaluation of x and such that φ is true according to the interpretation M and the variable assignment μ' . This formal definition captures the idea that $\exists x \phi(x)$ is true if and only if there is a way to choose a value for x such that $\varphi(x)$ is satisfied.

5. **Universal quantifiers.** A formula $\forall x \phi(x)$ is true according to M and μ if $\varphi(x)$ is true for every pair composed by the interpretation M and some variable assignment μ' that differs from μ only on the value of x. This captures the idea that $\forall x \phi(x)$ is true if every possible choice of a value for x causes $\varphi(x)$ to be true.

If a formula does not contain free variables, and so is a sentence, then the initial variable assignment does not affect its truth value. In other words, a sentence is true according to M and μ if and only if it is true according to M and every other variable assignment μ' .

There is a second common approach to defining truth values that does not rely on variable assignment functions. Instead, given an interpretation M, one first adds to the signature a collection of constant symbols, one for each element of the domain of discourse in M; say that for each d in the domain the constant symbol cd is fixed. The interpretation is extended so that each new constant symbol is assigned to its corresponding element of the domain. One now defines truth for quantified formulas syntactically, as follows:

1. **Existential quantifiers (alternate).** A formula $\exists x \phi(x)$ is true according to M if there is some d in the domain of discourse such that $\phi(c_d)$ holds. Here $\phi(c_d)$ is the result of substituting cd for every free occurrence of x in φ.

2. **Universal quantifiers (alternate).** A formula $\forall x \phi(x)$ is true according to M if, for every d in the domain of discourse, $\phi(c_d)$ is true according to M.

This alternate approach gives exactly the same truth values to all sentences as the approach via variable assignments.

24.3.3 Validity, satisfiability, and logical consequence

See also: Satisfiability

If a sentence φ evaluates to True under a given interpretation M, one says that M **satisfies** φ; this is denoted $M \vDash \varphi$. A sentence is **satisfiable** if there is some interpretation under which it is true.

Satisfiability of formulas with free variables is more complicated, because an interpretation on its own does not determine the truth value of such a formula. The most common convention is that a formula with free variables is said to be satisfied by an interpretation if the formula remains true regardless which individuals from the domain of discourse are assigned to its free variables. This has the same effect as saying that a formula is satisfied if and only if its universal closure is satisfied.

A formula is **logically valid** (or simply **valid**) if it is true in every interpretation. These formulas play a role similar to tautologies in propositional logic.

A formula φ is a **logical consequence** of a formula ψ if every interpretation that makes ψ true also makes φ true. In this case one says that φ is logically implied by ψ.

24.3.4 Algebraizations

An alternate approach to the semantics of first-order logic proceeds via abstract algebra. This approach generalizes the Lindenbaum–Tarski algebras of propositional logic. There are three ways of eliminating quantified variables from first-order logic that do not involve replacing quantifiers with other variable binding term operators:

- Cylindric algebra, by Alfred Tarski and his coworkers;
- Polyadic algebra, by Paul Halmos;
- Predicate functor logic, mainly due to Willard Quine.

These algebras are all lattices that properly extend the two-element Boolean algebra.

Tarski and Givant (1987) showed that the fragment of first-order logic that has no atomic sentence lying in the scope of more than three quantifiers has the same expressive power as relation algebra. This fragment is of great interest because it suffices for Peano arithmetic and most axiomatic set theory, including the canonical ZFC. They also prove that first-order logic with a primitive ordered pair is equivalent to a relation algebra with two ordered pair projection functions.

24.3.5 First-order theories, models, and elementary classes

A **first-order theory** of a particular signature is a set of axioms, which are sentences consisting of symbols from that signature. The set of axioms is often finite or recursively enumerable, in which case the theory is called **effective**. Some authors require theories to also include all logical consequences of the axioms. The axioms are considered to hold within the theory and from them other sentences that hold within the theory can be derived.

A first-order structure that satisfies all sentences in a given theory is said to be a **model** of the theory. An **elementary class** is the set of all structures satisfying a particular theory. These classes are a main subject of study in model theory.

Many theories have an **intended interpretation**, a certain model that is kept in mind when studying the theory. For example, the intended interpretation of Peano arithmetic consists of the usual natural numbers with their usual operations. However, the Löwenheim–Skolem theorem shows that most first-order theories will also have other, nonstandard models.

A theory is **consistent** if it is not possible to prove a contradiction from the axioms of the theory. A theory is **complete** if, for every formula in its signature, either that formula or its negation is a logical consequence of the axioms of the theory. Gödel's incompleteness theorem shows that effective first-order theories that include a sufficient portion of the theory of the natural numbers can never be both consistent and complete.

For more information on this subject see List of first-order theories and Theory (mathematical logic)

24.3.6   Empty domains

Main article: Empty domain

The definition above requires that the domain of discourse of any interpretation must be a nonempty set. There are settings, such as inclusive logic, where empty domains are permitted. Moreover, if a class of algebraic structures includes an empty structure (for example, there is an empty poset), that class can only be an elementary class in first-order logic if empty domains are permitted or the empty structure is removed from the class.

There are several difficulties with empty domains, however:

- Many common rules of inference are only valid when the domain of discourse is required to be nonempty. One example is the rule stating that $\phi \lor \exists x \psi$ implies $\exists x (\phi \lor \psi)$ when x is not a free variable in φ. This rule, which is used to put formulas into prenex normal form, is sound in nonempty domains, but unsound if the empty domain is permitted.

- The definition of truth in an interpretation that uses a variable assignment function cannot work with empty domains, because there are no variable assignment functions whose range is empty. (Similarly, one cannot assign interpretations to constant symbols.) This truth definition requires that one must select a variable assignment function (μ above) before truth values for even atomic formulas can be defined. Then the truth value of a sentence is defined to be its truth value under any variable assignment, and it is proved that this truth value does not depend on which assignment is chosen. This technique does not work if there are no assignment functions at all; it must be changed to accommodate empty domains.

Thus, when the empty domain is permitted, it must often be treated as a special case. Most authors, however, simply exclude the empty domain by definition.

24.4   Deductive systems

A **deductive system** is used to demonstrate, on a purely syntactic basis, that one formula is a logical consequence of another formula. There are many such systems for first-order logic, including Hilbert-style deductive systems, natural deduction, the sequent calculus, the tableaux method, and resolution. These share the common property that a deduction is a finite syntactic object; the format of this object, and the way it is constructed, vary widely. These finite deductions themselves are often called **derivations** in proof theory. They are also often called proofs, but are completely formalized unlike natural-language mathematical proofs.

A deductive system is **sound** if any formula that can be derived in the system is logically valid. Conversely, a deductive system is **complete** if every logically valid formula is derivable. All of the systems discussed in this article are both sound and complete. They also share the property that it is possible to effectively verify that a purportedly valid deduction is actually a deduction; such deduction systems are called **effective**.

A key property of deductive systems is that they are purely syntactic, so that derivations can be verified without considering any interpretation. Thus a sound argument is correct in every possible interpretation of the language, regardless whether that interpretation is about mathematics, economics, or some other area.

In general, logical consequence in first-order logic is only semidecidable: if a sentence A logically implies a sentence B then this can be discovered (for example, by searching for a proof until one is found, using some effective, sound, complete proof system). However, if A does not logically imply B, this does not mean that A logically implies the negation of B. There is no effective procedure that, given formulas A and B, always correctly decides whether A logically implies B.

24.4.1   Rules of inference

Further information: List of rules of inference

A **rule of inference** states that, given a particular formula (or set of formulas) with a certain property as a hypothesis, another specific formula (or set of formulas) can be derived as a conclusion. The rule is sound (or truth-preserving) if it preserves validity in the sense that whenever any interpretation satisfies the hypothesis, that interpretation also satisfies the conclusion.

For example, one common rule of inference is the **rule of substitution**. If t is a term and φ is a formula possibly containing the variable x, then $\varphi[t/x]$ (often denoted $\varphi[x/t]$) is the result of replacing all free instances of x by t in φ. The substitution rule states that for any φ and any term t, one can conclude $\varphi[t/x]$ from φ provided that no free variable of t becomes bound during the substitution process. (If some free variable of t becomes bound, then to substitute t for x it is first necessary to change the bound variables of φ to differ from the free variables of t.)

To see why the restriction on bound variables is necessary, consider the logically valid formula φ given by $\exists x(x = y)$, in the signature of $(0,1,+,\times,=)$ of arithmetic. If t is the term "x + 1", the formula $\varphi[t/y]$ is $\exists x(x = x + 1)$, which will be false in many interpretations. The problem is that the free variable x of t became bound during the substitution. The intended replacement can be obtained by renaming the bound variable x of φ to something else, say z, so that the formula after substitution is $\exists z(z = x + 1)$, which is again logically valid.

The substitution rule demonstrates several common aspects of rules of inference. It is entirely syntactical; one can tell whether it was correctly applied without appeal to any interpretation. It has (syntactically defined) limitations on when it can be applied, which must be respected to preserve the correctness of derivations. Moreover, as is often the case, these limitations are necessary because of interactions between free and bound variables that occur during syntactic manipulations of the formulas involved in the inference rule.

24.4.2 Hilbert-style systems and natural deduction

A deduction in a Hilbert-style deductive system is a list of formulas, each of which is a **logical axiom**, a hypothesis that has been assumed for the derivation at hand, or follows from previous formulas via a rule of inference. The logical axioms consist of several axiom schemas of logically valid formulas; these encompass a significant amount of propositional logic. The rules of inference enable the manipulation of quantifiers. Typical Hilbert-style systems have a small number of rules of inference, along with several infinite schemas of logical axioms. It is common to have only modus ponens and universal generalization as rules of inference.

Natural deduction systems resemble Hilbert-style systems in that a deduction is a finite list of formulas. However, natural deduction systems have no logical axioms; they compensate by adding additional rules of inference that can be used to manipulate the logical connectives in formulas in the proof.

24.4.3 Sequent calculus

Further information: Sequent calculus

The sequent calculus was developed to study the properties of natural deduction systems. Instead of working with one formula at a time, it uses **sequents**, which are expressions of the form

$$A_1, \ldots, A_n \vdash B_1, \ldots, B_k,$$

where A_1, ..., An, B_1, ..., Bk are formulas and the turnstile symbol $\vdash$ is used as punctuation to separate the two halves. Intuitively, a sequent expresses the idea that $(A_1 \wedge \cdots \wedge A_n)$ implies $(B_1 \vee \cdots \vee B_k)$.

24.4.4 Tableaux method

Further information: Method of analytic tableaux

Unlike the methods just described, the derivations in the tableaux method are not lists of formulas. Instead, a derivation is a tree of formulas. To show that a formula A is provable, the tableaux method attempts to demonstrate that the negation of A is unsatisfiable. The tree of the derivation has $\neg A$ at its root; the tree branches in a way that reflects the structure of the formula. For example, to show that $C \vee D$ is unsatisfiable requires showing that C and D are each unsatisfiable; this corresponds to a branching point in the tree with parent $C \vee D$ and children C and D.

24.4.5   Resolution

The resolution rule is a single rule of inference that, together with unification, is sound and complete for first-order logic. As with the tableaux method, a formula is proved by showing that the negation of the formula is unsatisfiable. Resolution is commonly used in automated theorem proving.

The resolution method works only with formulas that are disjunctions of atomic formulas; arbitrary formulas must first be converted to this form through Skolemization. The resolution rule states that from the hypotheses $A_1 \vee \cdots \vee A_k \vee C$ and $B_1 \vee \cdots \vee B_l \vee \neg C$, the conclusion $A_1 \vee \cdots \vee A_k \vee B_1 \vee \cdots \vee B_l$ can be obtained.

24.4.6   Provable identities

The following sentences can be called "identities" because the main connective in each is the biconditional.

$$\neg \forall x \, P(x) \Leftrightarrow \exists x \, \neg P(x)$$

$$\neg \exists x \, P(x) \Leftrightarrow \forall x \, \neg P(x)$$

$$\forall x \, \forall y \, P(x,y) \Leftrightarrow \forall y \, \forall x \, P(x,y)$$

$$\exists x \, \exists y \, P(x,y) \Leftrightarrow \exists y \, \exists x \, P(x,y)$$

$$\forall x \, P(x) \wedge \forall x \, Q(x) \Leftrightarrow \forall x \, (P(x) \wedge Q(x))$$

$$\exists x \, P(x) \vee \exists x \, Q(x) \Leftrightarrow \exists x \, (P(x) \vee Q(x))$$

$$P \wedge \exists x \, Q(x) \Leftrightarrow \exists x \, (P \wedge Q(x)) \text{ (where } x \text{ must not occur free in } P)$$

$$P \vee \forall x \, Q(x) \Leftrightarrow \forall x \, (P \vee Q(x)) \text{ (where } x \text{ must not occur free in } P)$$

24.5   Equality and its axioms

There are several different conventions for using equality (or identity) in first-order logic. The most common convention, known as **first-order logic with equality**, includes the equality symbol as a primitive logical symbol which is always interpreted as the real equality relation between members of the domain of discourse, such that the "two" given members are the same member. This approach also adds certain axioms about equality to the deductive system employed. These equality axioms are:

1. **Reflexivity**. For each variable x, $x = x$.

2. **Substitution for functions.** For all variables x and y, and any function symbol f,

    $x = y \rightarrow f(...,x,...) = f(...,y,...)$.

3. **Substitution for formulas.** For any variables x and y and any formula $\varphi(x)$, if φ' is obtained by replacing any number of free occurrences of x in φ with y, such that these remain free occurrences of y, then

    $x = y \rightarrow (\varphi \rightarrow \varphi')$.

These are axiom schemas, each of which specifies an infinite set of axioms. The third schema is known as **Leibniz's law**, "the principle of substitutivity", "the indiscernibility of identicals", or "the replacement property". The second schema, involving the function symbol f, is (equivalent to) a special case of the third schema, using the formula

$$x = y \rightarrow (f(...,x,...) = z \rightarrow f(...,y,...) = z).$$

Many other properties of equality are consequences of the axioms above, for example:

1. **Symmetry.** If $x = y$ then $y = x$.

2. **Transitivity.** If $x = y$ and $y = z$ then $x = z$.

24.5.1   First-order logic without equality

An alternate approach considers the equality relation to be a non-logical symbol. This convention is known as **first-order logic without equality**. If an equality relation is included in the signature, the axioms of equality must now be added to the theories under consideration, if desired, instead of being considered rules of logic. The main difference between this method and first-order logic with equality is that an interpretation may now interpret two distinct individuals as "equal" (although, by Leibniz's law, these will satisfy exactly the same formulas under any interpretation). That is, the equality relation may now be interpreted by an arbitrary equivalence relation on the domain of discourse that is congruent with respect to the functions and relations of the interpretation.

When this second convention is followed, the term **normal model** is used to refer to an interpretation where no distinct individuals a and b satisfy $a = b$. In first-order logic with equality, only normal models are considered, and so there is no term for a model other than a normal model. When first-order logic without equality is studied, it is necessary to amend the statements of results such as the Löwenheim–Skolem theorem so that only normal models are considered.

First-order logic without equality is often employed in the context of second-order arithmetic and other higher-order theories of arithmetic, where the equality relation between sets of natural numbers is usually omitted.

24.5.2   Defining equality within a theory

If a theory has a binary formula $A(x,y)$ which satisfies reflexivity and Leibniz's law, the theory is said to have equality, or to be a theory with equality. The theory may not have all instances of the above schemas as axioms, but rather as derivable theorems. For example, in theories with no function symbols and a finite number of relations, it is possible to define equality in terms of the relations, by defining the two terms s and t to be equal if any relation is unchanged by changing s to t in any argument.

Some theories allow other *ad hoc* definitions of equality:

- In the theory of partial orders with one relation symbol $\leq$, one could define $s = t$ to be an abbreviation for $s \leq t \wedge t \leq s$.

- In set theory with one relation $\in$, one may define $s = t$ to be an abbreviation for $\forall x \, (s \in x \leftrightarrow t \in x) \wedge \forall x \, (x \in s \leftrightarrow x \in t)$. This definition of equality then automatically satisfies the axioms for equality. In this case, one should replace the usual axiom of extensionality, $\forall x \forall y [\forall z (z \in x \Leftrightarrow z \in y) \Rightarrow x = y]$, by $\forall x \forall y [\forall z (z \in x \Leftrightarrow z \in y) \Rightarrow \forall z (x \in z \Leftrightarrow y \in z)]$, i.e. if x and y have the same elements, then they belong to the same sets.

24.6   Metalogical properties

One motivation for the use of first-order logic, rather than higher-order logic, is that first-order logic has many metalogical properties that stronger logics do not have. These results concern general properties of first-order logic itself, rather than properties of individual theories. They provide fundamental tools for the construction of models of first-order theories.

24.6.1 Completeness and undecidability

Gödel's completeness theorem, proved by Kurt Gödel in 1929, establishes that there are sound, complete, effective deductive systems for first-order logic, and thus the first-order logical consequence relation is captured by finite provability. Naively, the statement that a formula φ logically implies a formula ψ depends on every model of φ; these models will in general be of arbitrarily large cardinality, and so logical consequence cannot be effectively verified by checking every model. However, it is possible to enumerate all finite derivations and search for a derivation of ψ from φ. If ψ is logically implied by φ, such a derivation will eventually be found. Thus first-order logical consequence is semidecidable: it is possible to make an effective enumeration of all pairs of sentences (φ,ψ) such that ψ is a logical consequence of φ.

Unlike propositional logic, first-order logic is undecidable (although semidecidable), provided that the language has at least one predicate of arity at least 2 (other than equality). This means that there is no decision procedure that determines whether arbitrary formulas are logically valid. This result was established independently by Alonzo Church and Alan Turing in 1936 and 1937, respectively, giving a negative answer to the Entscheidungsproblem posed by David Hilbert in 1928. Their proofs demonstrate a connection between the unsolvability of the decision problem for first-order logic and the unsolvability of the halting problem.

There are systems weaker than full first-order logic for which the logical consequence relation is decidable. These include propositional logic and monadic predicate logic, which is first-order logic restricted to unary predicate symbols and no function symbols. Other logics with no function symbols which are decidable are the guarded fragment of first-order logic, as well as two-variable logic. The Bernays–Schönfinkel class of first-order formulas is also decidable. Decidable subsets of first-order logic are also studied in the framework of description logics.

24.6.2 The Löwenheim–Skolem theorem

The Löwenheim–Skolem theorem shows that if a first-order theory of cardinality λ has an infinite model, then it has models of every infinite cardinality greater than or equal to λ. One of the earliest results in model theory, it implies that it is not possible to characterize countability or uncountability in a first-order language. That is, there is no first-order formula $\varphi(x)$ such that an arbitrary structure M satisfies φ if and only if the domain of discourse of M is countable (or, in the second case, uncountable).

The Löwenheim–Skolem theorem implies that infinite structures cannot be categorically axiomatized in first-order logic. For example, there is no first-order theory whose only model is the real line: any first-order theory with an infinite model also has a model of cardinality larger than the continuum. Since the real line is infinite, any theory satisfied by the real line is also satisfied by some nonstandard models. When the Löwenheim–Skolem theorem is applied to first-order set theories, the nonintuitive consequences are known as Skolem's paradox.

24.6.3 The compactness theorem

The compactness theorem states that a set of first-order sentences has a model if and only if every finite subset of it has a model. This implies that if a formula is a logical consequence of an infinite set of first-order axioms, then it is a logical consequence of some finite number of those axioms. This theorem was proved first by Kurt Gödel as a consequence of the completeness theorem, but many additional proofs have been obtained over time. It is a central tool in model theory, providing a fundamental method for constructing models.

The compactness theorem has a limiting effect on which collections of first-order structures are elementary classes. For example, the compactness theorem implies that any theory that has arbitrarily large finite models has an infinite model. Thus the class of all finite graphs is not an elementary class (the same holds for many other algebraic structures).

There are also more subtle limitations of first-order logic that are implied by the compactness theorem. For example, in computer science, many situations can be modeled as a directed graph of states (nodes) and connections (directed edges). Validating such a system may require showing that no "bad" state can be reached from any "good" state. Thus one seeks to determine if the good and bad states are in different connected components of the graph. However, the compactness theorem can be used to show that connected graphs are not an elementary class in first-order logic, and there is no formula $\varphi(x,y)$ of first-order logic, in the logic of graphs, that expresses the idea that there is a path from x to y. Connectedness can be expressed in second-order logic, however, but not with only existential set quantifiers, as Σ_1^1 also enjoys compactness.

24.6.4 Lindström's theorem

Main article: Lindström's theorem

Per Lindström showed that the metalogical properties just discussed actually characterize first-order logic in the sense that no stronger logic can also have those properties (Ebbinghaus and Flum 1994, Chapter XIII). Lindström defined a class of abstract logical systems, and a rigorous definition of the relative strength of a member of this class. He established two theorems for systems of this type:

- A logical system satisfying Lindström's definition that contains first-order logic and satisfies both the Löwenheim–Skolem theorem and the compactness theorem must be equivalent to first-order logic.

- A logical system satisfying Lindström's definition that has a semidecidable logical consequence relation and satisfies the Löwenheim–Skolem theorem must be equivalent to first-order logic.

24.7 Limitations

Although first-order logic is sufficient for formalizing much of mathematics, and is commonly used in computer science and other fields, it has certain limitations. These include limitations on its expressiveness and limitations of the fragments of natural languages that it can describe.

For instance, first-order logic is undecidable, meaning a sound, complete and terminating decision algorithm is impossible. This has led to the study of interesting decidable fragments such as C_2, first-order logic with two variables and the counting quantifiers $\exists^{\geq n}$ and $\exists^{\leq n}$ (these quantifiers are, respectively, "there exists at least n" and "there exists at most n") (Horrocks 2010).

24.7.1 Expressiveness

The Löwenheim–Skolem theorem shows that if a first-order theory has any infinite model, then it has infinite models of every cardinality. In particular, no first-order theory with an infinite model can be categorical. Thus there is no first-order theory whose only model has the set of natural numbers as its domain, or whose only model has the set of real numbers as its domain. Many extensions of first-order logic, including infinitary logics and higher-order logics, are more expressive in the sense that they do permit categorical axiomatizations of the natural numbers or real numbers. This expressiveness comes at a metalogical cost, however: by Lindström's theorem, the compactness theorem and the downward Löwenheim–Skolem theorem cannot hold in any logic stronger than first-order.

24.7.2 Formalizing natural languages

First-order logic is able to formalize many simple quantifier constructions in natural language, such as "every person who lives in Perth lives in Australia". But there are many more complicated features of natural language that cannot be expressed in (single-sorted) first-order logic. "Any logical system which is appropriate as an instrument for the analysis of natural language needs a much richer structure than first-order predicate logic" (Gamut 1991, p. 75).

24.8 Restrictions, extensions, and variations

There are many variations of first-order logic. Some of these are inessential in the sense that they merely change notation without affecting the semantics. Others change the expressive power more significantly, by extending the semantics through additional quantifiers or other new logical symbols. For example, infinitary logics permit formulas of infinite size, and modal logics add symbols for possibility and necessity.

24.8.1   Restricted languages

First-order logic can be studied in languages with fewer logical symbols than were described above.

- Because $\exists x\phi(x)$ can be expressed as $\neg\forall x\neg\phi(x)$, and $\forall x\phi(x)$ can be expressed as $\neg\exists x\neg\phi(x)$, either of the two quantifiers $\exists$ and $\forall$ can be dropped.

- Since $\phi \vee \psi$ can be expressed as $\neg(\neg\phi \wedge \neg\psi)$ and $\phi \wedge \psi$ can be expressed as $\neg(\neg\phi \vee \neg\psi)$, either $\vee$ or $\wedge$ can be dropped. In other words, it is sufficient to have $\neg$ and $\vee$, or $\neg$ and $\wedge$, as the only logical connectives.

- Similarly, it is sufficient to have only $\neg$ and $\rightarrow$ as logical connectives, or to have only the Sheffer stroke (**NAND**) or the Peirce arrow (**NOR**) operator.

- It is possible to entirely avoid function symbols and constant symbols, rewriting them via predicate symbols in an appropriate way. For example, instead of using a constant symbol 0 one may use a predicate $0(x)$ (interpreted as $x = 0$), and replace every predicate such as $P(0,y)$ with $\forall x \ (0(x) \rightarrow P(x,y))$. A function such as $f(x_1, x_2, ..., x_n)$ will similarly be replaced by a predicate $F(x_1, x_2, ..., x_n, y)$ interpreted as $y = f(x_1, x_2, ..., x_n)$. This change requires adding additional axioms to the theory at hand, so that interpretations of the predicate symbols used have the correct semantics.

Restrictions such as these are useful as a technique to reduce the number of inference rules or axiom schemas in deductive systems, which leads to shorter proofs of metalogical results. The cost of the restrictions is that it becomes more difficult to express natural-language statements in the formal system at hand, because the logical connectives used in the natural language statements must be replaced by their (longer) definitions in terms of the restricted collection of logical connectives. Similarly, derivations in the limited systems may be longer than derivations in systems that include additional connectives. There is thus a trade-off between the ease of working within the formal system and the ease of proving results about the formal system.

It is also possible to restrict the arities of function symbols and predicate symbols, in sufficiently expressive theories. One can in principle dispense entirely with functions of arity greater than 2 and predicates of arity greater than 1 in theories that include a pairing function. This is a function of arity 2 that takes pairs of elements of the domain and returns an ordered pair containing them. It is also sufficient to have two predicate symbols of arity 2 that define projection functions from an ordered pair to its components. In either case it is necessary that the natural axioms for a pairing function and its projections are satisfied.

24.8.2   Many-sorted logic

Ordinary first-order interpretations have a single domain of discourse over which all quantifiers range. **Many-sorted first-order logic** allows variables to have different **sorts**, which have different domains. This is also called **typed first-order logic**, and the sorts called **types** (as in data type), but it is not the same as first-order type theory. Many-sorted first-order logic is often used in the study of second-order arithmetic.

When there are only finitely many sorts in a theory, many-sorted first-order logic can be reduced to single-sorted first-order logic. One introduces into the single-sorted theory a unary predicate symbol for each sort in the many-sorted theory, and adds an axiom saying that these unary predicates partition the domain of discourse. For example, if there are two sorts, one adds predicate symbols $P_1(x)$ and $P_2(x)$ and the axiom

$$\forall x(P_1(x) \vee P_2(x)) \wedge \neg\exists x(P_1(x) \wedge P_2(x))$$

Then the elements satisfying P_1 are thought of as elements of the first sort, and elements satisfying P_2 as elements of the second sort. One can quantify over each sort by using the corresponding predicate symbol to limit the range of quantification. For example, to say there is an element of the first sort satisfying formula $\varphi(x)$, one writes

$\exists x(P_1(x) \wedge \phi(x))$

24.8.3 Additional quantifiers

Additional quantifiers can be added to first-order logic.

- Sometimes it is useful to say that "*P(x)* holds for exactly one *x*", which can be expressed as $\exists!\, x\, P(x)$. This notation, called uniqueness quantification, may be taken to abbreviate a formula such as $\exists\, x\, (P(x) \wedge \forall\, y\, (P(y) \rightarrow (x = y)))$.

- **First-order logic with extra quantifiers** has new quantifiers $Qx,...$, with meanings such as "there are many *x* such that ...". Also see branching quantifiers and the plural quantifiers of George Boolos and others.

- **Bounded quantifiers** are often used in the study of set theory or arithmetic.

24.8.4 Infinitary logics

Main article: Infinitary logic

Infinitary logic allows infinitely long sentences. For example, one may allow a conjunction or disjunction of infinitely many formulas, or quantification over infinitely many variables. Infinitely long sentences arise in areas of mathematics including topology and model theory.

Infinitary logic generalizes first-order logic to allow formulas of infinite length. The most common way in which formulas can become infinite is through infinite conjunctions and disjunctions. However, it is also possible to admit generalized signatures in which function and relation symbols are allowed to have infinite arities, or in which quantifiers can bind infinitely many variables. Because an infinite formula cannot be represented by a finite string, it is necessary to choose some other representation of formulas; the usual representation in this context is a tree. Thus formulas are, essentially, identified with their parse trees, rather than with the strings being parsed.

The most commonly studied infinitary logics are denoted $L\alpha_\beta$, where α and β are each either cardinal numbers or the symbol ∞. In this notation, ordinary first-order logic is $L\omega\omega$. In the logic $L\infty\omega$, arbitrary conjunctions or disjunctions are allowed when building formulas, and there is an unlimited supply of variables. More generally, the logic that permits conjunctions or disjunctions with less than κ constituents is known as $L\kappa\omega$. For example, $L\omega_1\omega$ permits countable conjunctions and disjunctions.

The set of free variables in a formula of $L\kappa\omega$ can have any cardinality strictly less than κ, yet only finitely many of them can be in the scope of any quantifier when a formula appears as a subformula of another.[8] In other infinitary logics, a subformula may be in the scope of infinitely many quantifiers. For example, in $L\kappa\infty$, a single universal or existential quantifier may bind arbitrarily many variables simultaneously. Similarly, the logic $L\kappa\lambda$ permits simultaneous quantification over fewer than λ variables, as well as conjunctions and disjunctions of size less than κ.

24.8.5 Non-classical and modal logics

- **Intuitionistic first-order logic** uses intuitionistic rather than classical propositional calculus; for example, $\neg\neg\varphi$ need not be equivalent to φ.

- First-order **modal logic** allows one to describe other possible worlds as well as this contingently true world which we inhabit. In some versions, the set of possible worlds varies depending on which possible world one inhabits. Modal logic has extra *modal operators* with meanings which can be characterized informally as, for example "it is necessary that φ" (true in all possible worlds) and "it is possible that φ" (true in some possible world). With standard first-order logic we have a single domain and each predicate is assigned one extension. With first-order modal logic we have a *domain function* that assigns each possible world its own domain, so that each predicate gets an extension only relative to these possible worlds. This allows us to model cases where, for example, Alex is a Philosopher, but might have been a Mathematician, and might not have existed at all. In the first possible world $P(a)$ is true, in the second $P(a)$ is false, and in the third possible world there is no a in the domain at all.

- **first-order fuzzy logics** are first-order extensions of propositional fuzzy logics rather than classical propositional calculus.

24.8.6 Fixpoint logic

Fixpoint logic extends first-order logic by adding the closure under the least fixed points of positive operators.[9]

24.8.7 Higher-order logics

Main article: Higher-order logic

The characteristic feature of first-order logic is that individuals can be quantified, but not predicates. Thus

$$\exists a(\text{Phil}(a))$$

is a legal first-order formula, but

$$\exists \text{Phil}(\text{Phil}(a))$$

is not, in most formalizations of first-order logic. Second-order logic extends first-order logic by adding the latter type of quantification. Other higher-order logics allow quantification over even higher types than second-order logic permits. These higher types include relations between relations, functions from relations to relations between relations, and other higher-type objects. Thus the "first" in first-order logic describes the type of objects that can be quantified.

Unlike first-order logic, for which only one semantics is studied, there are several possible semantics for second-order logic. The most commonly employed semantics for second-order and higher-order logic is known as **full semantics**. The combination of additional quantifiers and the full semantics for these quantifiers makes higher-order logic stronger than first-order logic. In particular, the (semantic) logical consequence relation for second-order and higher-order logic is not semidecidable; there is no effective deduction system for second-order logic that is sound and complete under full semantics.

Second-order logic with full semantics is more expressive than first-order logic. For example, it is possible to create axiom systems in second-order logic that uniquely characterize the natural numbers and the real line. The cost of this expressiveness is that second-order and higher-order logics have fewer attractive metalogical properties than first-order logic. For example, the Löwenheim–Skolem theorem and compactness theorem of first-order logic become false when generalized to higher-order logics with full semantics.

24.9 Automated theorem proving and formal methods

Further information: First-order theorem proving

Automated theorem proving refers to the development of computer programs that search and find derivations (formal proofs) of mathematical theorems. Finding derivations is a difficult task because the search space can be very large; an exhaustive search of every possible derivation is theoretically possible but computationally infeasible for many systems of interest in mathematics. Thus complicated heuristic functions are developed to attempt to find a derivation in less time than a blind search.

The related area of automated proof verification uses computer programs to check that human-created proofs are correct. Unlike complicated automated theorem provers, verification systems may be small enough that their correctness can be checked both by hand and through automated software verification. This validation of the proof verifier is needed to give confidence that any derivation labeled as "correct" is actually correct.

Some proof verifiers, such as Metamath, insist on having a complete derivation as input. Others, such as Mizar and Isabelle, take a well-formatted proof sketch (which may still be very long and detailed) and fill in the missing pieces by doing simple proof searches or applying known decision procedures: the resulting derivation is then verified by a small, core "kernel". Many such systems are primarily intended for interactive use by human mathematicians: these are known as proof assistants. They may also use formal logics that are stronger than first-order logic, such as type theory. Because a full derivation of any nontrivial result in a first-order deductive system will be extremely long for a human to write,[10] results are often formalized as a series of lemmas, for which derivations can be constructed separately.

Automated theorem provers are also used to implement formal verification in computer science. In this setting, theorem provers are used to verify the correctness of programs and of hardware such as processors with respect to a formal specification. Because such analysis is time-consuming and thus expensive, it is usually reserved for projects in which a malfunction would have grave human or financial consequences.

24.10   See also

- ACL2 — A Computational Logic for Applicative Common Lisp.

- Equiconsistency

- Extension by definitions

- Herbrandization

- Higher-order logic

- List of logic symbols

- Löwenheim number

- Prenex normal form

- Relational algebra

- Relational model

- Second-order logic

- Skolem normal form

- Tarski's World

- Truth table

- Type (model theory)

24.11   Notes

[1] Mendelson, Elliott (1964). *Introduction to Mathematical Logic*. Van Nostrand Reinhold. p. 56.

[2] The word *language* is sometimes used as a synonym for signature, but this can be confusing because "language" can also refer to the set of formulas.

[3] More precisely, there is only one language of each variant of one-sorted first-order logic: with or without equality, with or without functions, with or without propositional variables, ...

[4] Some authors who use the term "well-formed formula" use "formula" to mean any string of symbols from the alphabet. However, most authors in mathematical logic use "formula" to mean "well-formed formula" and have no term for non-well-formed formulas. In every context, it is only the well-formed formulas that are of interest.

[5] The SMT-LIB Standard: Version 2.0, by Clark Barrett, Aaron Stump, and Cesare Tinelli. http://smtlib.cs.uiowa.edu/language.shtml

[6] e.g. the matrix shown at 4

[7] e.g. the matrix shown at 2

[8] Some authors only admit formulas with finitely many free variables in $L\kappa\omega$, and more generally only formulas with $< \lambda$ free variables in $L\kappa\lambda$.

[9] Bosse, Uwe (1993). "An Ehrenfeucht–Fraïssé game for fixpoint logic and stratified fixpoint logic". In Börger, Egon. *Computer Science Logic: 6th Workshop, CSL'92, San Miniato, Italy, September 28 - October 2, 1992. Selected Papers.* Lecture Notes in Computer Science **702**. Springer-Verlag. pp. 100–114. ISBN 3-540-56992-8. Zbl 0808.03024.

[10] Avigad *et al.* (2007) discuss the process of formally verifying a proof of the prime number theorem. The formalized proof required approximately 30,000 lines of input to the Isabelle proof verifier.

24.12   References

- Andrews, Peter B. (2002); *An Introduction to Mathematical Logic and Type Theory: To Truth Through Proof*, 2nd ed., Berlin: Kluwer Academic Publishers. Available from Springer.

- Avigad, Jeremy; Donnelly, Kevin; Gray, David; and Raff, Paul (2007); "A formally verified proof of the prime number theorem", *ACM Transactions on Computational Logic*, vol. 9 no. 1 doi:10.1145/1297658.1297660

- Barwise, Jon (1977); "An Introduction to First-Order Logic", in Barwise, Jon, ed. (1982). *Handbook of Mathematical Logic*. Studies in Logic and the Foundations of Mathematics. Amsterdam, NL: North-Holland. ISBN 978-0-444-86388-1.

- Barwise, Jon; and Etchemendy, John (2000); *Language Proof and Logic*, Stanford, CA: CSLI Publications (Distributed by the University of Chicago Press)

- Bocheński, Józef Maria (2007); *A Précis of Mathematical Logic*, Dordrecht, NL: D. Reidel, translated from the French and German editions by Otto Bird

- Ferreirós, José (2001); *The Road to Modern Logic — An Interpretation*, Bulletin of Symbolic Logic, Volume 7, Issue 4, 2001, pp. 441–484, DOI 10.2307/2687794, JStor

- Gamut, L. T. F. (1991); *Logic, Language, and Meaning, Volume 2: Intensional Logic and Logical Grammar*, Chicago, IL: University of Chicago Press, ISBN 0-226-28088-8

- Hilbert, David; and Ackermann, Wilhelm (1950); *Principles of Mathematical Logic*, Chelsea (English translation of *Grundzüge der theoretischen Logik*, 1928 German first edition)

- Hodges, Wilfrid (2001); "Classical Logic I: First Order Logic", in Goble, Lou (ed.); *The Blackwell Guide to Philosophical Logic*, Blackwell

- Ebbinghaus, Heinz-Dieter; Flum, Jörg; and Thomas, Wolfgang (1994); *Mathematical Logic*, Undergraduate Texts in Mathematics, Berlin, DE/New York, NY: Springer-Verlag, Second Edition, ISBN 978-0-387-94258-2

- Rautenberg, Wolfgang (2010), *A Concise Introduction to Mathematical Logic* (3rd ed.), New York, NY: Springer Science+Business Media, doi:10.1007/978-1-4419-1221-3, ISBN 978-1-4419-1220-6

- Tarski, Alfred and Givant, Steven (1987); *A Formalization of Set Theory without Variables*. Vol.41 of American Mathematical Society colloquium publications, Providence RI: American Mathematical Society, ISBN 978-0821810415.

24.13 External links

- Hazewinkel, Michiel, ed. (2001), "Predicate calculus", *Encyclopedia of Mathematics*, Springer, ISBN 978-1-55608-010-4

- Stanford Encyclopedia of Philosophy: Shapiro, Stewart; "Classical Logic". Covers syntax, model theory, and metatheory for first-order logic in the natural deduction style.

- Magnus, P. D.; *forall x: an introduction to formal logic.* Covers formal semantics and proof theory for first-order logic.

- Metamath: an ongoing online project to reconstruct mathematics as a huge first-order theory, using first-order logic and the axiomatic set theory ZFC. *Principia Mathematica* modernized.

- Podnieks, Karl; *Introduction to mathematical logic*

- *Cambridge Mathematics Tripos Notes* (typeset by John Fremlin). These notes cover part of a past Cambridge Mathematics Tripos course taught to undergraduates students (usually) within their third year. The course is entitled "Logic, Computation and Set Theory" and covers Ordinals and cardinals, Posets and Zorn's Lemma, Propositional logic, Predicate logic, Set theory and Consistency issues related to ZFC and other set theories.

- Tree Proof Generator can validate or invalidate formulas of FOL through the semantic tableaux method.

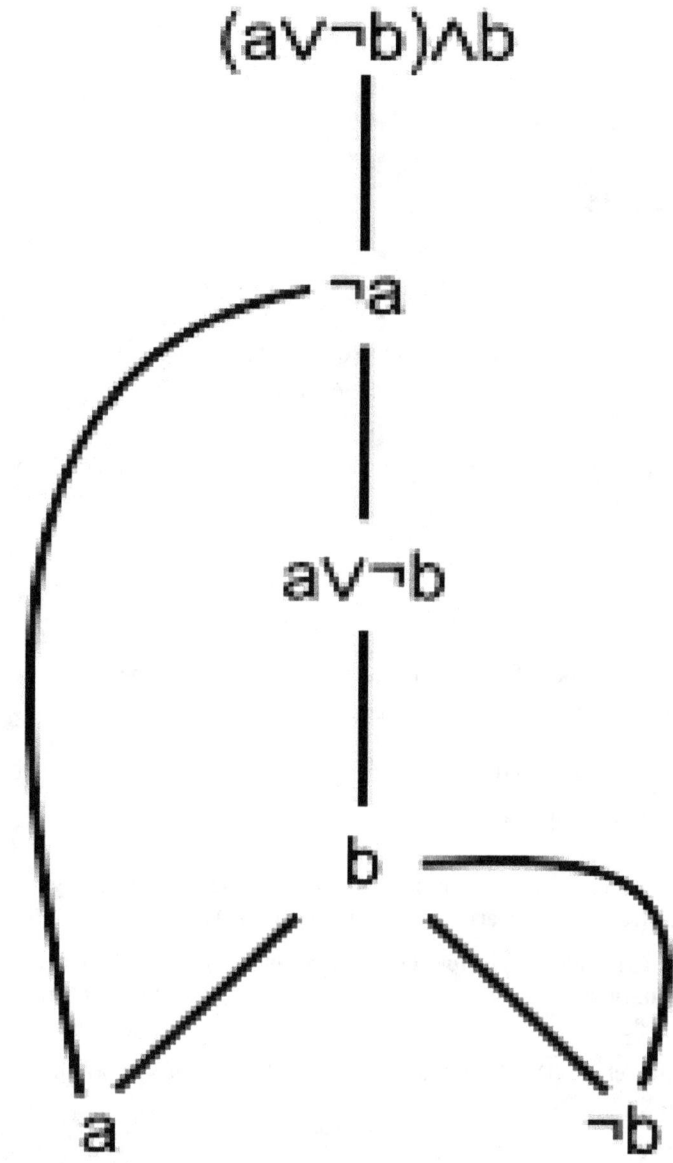

A tableaux proof for the propositional formula ((a ∨ ~b) & b) → a.

Chapter 25

Signature (logic)

In logic, especially mathematical logic, a **signature** lists and describes the non-logical symbols of a formal language. In universal algebra, a signature lists the operations that characterize an algebraic structure. In model theory, signatures are used for both purposes.

Signatures play the same role in mathematics as type signatures in computer programming. They are rarely made explicit in more philosophical treatments of logic.

25.1 Definition

Formally, a (single-sorted) **signature** can be defined as a triple $\sigma = (S_{\text{func}}, S_{\text{rel}}, \text{ar})$, where S_{func} and S_{rel} are disjoint sets not containing any other basic logical symbols, called respectively

- *function symbols* (examples: $+$, $\times$, 0, 1) and

- *relation symbols* or *predicates* (examples: $\leq$, $\in$),

and a function ar: $S_{\text{func}} \cup S_{\text{rel}} \to \mathbb{N}_0$ which assigns a non-negative integer called *arity* to every function or relation symbol. A function or relation symbol is called *n-ary* if its arity is n. A nullary (*0-ary*) function symbol is called a *constant symbol*.

A signature with no function symbols is called a **relational signature**, and a signature with no relation symbols is called an **algebraic signature**. A **finite signature** is a signature such that S_{func} and S_{rel} are finite. More generally, the **cardinality** of a signature $\sigma = (S_{\text{func}}, S_{\text{rel}}, \text{ar})$ is defined as $|\sigma| = |S_{\text{func}}| + |S_{\text{rel}}|$.

The **language of a signature** is the set of all well formed sentences built from the symbols in that signature together with the symbols in the logical system.

25.2 Other conventions

In universal algebra the word **type** or **similarity type** is often used as a synonym for "signature". In model theory, a signature σ is often called **vocabulary**, or identified with the (first-order) language L to which it provides the non-logical symbols. However, the cardinality of the language L will always be infinite; if σ is finite then $|L|$ will be $\aleph_0$.

As the formal definition is inconvenient for everyday use, the definition of a specific signature is often abbreviated in an informal way, as in:

> "The standard signature for abelian groups is $\sigma = (+,-,0)$, where $-$ is a unary operator."

Sometimes an algebraic signature is regarded as just a list of arities, as in:

225

"The similarity type for abelian groups is $\sigma = (2,1,0)$."

Formally this would define the function symbols of the signature as something like f_0 (nullary), f_1 (unary) and f_2 (binary), but in reality the usual names are used even in connection with this convention.

In mathematical logic, very often symbols are not allowed to be nullary, so that constant symbols must be treated separately rather than as nullary function symbols. They form a set S_{const} disjoint from S_{func}, on which the arity function ar is not defined. However, this only serves to complicate matters, especially in proofs by induction over the structure of a formula, where an additional case must be considered. Any nullary relation symbol, which is also not allowed under such a definition, can be emulated by a unary relation symbol together with a sentence expressing that its value is the same for all elements. This translation fails only for empty structures (which are often excluded by convention). If nullary symbols are allowed, then every formula of propositional logic is also a formula of first-order logic.

25.3   Use of signatures in logic and algebra

In the context of first-order logic, the symbols in a signature are also known as the non-logical symbols, because together with the logical symbols they form the underlying alphabet over which two formal languages are inductively defined: The set of *terms* over the signature and the set of (well-formed) *formulas* over the signature.

In a structure, an *interpretation* ties the function and relation symbols to mathematical objects that justify their names: The interpretation of an n-ary function symbol f in a structure A with *domain* A is a function $f^A: A^n \to A$, and the interpretation of an n-ary relation symbol is a relation $R^A \subseteq A^n$. Here $A^n = A \times A \times \ldots \times A$ denotes the n-fold cartesian product of the domain A with itself, and so f is in fact an n-ary function, and R an n-ary relation.

25.4   Many-sorted signatures

For many-sorted logic and for many-sorted structures signatures must encode information about the sorts. The most straightforward way of doing this is via **symbol types** that play the role of generalized arities.[1]

Symbol types

Let S be a set (of sorts) not containing the symbols $\times$ or $\to$.

The symbol types over S are certain words over the alphabet $S \cup \{\times, \to\}$: the relational symbol types $s_1 \times \ldots \times sn$, and the functional symbol types $s_1 \times \ldots \times sn \to s'$, for non-negative integers n and $s_1, s_2, \ldots, sn, s' \in S$. (For $n = 0$, the expression $s_1 \times \ldots \times sn$ denotes the empty word.)

Signature

A (many-sorted) signature is a triple $(S, P, type)$ consisting of

- a set S of sorts,

- a set P of symbols, and

- a map type which associates to every symbol in P a symbol type over S.

25.5   Notes

[1] Many-Sorted Logic, the first chapter in Lecture notes on Decision Procedures, written by Calogero G. Zarba.

25.6   References

- Burris, Stanley N.; Sankappanavar, H.P. (1981). *A Course in Universal Algebra*. Springer. ISBN 3-540-90578-2. Free online edition.

- Hodges, Wilfrid (1997). *A shorter model theory*. Cambridge University Press. ISBN 0-521-58713-1.

25.7   External links

- Stanford Encyclopedia of Philosophy: "Model theory"—by Wilfred Hodges.

- PlanetMath: Entry "Signature" describes the concept for the case when no sorts are introduced.

- Baillie, Jean, "An Introduction to the Algebraic Specification of Abstract Data Types."

Chapter 26

Quasiidentity

In universal algebra, a **quasiidentity** is an implication of the form

$$s_1 = t_1 \wedge \ldots \wedge sn = tn \rightarrow s = t$$

where s_1, ..., sn, s and t_1, ..., tn, t are terms built up from variables using the operation symbols of the specified signature.

Quasiidentities amount to conditional equations for which the conditions themselves are equations. A quasiidentity for which $n = 0$ is an ordinary identity or equation, whence quasiidentities are a generalization of identities. Quasiidentities are special type of Horn clauses.

26.1 See also

Quasivariety

26.2 References

- Burris, Stanley N.; H.P. Sankappanavar (1981). *A Course in Universal Algebra*. Springer. ISBN 3-540-90578-2. Free online edition.

Chapter 27

Atomic formula

In mathematical logic, an **atomic formula** (also known simply as an **atom**) is a formula with no deeper propositional structure, that is, a formula that contains no logical connectives or equivalently a formula that has no strict subformulas. Atoms are thus the simplest well-formed formulas of the logic. Compound formulas are formed by combining the atomic formulas using the logical connectives.

The precise form of atomic formulas depends on the logic under consideration; for propositional logic, for example, the atomic formulas are the propositional variables. For predicate logic, the atoms are predicate symbols together with their arguments, each argument being a term. In model theory, atomic formula are merely strings of symbols with a given signature, which may or may not be satisfiable with respect to a given model.[1]

27.1   Atomic formula in first-order logic

The well-formed terms and propositions of ordinary first-order logic have the following syntax:

Terms:

- $t \equiv c \mid x \mid f(t_1, ..., t_n)$,

that is, a term is recursively defined to be a constant c (a named object from the domain of discourse), or a variable x (ranging over the objects in the domain of discourse), or an n-ary function f whose arguments are terms tk. Functions map tuples of objects to objects.

Propositions:

- $A, B, ... \equiv P(t_1, ..., t_n) \mid A \wedge B \mid \top \mid A \vee B \mid \bot \mid A \supset B \mid \forall x.\, A \mid \exists x.\, A$,

that is, a proposition is recursively defined to be an n-ary predicate P whose arguments are terms tk, or an expression composed of logical connectives (and, or) and quantifiers (for-all, there-exists) used with other propositions.

An **atomic formula** or **atom** is simply a predicate applied to a tuple of terms; that is, an atomic formula is a formula of the form $P(t_1, ..., tn)$ for P a predicate, and the tn terms.

All other well-formed formulae are obtained by composing atoms with logical connectives and quantifiers.

For example, the formula $\forall x.\, P(x) \wedge \exists y.\, Q(y, f(x)) \vee \exists z.\, R(z)$ contains the atoms

- $P(x)$

- $Q(y, f(x))$

- $R(z)$

When all of the terms in an atom are ground terms, then the atom is called a ground atom or *ground predicate*.

27.2 See also

- In model theory, structures assign an interpretation to the atomic formulas.

- In proof theory, polarity assignment for atomic formulas is an essential component of focusing.

- Atomic sentence

27.3 References

[1] Wilfrid Hodges (1997). *A Shorter Model Theory*. Cambridge University Press. pp. 11–14. ISBN 0-521-58713-1.

27.4 Further reading

- Hinman, P. (2005). *Fundamentals of Mathematical Logic*. A K Peters. ISBN 1-56881-262-0.

Chapter 28

List of logic symbols

In logic, a set of symbols is commonly used to express logical representation. As logicians are familiar with these symbols, they are not explained each time they are used. So, for students of logic, the following table lists many common symbols together with their name, pronunciation, and the related field of mathematics. Additionally, the third column contains an informal definition, the fourth column gives a short example, the fifth and sixth give the unicode location and name for use in HTML documents.[1] The last column provides the LaTeX symbol.

Be aware that, outside of logic, different symbols have the same meaning, and the same symbol has, depending on the context, different meanings.

28.1 Basic logic symbols

28.2 Advanced and rarely used logical symbols

These symbols are sorted by their Unicode value:

- U+00B7 · middle dot, an outdated way for denoting AND,[3] still in use in electronics; for example "A·B" is the same as "A&B"

- ·: Center dot with a line above it. Outdated way for denoting NAND, for example "A·B" is the same as "A NAND B" or "A|B" or "¬(A & B)". See also Unicode U+22C5 · dot operator.

- U+0305‾ combining overline, used as abbreviation for standard numerals (Typographical Number Theory). For example, using HTML style "4̅" is a shorthand for the standard numeral "SSSS0".

- Overline, is also a rarely used format for denoting Gödel numbers, for example "AVB" says the Gödel number of "(AVB)"

- Overline is also an outdated way for denoting negation, still in use in electronics; for example "AVB" is the same as "¬(AVB)"

- U+2191 ↑ upwards arrow or U+007C | vertical line: Sheffer stroke, the sign for the NAND operator.

- U+2201 ∁ complement

- U+2204 ∄ there does not exist: strike out existential quantifier same as "¬∃"

- U+2234 ∴ therefore

- U+2235 ∵ because

- U+22A7 ⊧ models: is a model of

- U+22A8 ⊨ true: is true of

- U+22AC ⊬ does not prove: negated ⊢, the sign for "does not prove", for example $T \nvdash P$ says "P is not a theorem of T"

- U+22AD ⊭ not true: is not true of

- U+22BC ⊼ nand: another NAND operator, can also be rendered as ∧

- U+22BD ⊽ nor: another NOR operator, can also be rendered as V

- U+22C4 ◇ diamond operator: modal operator for "it is possible that", "it is not necessarily not" or rarely "it is not provable not" (in most modal logics it is defined as "¬□¬")

- U+22C6 ⋆ star operator: usually used for ad-hoc operators

- U+22A5 ⊥ up tack or U+2193 ↓ downwards arrow: Webb-operator or Peirce arrow, the sign for NOR. Confusingly, "⊥" is also the sign for contradiction or absurdity.

- U+2310 ⌐ reversed not sign

- U+231C ⌜ top left corner and U+231D ⌝ top right corner: corner quotes, also called "Quine quotes"; for quasi-quotation, i.e. quoting specific context of unspecified ("variable") expressions;[4] also used for denoting Gödel number;[5] for example "⌜G⌝" denotes the Gödel number of G. (Typographical note: although the quotes appears as a "pair" in unicode (231C and 231D), they are not symmetrical in some fonts. And in some fonts (for example Arial) they are only symmetrical in certain sizes. Alternatively the quotes can be rendered as ⌈ and ⌉ (U+2308 and U+2309) or by using a negation symbol and a reversed negation symbol ⌐ ¬ in superscript mode.)

- U+25FB □ white medium square or U+25A1 □ white square: modal operator for "it is necessary that" (in modal logic), or "it is provable that" (in provability logic), or "it is obligatory that" (in deontic logic), or "it is believed that" (in doxastic logic).

Note that the following operators are rarely supported by natively installed fonts. If you wish to use these in a web page, you should always embed the necessary fonts so the page viewer can see the web page without having the necessary fonts installed in their computer.

- U+27E1 ◇ white concave-sided diamond

- U+27E2 ◇ white concave-sided diamond with leftwards tick: modal operator for was never

- U+27E3 ◇ white concave-sided diamond with rightwards tick: modal operator for will never be

- U+27E4 ◻ white square with leftwards tick: modal operator for was always

- U+27E5 ◻ white square with rightwards tick: modal operator for will always be

- U+297D ⥽ right fish tail: sometimes used for "relation", also used for denoting various ad hoc relations (for example, for denoting "witnessing" in the context of Rosser's trick) The fish hook is also used as strict implication by C.I.Lewis $p \,⥽\, q \equiv \Box(p \to q)$, the corresponding LaTeX macro is \strictif. See here for an image of glyph. Added to Unicode 3.2.0.

28.2.1   Poland and Germany

As of 2014 in Poland, the universal quantifier is sometimes written ∧ and the existential quantifier as ∨ . The same applies for Germany.

28.3   See also

- Józef Maria Bocheński
- List of notation used in Principia Mathematica
- List of mathematical symbols
- Logic alphabet, a suggested set of logical symbols
- Logical connective
- Mathematical operators and symbols in Unicode
- Polish notation
- Truth function
- Truth table

28.4   References

[1] "Named character references". *HTML 5.1 Nightly.* W3C. Retrieved 9 September 2015.

[2] Although this character is available in LaTeX, the MediaWiki TeX system doesn't support this character.

[3] Brody, Baruch A. (1973), *Logic: theoretical and applied*, Prentice-Hall, p. 93, ISBN 9780135401460, We turn now to the second of our connective symbols, the centered dot, which is called the conjunction sign.

[4] Quine, W.V. (1981): *Mathematical Logic*, §6

[5] Hintikka, Jaakko (1998), *The Principles of Mathematics Revisited*, Cambridge University Press, p. 113, ISBN 9780521624985.

28.5   Further reading

Józef Maria Bocheński (1959), *A Précis of Mathematical Logic*, trans., Otto Bird, from the French and German editions, Dordrecht, South Holland: D. Reidel.

28.6   External links

- Named character entities in HTML 4.0

Chapter 29

T-schema

The **T-schema** or truth schema (not to be confused with 'Convention T') is used to give an inductive definition of truth which lies at the heart of any realisation of Alfred Tarski's semantic theory of truth. Some authors refer to it as the "Equivalence Schema", a synonym introduced by Michael Dummett.[1]

The T-schema is often expressed in natural language, but it can be formalized in many-sorted predicate logic or modal logic; such a formalisation is called a **T-theory**. T-theories form the basis of much fundamental work in philosophical logic, where they are applied in several important controversies in analytic philosophy.

As expressed in semi-natural language (where 'S' is the name of the sentence abbreviated to S): 'S' is true if and only if S

Example: 'snow is white' is true if and only if snow is white.

29.1 The inductive definition

By using the schema one can give an inductive definition for the truth of compound sentences. Atomic sentences are assigned truth values disquotationally. For example, the sentence "'Snow is white' is true" becomes materially equivalent with the sentence "snow is white", i.e. 'snow is white' is true if and only if snow is white. The truth of more complex sentences is defined in terms of the components of the sentence:

- A sentence of the form "A and B" is true if and only if A is true and B is true

- A sentence of the form "A or B" is true if and only if A is true or B is true

- A sentence of the form "if A then B" is true if and only if A is false or B is true; see material implication.

- A sentence of the form "not A" is true if and only if A is false

- A sentence of the form "for all x, $A(x)$" is true if and only if, for every possible value of x, $A(x)$ is true.

- A sentence of the form "for some x, $A(x)$" is true if and only if, for some possible value of x, $A(x)$ is true.

29.2 Natural languages

Joseph Heath points out[2] that "The analysis of the truth predicate provided by Tarski's Schema T is not capable of handling all occurrences of the truth predicate in natural language. In particular, Schema T treats only "freestanding" uses of the predicate—cases when it is applied to complete sentences." He gives as "obvious problem" the sentence:

- Everything that Bill believes is true.

Heath argues that analyzing this sentence using T-schema generates the sentence fragment—"everything that Bill believes"—on the righthand side of the Logical biconditional.

29.3 See also

- Principle of bivalence
- Law of excluded middle

29.4 References

[1] Wolfgang Künne (2003). *Conceptions of truth*. Clarendon Press. p. 18. ISBN 978-0-19-928019-3.

[2] Joseph Heath (2001). *Communicative action and rational choice*. MIT Press. p. 186. ISBN 978-0-262-08291-4.

29.5 External links

- Tarski's Truth Definitions entry in the *Stanford Encyclopedia of Philosophy*
- Consequences of the Semantic Paradoxes entry in the *Stanford Encyclopedia of Philosophy*

Chapter 30

Theory (mathematical logic)

In mathematical logic, a **theory** (also called a **formal theory**) is a set of sentences in a formal language. Usually a deductive system is understood from context. An element $\phi \in T$ of a theory T is then called an axiom of the theory, and any sentence that follows from the axioms ($T \vdash \phi$) is called a theorem of the theory. Every axiom is also a theorem. A **first-order theory** is a set of first-order sentences.

30.1 Theories expressed in formal language generally

When defining theories for foundational purposes, additional care must be taken and normal set-theoretic language may not be appropriate.

The construction of a theory begins by specifying a definite non-empty *conceptual class* $\mathcal{E}$, the elements of which are called *statements*. These initial statements are often called the *primitive elements* or *elementary* statements of the theory, to distinguish them from other statements which may be derived from them.

A theory $\mathcal{T}$ is a conceptual class consisting of certain of these elementary statements. The elementary statements which belong to $\mathcal{T}$ are called the *elementary theorems* of $\mathcal{T}$ and said to be *true*. In this way, a theory is a way of designating a subset of $\mathcal{E}$ which consists entirely of true statements.

This general way of designating a theory stipulates that the truth of any of its elementary statements is not known without reference to $\mathcal{T}$. Thus the same elementary statement may be true with respect to one theory, and not true with respect to another. This is as in ordinary language, where statements such as "He is a terrible person." cannot be judged to be true or false without reference to some interpretation of who "He" is and for that matter what a "terrible person" is under this theory.[1]

30.1.1 Subtheories and extensions

A theory S is a **subtheory** of a theory T if S is a subset of T. If T is a subset of S then S is an **extension** or **supertheory** of T

30.1.2 Deductive theories

A theory is said to be a *deductive theory* if $\mathcal{T}$ is an inductive class. That is, that its content is based on some formal deductive system and that some of its elementary statements are taken as axioms. In a deductive theory, any sentence which is a logical consequence of one or more of the axioms is also a sentence of that theory.[1]

236

30.1.3 Consistency and completeness

Main articles: Consistency and Completeness (logic)

A **syntactically consistent theory** is a theory from which not every sentence in the underlying language can be proven (with respect to some deductive system which is usually clear from context). In a deductive system (such as first-order logic) that satisfies the principle of explosion, this is equivalent to requiring that there is no sentence φ such that both φ and its negation can be proven from the theory.

A **satisfiable theory** is a theory that has a model. This means there is a structure M that satisfies every sentence in the theory. Any satisfiable theory is syntactically consistent, because the structure satisfying the theory will satisfy exactly one of φ and the negation of φ, for each sentence φ.

A **consistent theory** is sometimes defined to be a syntactically consistent theory, and sometimes defined to be a satisfiable theory. For first-order logic, the most important case, it follows from the completeness theorem that the two meanings coincide. In other logics, such as second-order logic, there are syntactically consistent theories that are not satisfiable, such as ω-inconsistent theories.

A complete consistent theory (or just a **complete theory**) is a consistent theory T such that for every sentence φ in its language, either φ is provable from T or $T \cup \{\varphi\}$ is inconsistent. For theories closed under logical consequence, this means that for every sentence φ, either φ or its negation is contained in the theory. An **incomplete theory** is a consistent theory that is not complete.

See also **ω-consistent theory** for a stronger notion of consistency.

30.1.4 Interpretation of a theory

Main article: Interpretation (logic)

An **interpretation of a theory** is the relationship between a theory and some contensive subject matter when there is a many-to-one correspondence between certain elementary statements of the theory, and certain contensive statements related to the subject matter. If every elementary statement in the theory has a contensive correspondent it is called a *full interpretation*, otherwise it is called a *partial interpretation*.[2]

30.1.5 Theories associated with a structure

Each structure has several associated theories. The **complete theory** of a structure A is the set of all first-order sentences over the signature of A which are satisfied by A. It is denoted by $\mathrm{Th}(A)$. More generally, the **theory** of K, a class of σ-structures, is the set of all first-order σ-sentences that are satisfied by all structures in K, and is denoted by $\mathrm{Th}(K)$. Clearly $\mathrm{Th}(A) = \mathrm{Th}(\{A\})$. These notions can also be defined with respect to other logics.

For each σ-structure A, there are several associated theories in a larger signature σ' that extends σ by adding one new constant symbol for each element of the domain of A. (If the new constant symbols are identified with the elements of A which they represent, σ' can be taken to be $\sigma \cup A$.) The cardinality of σ' is thus the larger of the cardinality of σ and the cardinality of A.

The **diagram** of A consists of all atomic or negated atomic σ'-sentences that are satisfied by A and is denoted by $\mathrm{diag}A$. The **positive diagram** of A is the set of all atomic σ'-sentences which A satisfies. It is denoted by diag^+A. The **elementary diagram** of A is the set $\mathrm{eldiag}A$ of *all* first-order σ'-sentences that are satisfied by A or, equivalently, the complete (first-order) theory of the natural expansion of A to the signature σ'.

30.2   First-order theories

Further information: List of first-order theories

A first-order theory $\mathcal{QS}$ is a set of sentences in a first-order formal language $\mathcal{Q}$.

30.2.1   Derivation in a first order theory

Main article: First order logic § Deductive systems

There are many formal derivation ("proof") systems for first-order logic.

30.2.2   Syntactic consequence in a first order theory

Main article: First-order logic § Validity, satisfiability, and logical consequence

A formula A is a **syntactic consequence** of a first-order theory $\mathcal{QS}$ if there is a derivation of A using only formulas in $\mathcal{QS}$ as non-logical axioms. Such a formula A is also called a theorem of $\mathcal{QS}$. The notation " $\mathcal{QS} \vdash A$ " indicates A is a theorem of $\mathcal{QS}$

30.2.3   Interpretation of a first order theory

Main article: Structure (mathematical logic)

An **interpretation** of a first-order theory provides a semantics for the formulas of the theory. An interpretation is said to satisfy a formula if the formula is true according to the interpretation. A **model** of a first order theory $\mathcal{QS}$ is an interpretation in which every formula of $\mathcal{QS}$ is satisfied.

30.2.4   First order theories with identity

Main article: First order logic § Equality and its axioms

A first order theory $\mathcal{QS}$ is a first-order theory with identity if $\mathcal{QS}$ includes the identity relation symbol "=" and the reflexivity and substitution axiom schemes for this symbol.

30.2.5   Topics related to first order theories

- Compactness theorem

- Consistent set

- Deduction theorem

- Enumeration theorem

- Lindenbaum's lemma

- Löwenheim–Skolem theorem

30.3 Examples

One way to specify a theory is to define a set of axioms in a particular language. The theory can be taken to include just those axioms, or their logical or provable consequences, as desired. Theories obtained this way include ZFC and Peano arithmetic.

A second way to specify a theory is to begin with a structure and then let the theory be the set of sentences that are satisfied by the structure. This is one method for producing complete theories, described below. Examples of theories of this sort include the sets of true sentences in the structures $(\mathbf{N}, +, \times, 0, 1, =)$ and $(\mathbf{R}, +, \times, 0, 1, =)$, where $\mathbf{N}$ is the set of natural numbers and $\mathbf{R}$ is the set of real numbers. The first of these, called the theory of true arithmetic, cannot be written as the set of logical consequences of any enumerable set of axioms. The theory of $(\mathbf{R}, +, \times, 0, 1, =)$ was shown by Tarski to be decidable; it is the theory of real closed fields.

30.4 See also

- Axiomatic system
- List of first-order theories

30.5 References

[1] Curry, Haskell, *Foundations of Mathematical Logic*

[2] Curry, Haskell, *Foundations of Mathematical Logic* p.48

30.6 Further reading

- Hodges, Wilfrid (1997). *A shorter model theory*. Cambridge University Press. ISBN 0-521-58713-1.

Chapter 31

Consistency

For other uses, see Consistency (disambiguation).

In classical deductive logic, a **consistent** theory is one that does not contain a contradiction.[1][2] The lack of contradiction can be defined in either semantic or syntactic terms. The semantic definition states that a theory is consistent if and only if it has a model, i.e. there exists an interpretation under which all formulas in the theory are true. This is the sense used in traditional Aristotelian logic, although in contemporary mathematical logic the term **satisfiable** is used instead. The syntactic definition states that a theory is consistent if and only if there is no formula P such that both P and its negation are provable from the axioms of the theory under its associated deductive system.

If these semantic and syntactic definitions are equivalent for any theory formulated using a particular deductive logic, the logic is called **complete**. The completeness of the sentential calculus was proved by Paul Bernays in 1918[3] and Emil Post in 1921,[4] while the completeness of predicate calculus was proved by Kurt Gödel in 1930,[5] and consistency proofs for arithmetics restricted with respect to the induction axiom schema were proved by Ackermann (1924), von Neumann (1927) and Herbrand (1931).[6] Stronger logics, such as second-order logic, are not complete.

A **consistency proof** is a mathematical proof that a particular theory is consistent. The early development of mathematical proof theory was driven by the desire to provide finitary consistency proofs for all of mathematics as part of Hilbert's program. Hilbert's program was strongly impacted by incompleteness theorems, which showed that sufficiently strong proof theories cannot prove their own consistency (provided that they are in fact consistent).

Although consistency can be proved by means of model theory, it is often done in a purely syntactical way, without any need to reference some model of the logic. The cut-elimination (or equivalently the normalization of the underlying calculus if there is one) implies the consistency of the calculus: since there is obviously no cut-free proof of falsity, there is no contradiction in general.

31.1 Consistency and completeness in arithmetic and set theory

In theories of arithmetic, such as Peano arithmetic, there is an intricate relationship between the consistency of the theory and its completeness. A theory is complete if, for every formula φ in its language, at least one of φ or $\neg\,\varphi$ is a logical consequence of the theory.

Presburger arithmetic is an axiom system for the natural numbers under addition. It is both consistent and complete.

Gödel's incompleteness theorems show that any sufficiently strong effective theory of arithmetic cannot be both complete and consistent. Gödel's theorem applies to the theories of Peano arithmetic (PA) and Primitive recursive arithmetic (PRA), but not to Presburger arithmetic.

Moreover, Gödel's second incompleteness theorem shows that the consistency of sufficiently strong effective theories of arithmetic can be tested in a particular way. Such a theory is consistent if and only if it does *not* prove a particular sentence, called the Gödel sentence of the theory, which is a formalized statement of the claim that the theory is indeed consistent. Thus the consistency of a sufficiently strong, effective, consistent theory of arithmetic can never be proven in that system itself. The same result is true for effective theories that can describe a strong enough fragment of arithmetic – including set theories such as Zermelo–Fraenkel set theory. These set theories cannot prove their own Gödel sentences – provided that they are consistent, which is generally believed.

Because consistency of ZF is not provable in ZF, the weaker notion **relative consistency** is interesting in set theory (and in other sufficiently expressive axiomatic systems). If T is a theory and A is an additional axiom, $T + A$ is said to be consistent relative to T (or simply that A is consistent with T) if it can be proved that if T is consistent then $T + A$ is consistent. If both A and $\neg A$ are consistent with T, then A is said to be independent of T.

31.2 First-order logic

31.2.1 Notation

$\vdash$ (Turnstile symbol) in the following context of Mathematical logic, means "provable from". That is, a $\vdash$ b reads: b is provable from a (in some specified formal system) -- see List of logic symbols) . In other cases, the turnstile symbol may stand to mean infers; derived from. See: List of mathematical symbols.

31.2.2 Definition

A set of formulas Φ in first-order logic is **consistent** (written Con Φ) if and only if there is no formula ϕ such that $\Phi \vdash \phi$ and $\Phi \vdash \neg\phi$. Otherwise Φ is **inconsistent** and is written Inc Φ .

Φ is said to be **simply consistent** if and only if for no formula ϕ of Φ , both ϕ and the negation of ϕ are theorems of Φ .

Φ is said to be **absolutely consistent** or **Post consistent** if and only if at least one formula of Φ is not a theorem of Φ .

Φ is said to be **maximally consistent** if and only if for every formula ϕ , if Con ($\Phi \cup \phi$) then $\phi \in \Phi$.

Φ is said to **contain witnesses** if and only if for every formula of the form $\exists x \phi$ there exists a term t such that $(\exists x \phi \to \phi\frac{t}{x}) \in \Phi$. See First-order logic.

31.2.3 Basic results

1. The following are equivalent:

   (a) Inc Φ

   (b) For all ϕ, $\Phi \vdash \phi$.

2. Every satisfiable set of formulas is consistent, where a set of formulas Φ is satisfiable if and only if there exists a model $\Im$ such that $\Im \models \Phi$.

3. For all Φ and ϕ :

   (a) if not $\Phi \vdash \phi$, then Con $(\Phi \cup \{\neg\phi\})$;

   (b) if Con Φ and $\Phi \vdash \phi$, then Con $(\Phi \cup \{\phi\})$;

   (c) if Con Φ , then Con $(\Phi \cup \{\phi\})$ or Con $(\Phi \cup \{\neg\phi\})$.

4. Let Φ be a maximally consistent set of formulas and contain witnesses. For all ϕ and ψ :

   (a) if $\Phi \vdash \phi$, then $\phi \in \Phi$,

   (b) either $\phi \in \Phi$ or $\neg\phi \in \Phi$,

   (c) $(\phi \lor \psi) \in \Phi$ if and only if $\phi \in \Phi$ or $\psi \in \Phi$,

   (d) if $(\phi \to \psi) \in \Phi$ and $\phi \in \Phi$, then $\psi \in \Phi$,

   (e) $\exists x \phi \in \Phi$ if and only if there is a term t such that $\phi\frac{t}{x} \in \Phi$.

31.2.4 Henkin's theorem

Let Φ be a maximally consistent set of S-formulas containing witnesses.

Define a binary relation $\sim$ on the set of S-terms such that $t_0 \sim t_1$ if and only if $t_0 \equiv t_1 \in \Phi$; and let $\bar{t}$ denote the equivalence class of terms containing t; and let $T_\Phi := \{\, \bar{t} \mid t \in T^S \,\}$ where T^S is the set of terms based on the symbol set S.

Define the S-structure $\mathfrak{T}_\Phi$ over T_Φ the **term-structure** corresponding to Φ by:

1. for n-ary $R \in S$, $R^{\mathfrak{T}_\Phi}\overline{t_0}\ldots\overline{t_{n-1}}$ if and only if $Rt_0\ldots t_{n-1} \in \Phi$;

2. for n-ary $f \in S$, $f^{\mathfrak{T}_\Phi}(\overline{t_0}\ldots\overline{t_{n-1}}) := \overline{ft_0\ldots t_{n-1}}$;

3. for $c \in S$, $c^{\mathfrak{T}_\Phi} := \bar{c}$.

Let $\mathfrak{I}_\Phi := (\mathfrak{T}_\Phi, \beta_\Phi)$ be the **term interpretation** associated with Φ, where $\beta_\Phi(x) := \bar{x}$.

$$\text{For all } \phi, \ \mathfrak{I}_\Phi \vDash \phi \text{ if and only if } \phi \in \Phi.$$

31.2.5 Sketch of proof

There are several things to verify. First, that $\sim$ is an equivalence relation. Then, it needs to be verified that (1), (2), and (3) are well defined. This falls out of the fact that $\sim$ is an equivalence relation and also requires a proof that (1) and (2) are independent of the choice of $t_0, \ldots, t_{n-1}$ class representatives. Finally, $\mathfrak{I}_\Phi \vDash \Phi$ can be verified by induction on formulas.

31.3 See also

- Equiconsistency

- Hilbert's problems

- Hilbert's second problem

- Jan Łukasiewicz

- Paraconsistent logic

- ω-consistency

- Gentzen's consistency proof

31.4 Footnotes

[1] Tarski 1946 states it this way: "A deductive theory is called CONSISTENT or NON-CONTRADICTORY if no two asserted statements of this theory contradict each other, or in other words, if of any two contradictory sentences . . . at least one cannot be proved," (p. 135) where Tarski defines *contradictory* as follows: "With the help of the word *not* one forms the NEGATION of any sentence; two sentences, of which the first is a negation of the second, are called CONTRADICTORY SENTENCES" (p. 20). This definition requires a notion of "proof". Gödel in his 1931 defines the notion this way: "The class of *provable formulas* is defined to be the smallest class of formulas that contains the axioms and is closed under the relation "immediate consequence", i.e. formula c of a and b is defined as an *immediate consequence* in terms of *modus ponens* or substitution; cf Gödel 1931 van Heijenoort 1967:601. Tarski defines "proof" informally as "statements follow one another in a definite order according to certain principles . . . and accompanied by considerations intended to establish their validity[true conclusion for all true premises -- Reichenbach 1947:68]" cf Tarski 1946:3. Kleene 1952 defines the notion with respect to either an induction

or as to paraphrase) a finite sequence of formulas such that each formula in the sequence is either an axiom or an "immediate consequence" of the preceding formulas; "A *proof is said to be a proof* of *its last formula, and this formula is said to be* (formally) provable *or be a* (formal) theorem" cf Kleene 1952:83.

[2] Paraconsistent logic *tolerates* contradictions, but toleration of contradiction does not entail consistency.

[3] van Heijenoort 1967:265 states that Bernays determined the *independence* of the axioms of *Principia Mathematica*, a result not published until 1926, but he says nothing about Bernays proving their *consistency*.

[4] Post proves both consistency and completeness of the propositional calculus of PM, cf van Heijenoort's commentary and Post's 1931 *Introduction to a general theory of elementary propositons* in van Heijenoort 1967:264ff. Also Tarski 1946:134ff.

[5] cf van Heijenoort's commentary and Gödel's 1930 *The completeness of the axioms of the functional calculus of logic* in van Heijenoort 1967:582ff

[6] cf van Heijenoort's commentary and Herbrand's 1930 *On the consistency of arithmetic* in van Heijenoort 1967:618ff.

31.5 References

- Stephen Kleene, 1952 10th impression 1991, *Introduction to Metamathematics*, North-Holland Publishing Company, Amsterday, New York, ISBN 0-7204-2103-9.

- Hans Reichenbach, 1947, *Elements of Symbolic Logic*, Dover Publications, Inc. New York, ISBN 0-486-24004-5,

- Alfred Tarski, 1946, *Introduction to Logic and to the Methodology of Deductive Sciences, Second Edition*, Dover Publications, Inc., New York, ISBN 0-486-28462-X.

- Jean van Heijenoort, 1967, *From Frege to Gödel: A Source Book in Mathematical Logic*, Harvard University Press, Cambridge, MA, ISBN 0-674-32449-8 (pbk.)

- The Cambridge Dictionary of Philosophy, *consistency*

- H.D. Ebbinghaus, J. Flum, W. Thomas, **Mathematical Logic**

- Jevons, W.S., 1870, *Elementary Lessons in Logic*

31.6 External links

- Chris Mortensen, Inconsistent Mathematics, Stanford Encyclopedia of Philosophy

Chapter 32

Löwenheim–Skolem theorem

In mathematical logic, the **Löwenheim–Skolem theorem**, named for Leopold Löwenheim and Thoralf Skolem, states that if a countable first-order theory has an infinite model, then for every infinite cardinal number κ it has a model of size κ. The result implies that first-order theories are unable to control the cardinality of their infinite models, and that no first-order theory with an infinite model can have a unique model up to isomorphism.

The (downward) Löwenheim–Skolem theorem is one of the two key properties, along with the compactness theorem, that are used in Lindström's theorem to characterize first-order logic. In general, the Löwenheim–Skolem theorem does not hold in stronger logics such as second-order logic.

32.1 Background

A signature consists of a set of function symbols S_{func}, a set of relation symbols S_{rel}, and a function ar $: S_{func} \cup S_{rel} \to \mathbb{N}_0$ representing the arity of function and relation symbols. (A nullary function symbol is called a constant symbol.) In the context of first-order logic, a signature is sometimes called a **language**. It is called countable if the set of function and relation symbols in it is countable, and in general the cardinality of a signature is the cardinality of the set of all the symbols it contains.

A first-order **theory** consists of a fixed signature and a fixed set of sentences (formulas with no free variables) in that signature. Theories are often specified by giving a list of axioms that generate the theory, or by giving a structure and taking the theory to consist of the sentences satisfied by the structure.

Given a signature σ, a σ-structure M is a concrete interpretation of the symbols in σ. It consists of an underlying set (often also denoted by "M") together with an interpretation of the function and relation symbols of σ. An interpretation of a constant symbol of σ in M is simply an element of M. More generally, an interpretation of an n-ary function symbol f is a function from M^n to M. Similarly, an interpretation of a relation symbol R is an n-ary relation on M, i.e. a subset of M^n.

A **substructure** of a σ-structure M is obtained by taking a subset N of M which is closed under the interpretations of all the function symbols in σ (hence includes the interpretations of all constant symbols in σ), and then restricting the interpretations of the relation symbols to N. An elementary substructure is a very special case of this; in particular an elementary substructure satisfies exactly the same first-order sentences as the original structure (its **elementary extension**).

32.2 Precise statement

The modern statement of the theorem is both more general and stronger than the version for countable signatures stated in the introduction.

In its general form, the **Löwenheim–Skolem Theorem** states that for every signature σ, every infinite σ-structure M and every infinite cardinal number $\kappa \geq |\sigma|$, there is a σ-structure N such that $|N| = \kappa$ and

- if $\kappa < |M|$ then N is an elementary substructure of M;

- if $\kappa > |M|$ then N is an elementary extension of M.

The theorem is often divided into two parts corresponding to the two bullets above. The part of the theorem asserting that a structure has elementary substructures of all smaller infinite cardinalities is known as the **downward Löwenheim–Skolem Theorem**. The part of the theorem asserting that a structure has elementary extensions of all larger cardinalities is known as the **upward Löwenheim–Skolem Theorem**.

The statement given in the introduction follows immediately by taking M to be an infinite model of the theory. The proof of the upward part of the theorem also shows that a theory with arbitrarily large finite models must have an infinite model; sometimes this is considered to be part of the theorem. For historical variants of the theorem, see the notes below.

32.3   Examples and consequences

Let **N** denote the natural numbers and **R** the reals. It follows from the theorem that the theory of (**N**, +, ×, 0, 1) (the theory of true first-order arithmetic) has uncountable models, and that the theory of (**R**, +, ×, 0, 1) (the theory of real closed fields) has a countable model. There are, of course, axiomatizations characterizing (**N**, +, ×, 0, 1) and (**R**, +, ×, 0, 1) up to isomorphism. The Löwenheim–Skolem theorem shows that these axiomatizations cannot be first-order. For example, the completeness of a linear order, which is used to characterize the real numbers as a complete ordered field, is a non-first-order property.

A theory is called **categorical** if it has only one model, up to isomorphism. This term was introduced by Veblen (1904), and for some time thereafter mathematicians hoped they could put mathematics on a solid foundation by describing a categorical first-order theory of some version of set theory. The Löwenheim–Skolem theorem dealt a first blow to this hope, as it implies that a first-order theory which has an infinite model cannot be categorical. Later, in 1931, the hope was shattered completely by Gödel's incompleteness theorem.

Many consequences of the Löwenheim–Skolem theorem seemed counterintuitive to logicians in the early 20th century, as the distinction between first-order and non-first-order properties was not yet understood. One such consequence is the existence of uncountable models of true arithmetic, which satisfy every first-order induction axiom but have non-inductive subsets. Another consequence that was considered particularly troubling is the existence of a countable model of set theory, which nevertheless must satisfy the sentence saying the real numbers are uncountable. This counterintuitive situation came to be known as Skolem's paradox; it shows that the notion of countability is not absolute.

32.4   Proof sketch

32.4.1   Downward part

For each first-order σ -formula $\varphi(y, x_1, \ldots, x_n)$, the axiom of choice implies the existence of a function

$$f_\varphi : M^n \to M$$

such that, for all $a_1, \ldots, a_n \in M$, either

$$M \models \varphi(f_\varphi(a_1, \ldots, a_n), a_1, \ldots, a_n)$$

or

$$M \models \neg \exists y \varphi(y, a_1, \ldots, a_n).$$

Applying the axiom of choice again we get a function from the first order formulas φ to such functions f_φ.

The family of functions f_φ gives rise to a preclosure operator F on the power set of M

$$F(A) = \{b \in M \mid b = f_\varphi(a_1, \ldots, a_n); \ \varphi \in \sigma; \ a_1, \ldots, a_n \in A\}$$

for $A \subseteq M$.

Iterating F countably many times results in a closure operator F^ω. Taking an arbitrary subset $A \subseteq M$ such that $|A| = \kappa$, and having defined $N = F^\omega(A)$, one can see that also $|N| = \kappa$. N is an elementary substructure of M by the Tarski–Vaught test.

The trick used in this proof is essentially due to Skolem, who introduced function symbols for the Skolem functions f_φ into the language. One could also define the f_φ as partial functions such that f_φ is defined if and only if $M \models \exists y \varphi(y, a_1, \ldots, a_n)$. The only important point is that F is a preclosure operator such that $F(A)$ contains a solution for every formula with parameters in A which has a solution in M and that

$$|F(A)| \leq |A| + |\sigma| + \aleph_0.$$

32.4.2 Upward part

First, one extends the signature by adding a new constant symbol for every element of M. The complete theory of M for the extended signature σ' is called the *elementary diagram* of M. In the next step one adds κ many new constant symbols to the signature and adds to the elementary diagram of M the sentences $c \neq c'$ for any two distinct new constant symbols c and c'. Using the compactness theorem, the resulting theory is easily seen to be consistent. Since its models must have cardinality at least κ, the downward part of this theorem guarantees the existence of a model N which has cardinality exactly κ. It contains an isomorphic copy of M as an elementary substructure.

32.5 Historical notes

This account is based mainly on Dawson (1993). To understand the early history of model theory one must distinguish between *syntactical consistency* (no contradiction can be derived using the deduction rules for first-order logic) and *satisfiability* (there is a model). Somewhat surprisingly, even before the completeness theorem made the distinction unnecessary, the term *consistent* was used sometimes in one sense and sometimes in the other.

The first significant result in what later became model theory was *Löwenheim's theorem* in Leopold Löwenheim's publication "Über Möglichkeiten im Relativkalkül" (1915):

> For every countable signature σ, every σ-sentence which is satisfiable is satisfiable in a countable model.

Löwenheim's paper was actually concerned with the more general Peirce–Schröder calculus of relatives (relation algebra with quantifiers). He also used the now antiquated notations of Ernst Schröder. For a summary of the paper in English and using modern notations see Brady (2000, chapter 8).

According to the received historical view, Löwenheim's proof was faulty because it implicitly used König's lemma without proving it, although the lemma was not yet a published result at the time. In a revisionist account, Badesa (2004) considers that Löwenheim's proof was complete.

Skolem (1920) gave a (correct) proof using formulas in what would later be called *Skolem normal form* and relying on the axiom of choice:

> Every countable theory which is satisfiable in a model *M*, is satisfiable in a countable substructure of *M*.

Skolem (1923) also proved the following weaker version without the axiom of choice:

> Every countable theory which is satisfiable in a model is also satisfiable in a countable model.

Skolem (1929) simplified Skolem (1920). Finally, Anatoly Ivanovich Maltsev (Анато́лий Ива́нович Ма́льцев, 1936) proved the Löwenheim–Skolem theorem in its full generality (Maltsev 1936). He cited a note by Skolem, according to which the theorem had been proved by Alfred Tarski in a seminar in 1928. Therefore the general theorem is sometimes known as the *Löwenheim–Skolem–Tarski theorem*. But Tarski did not remember his proof, and it remains a mystery how he could do it without the compactness theorem.

It is somewhat ironic that Skolem's name is connected with the upward direction of the theorem as well as with the downward direction:

> *"I follow custom in calling Corollary 6.1.4 the upward Löwenheim-Skolem theorem. But in fact Skolem didn't even believe it, because he didn't believe in the existence of uncountable sets."* – Hodges (1993).

> *"Skolem [...] rejected the result as meaningless; Tarski [...] very reasonably responded that Skolem's formalist viewpoint ought to reckon the downward Löwenheim-Skolem theorem meaningless just like the upward."* – Hodges (1993).

> *"Legend has it that Thoralf Skolem, up until the end of his life, was scandalized by the association of his name to a result of this type, which he considered an absurdity, nondenumerable sets being, for him, fictions without real existence."* – Poizat (2000).

32.6 References

The Löwenheim–Skolem theorem is treated in all introductory texts on model theory or mathematical logic.

32.6.1 Historical publications

- Löwenheim, Leopold (1915), "Über Möglichkeiten im Relativkalkül" (PDF), *Mathematische Annalen* **76** (4): 447–470, doi:10.1007/BF01458217, ISSN 0025-5831

    - Löwenheim, Leopold (1977), "On possibilities in the calculus of relatives", *From Frege to Gödel: A Source Book in Mathematical Logic, 1879-1931* (3rd ed.), Cambridge, Massachusetts: Harvard University Press, pp. 228–251, ISBN 0-674-32449-8 (*online copy*, p. 228, at Google Books)

- Maltsev, Anatoly Ivanovich (1936), "Untersuchungen aus dem Gebiete der mathematischen Logik", *Matematicheskii Sbornik, n.s.* **1**: 323–336

- Skolem, Thoralf (1920), "Logisch-kombinatorische Untersuchungen über die Erfüllbarkeit oder Beweisbarkeit mathematischer Sätze nebst einem Theoreme über dichte Mengen", *Videnskapsselskapet Skrifter, I. Matematisk-naturvidenskabelig Klasse* **4**: 1–36

    - Skolem, Thoralf (1977), "Logico-combinatorical investigations in the satisfiability or provabilitiy of mathematical propositions: A simplified proof of a theorem by L. Löwenheim and generalizations of the theorem", *From Frege to Gödel: A Source Book in Mathematical Logic, 1879-1931* (3rd ed.), Cambridge, Massachusetts: Harvard University Press, pp. 252–263, ISBN 0-674-32449-8 (*online copy*, p. 252, at Google Books)

- Skolem, Thoralf (1922), "Einige Bemerkungen zu axiomatischen Begründung der Mengenlehre", *Mathematikerkongressen i Helsingfors den 4–7 Juli 1922, Den femte skandinaviska matematikerkongressen, Redogörelse*: 217–232

- Skolem, Thoralf (1977), "Some remarks on axiomatized set theory", *From Frege to Gödel: A Source Book in Mathematical Logic, 1879-1931* (3rd ed.), Cambridge, Massachusetts: Harvard University Press, pp. 290–301, ISBN 0-674-32449-8 (*online copy*, p. 290, at Google Books)

- Skolem, Thoralf (1929), "Über einige Grundlagenfragen der Mathematik", *Skrifter utgitt av det Norske Videnskaps-Akademi i Oslo, I. Matematisk-naturvidenskabelig Klasse* **7**: 1–49

- Veblen, Oswald (1904), "A System of Axioms for Geometry", *Transactions of the American Mathematical Society* **5** (3): 343–384, doi:10.2307/1986462, ISSN 0002-9947, JSTOR 1986462

32.6.2 Secondary sources

- Badesa, Calixto (2004), *The Birth of Model Theory: Löwenheim's Theorem in the Frame of the Theory of Relatives*, Princeton, NJ: Princeton University Press, ISBN 978-0-691-05853-5; A more concise account appears in chapter 9 of Leila Haaparanta, ed. (2009), *The Development of Modern Logic*, Oxford University Press, ISBN 978-0-19-513731-6

- Brady, Geraldine (2000), *From Peirce to Skolem: A Neglected Chapter in the History of Logic*, Elsevier, ISBN 978-0-444-50334-3

- Crossley, J.N.; Ash, C.J.; Brickhill, C.J.; Stillwell, J.C.; Williams, N.H. (1972), *What is mathematical logic?*, London-Oxford-New York: Oxford University Press, pp. 59–60, ISBN 0-19-888087-1, Zbl 0251.02001

- Dawson, John W., Jr. (1993), "The compactness of First-Order Logic: From Gödel to Lindström", *History and Philosophy of Logic* **14**: 15–37, doi:10.1080/01445349308837208

- Hodges, Wilfrid (1993), *Model theory*, Cambridge: Cambridge Univ. Pr., ISBN 978-0-521-30442-9

- Poizat, Bruno (2000), *A Course in Model Theory: An Introduction to Contemporary Mathematical Logic*, Berlin, New York: Springer, ISBN 978-0-387-98655-5

32.7 External links

- Sakharov, Alex and Weisstein, Eric W., "Löwenheim-Skolem Theorem", *MathWorld*.

- Burris, Stanley N., Contributions of the Logicians, Part II, From Richard Dedekind to Gerhard Gentzen

- Burris, Stanley N., Downward Löwenheim–Skolem theorem

- Simpson, Stephen G. (1998), Model Theory

Chapter 33

Cardinal number

This article is about the mathematical concept. For number words indicating quantity ("three" apples, "four" birds, etc.), see Cardinal number (linguistics).

In mathematics, **cardinal numbers**, or **cardinals** for short, are a generalization of the natural numbers used to measure the cardinality (size) of sets. The cardinality of a finite set is a natural number: the number of elements in the set. The *transfinite* cardinal numbers describe the sizes of infinite sets.

Cardinality is defined in terms of bijective functions. Two sets have the same cardinality if, and only if, there is a one-to-one correspondence (bijection) between the elements of the two sets. In the case of finite sets, this agrees with the intuitive notion of size. In the case of infinite sets, the behavior is more complex. A fundamental theorem due to Georg Cantor shows that it is possible for infinite sets to have different cardinalities, and in particular the cardinality of the set of real numbers is greater than the cardinality of the set of natural numbers. It is also possible for a proper subset of an infinite set to have the same cardinality as the original set, something that cannot happen with proper subsets of finite sets.

There is a transfinite sequence of cardinal numbers:

$$0, 1, 2, 3, \ldots, n, \ldots; \aleph_0, \aleph_1, \aleph_2, \ldots, \aleph_\alpha, \ldots.$$

This sequence starts with the natural numbers including zero (finite cardinals), which are followed by the aleph numbers (infinite cardinals of well-ordered sets). The aleph numbers are indexed by ordinal numbers. Under the assumption of the axiom of choice, this transfinite sequence includes every cardinal number. If one rejects that axiom, the situation is more complicated, with additional infinite cardinals that are not alephs.

Cardinality is studied for its own sake as part of set theory. It is also a tool used in branches of mathematics including model theory, combinatorics, abstract algebra, and mathematical analysis. In category theory, the cardinal numbers form a skeleton of the category of sets.

33.1 History

The notion of cardinality, as now understood, was formulated by Georg Cantor, the originator of set theory, in 1874–1884. Cardinality can be used to compare an aspect of finite sets; e.g. the sets {1,2,3} and {4,5,6} are not *equal*, but have the *same cardinality*, namely three (this is established by the existence of a bijection, i.e. a one-to-one correspondence, between the two sets; e.g. {1->4, 2->5, 3->6}).

Cantor applied his concept of bijection to infinite sets;[1] e.g. the set of natural numbers $\mathbf{N} = \{0, 1, 2, 3, \ldots\}$. Thus, all sets having a bijection with **N** he called denumerable (countably infinite) sets and they all have the same cardinal number. This cardinal number is called $\aleph_0$, aleph-null. He called the cardinal numbers of these infinite sets, transfinite cardinal numbers.

Cantor proved that any unbounded subset of **N** has the same cardinality as **N**, even though this might appear to run contrary

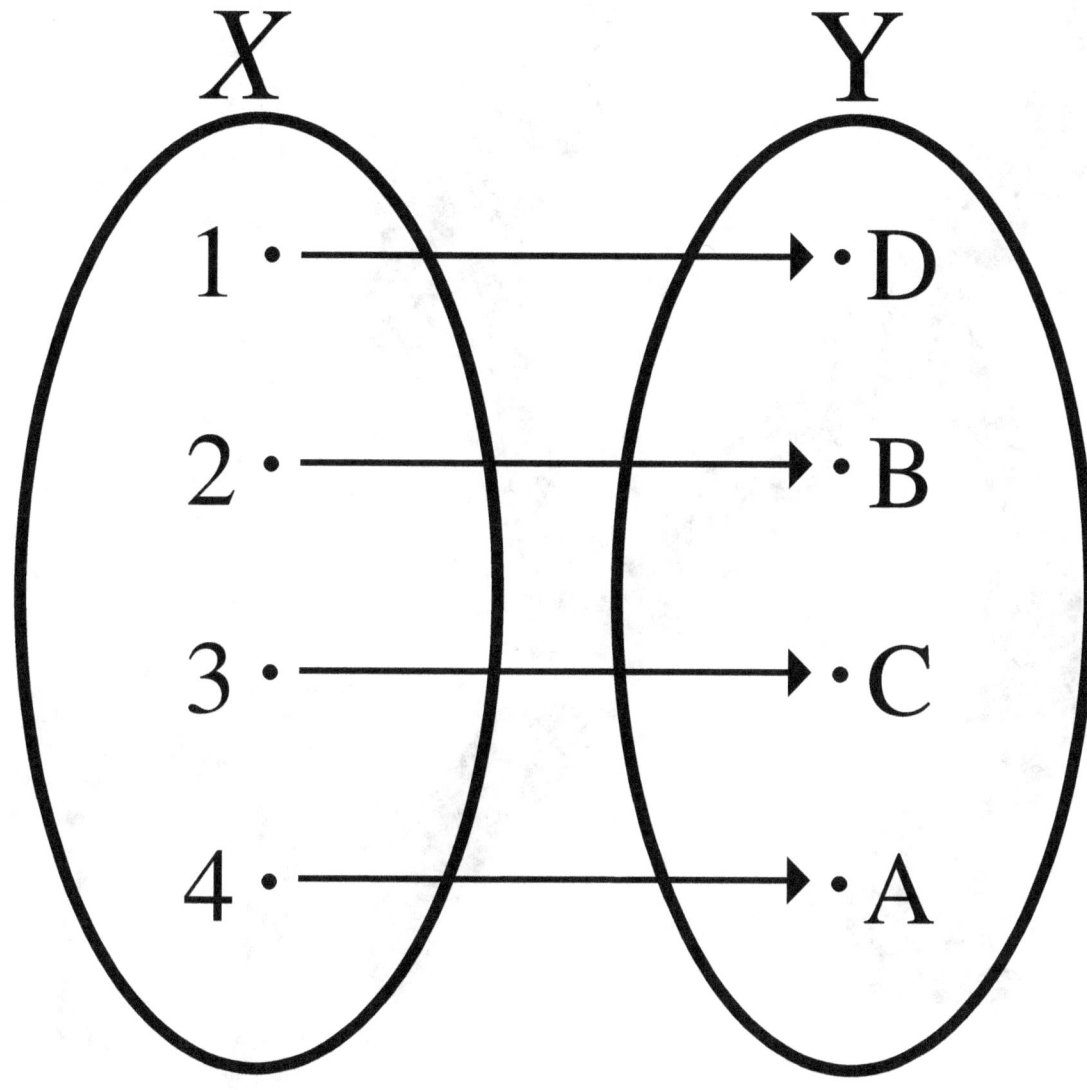

A bijective function, f: X → Y, *from set* X *to set* Y *demonstrates that the sets have the same cardinality, in this case equal to the cardinal number 4.*

to intuition. He also proved that the set of all ordered pairs of natural numbers is denumerable (which implies that the set of all rational numbers is denumerable), and later proved that the set of all algebraic numbers is also denumerable. Each algebraic number z may be encoded as a finite sequence of integers which are the coefficients in the polynomial equation of which it is the solution, i.e. the ordered n-tuple $(a_0, a_1, ..., an)$, $ai \in \mathbf{Z}$ together with a pair of rationals (b_0, b_1) such that z is the unique root of the polynomial with coefficients $(a_0, a_1, ..., an)$ that lies in the interval (b_0, b_1).

In his 1874 paper, Cantor proved that there exist higher-order cardinal numbers by showing that the set of real numbers has cardinality greater than that of $\mathbf{N}$. His original presentation used a complex argument with nested intervals, but in an 1891 paper he proved the same result using his ingenious but simple diagonal argument. The new cardinal number of the set of real numbers is called the cardinality of the continuum and Cantor used the symbol $\mathfrak{c}$ for it.

Cantor also developed a large portion of the general theory of cardinal numbers; he proved that there is a smallest transfinite cardinal number ($\aleph_0$, aleph-null) and that for every cardinal number, there is a next-larger cardinal

Aleph null, the smallest infinite cardinal

$(\aleph_1, \aleph_2, \aleph_3, \cdots)$.

His continuum hypothesis is the proposition that $\mathfrak{c}$ is the same as $\aleph_1$. This hypothesis has been found to be independent of the standard axioms of mathematical set theory; it can neither be proved nor disproved from the standard assumptions.

33.2 Motivation

In informal use, a **cardinal number** is what is normally referred to as a *counting number*, provided that 0 is included: 0, 1, 2, ... They may be identified with the natural numbers beginning with 0. The counting numbers are exactly what can be defined formally as the finite cardinal numbers. Infinite cardinals only occur in higher-level mathematics and logic.

More formally, a non-zero number can be used for two purposes: to describe the size of a set, or to describe the position of an element in a sequence. For finite sets and sequences it is easy to see that these two notions coincide, since for every number describing a position in a sequence we can construct a set which has exactly the right size, e.g. 3 describes the position of 'c' in the sequence <'a','b','c','d',...>, and we can construct the set {a,b,c} which has 3 elements. However when dealing with infinite sets it is essential to distinguish between the two — the two notions are in fact different for infinite sets. Considering the position aspect leads to ordinal numbers, while the size aspect is generalized by the **cardinal numbers** described here.

The intuition behind the formal definition of cardinal is the construction of a notion of the relative size or "bigness" of a set without reference to the kind of members which it has. For finite sets this is easy; one simply counts the number of elements a set has. In order to compare the sizes of larger sets, it is necessary to appeal to more subtle notions.

A set Y is at least as big as a set X if there is an injective mapping from the elements of X to the elements of Y. An injective mapping identifies each element of the set X with a unique element of the set Y. This is most easily understood by an example; suppose we have the sets $X = \{1,2,3\}$ and $Y = \{a,b,c,d\}$, then using this notion of size we would observe that there is a mapping:

$1 \rightarrow a$

$2 \rightarrow b$

$3 \rightarrow c$

which is injective, and hence conclude that Y has cardinality greater than or equal to X. Note the element d has no element mapping to it, but this is permitted as we only require an injective mapping, and not necessarily an injective and onto mapping. The advantage of this notion is that it can be extended to infinite sets.

We can then extend this to an equality-style relation. Two sets X and Y are said to have the same **cardinality** if there exists a bijection between X and Y. By the Schroeder–Bernstein theorem, this is equivalent to there being *both* an injective mapping from X to Y *and* an injective mapping from Y to X. We then write $|X| = |Y|$. The cardinal number of X itself is often defined as the least ordinal a with $|a| = |X|$. This is called the von Neumann cardinal assignment; for this definition to make sense, it must be proved that every set has the same cardinality as *some* ordinal; this statement is the well-ordering principle. It is however possible to discuss the relative cardinality of sets without explicitly assigning names to objects.

The classic example used is that of the infinite hotel paradox, also called Hilbert's paradox of the Grand Hotel. Suppose you are an innkeeper at a hotel with an infinite number of rooms. The hotel is full, and then a new guest arrives. It is possible to fit the extra guest in by asking the guest who was in room 1 to move to room 2, the guest in room 2 to move to room 3, and so on, leaving room 1 vacant. We can explicitly write a segment of this mapping:

$1 \rightarrow 2$

$2 \rightarrow 3$

$3 \rightarrow 4$

...

$n \rightarrow n + 1$

...

In this way we can see that the set $\{1,2,3,...\}$ has the same cardinality as the set $\{2,3,4,...\}$ since a bijection between the first and the second has been shown. This motivates the definition of an infinite set being any set which has a proper subset of the same cardinality; in this case $\{2,3,4,...\}$ is a proper subset of $\{1,2,3,...\}$.

When considering these large objects, we might also want to see if the notion of counting order coincides with that of cardinal defined above for these infinite sets. It happens that it doesn't; by considering the above example we can see that if some object "one greater than infinity" exists, then it must have the same cardinality as the infinite set we started out with. It is possible to use a different formal notion for number, called ordinals, based on the ideas of counting and considering each number in turn, and we discover that the notions of cardinality and ordinality are divergent once we move out of the finite numbers.

It can be proved that the cardinality of the real numbers is greater than that of the natural numbers just described. This can be visualized using Cantor's diagonal argument; classic questions of cardinality (for instance the continuum hypothesis) are concerned with discovering whether there is some cardinal between some pair of other infinite cardinals. In more recent times mathematicians have been describing the properties of larger and larger cardinals.

Since cardinality is such a common concept in mathematics, a variety of names are in use. Sameness of cardinality is sometimes referred to as **equipotence**, **equipollence**, or **equinumerosity**. It is thus said that two sets with the same cardinality are, respectively, **equipotent**, **equipollent**, or **equinumerous**.

33.3 Formal definition

Formally, assuming the axiom of choice, the cardinality of a set X is the least ordinal α such that there is a bijection between X and α. This definition is known as the von Neumann cardinal assignment. If the axiom of choice is not assumed we need to do something different. The oldest definition of the cardinality of a set X (implicit in Cantor and explicit in Frege and Principia Mathematica) is as the class $[X]$ of all sets that are equinumerous with X. This does not work in ZFC or other related systems of axiomatic set theory because if X is non-empty, this collection is too large to be a set. In fact, for $X \neq \varnothing$ there is an injection from the universe into $[X]$ by mapping a set m to $\{m\} \times X$ and so by the axiom of limitation of size, $[X]$ is a proper class. The definition does work however in type theory and in New Foundations and related systems. However, if we restrict from this class to those equinumerous with X that have the least rank, then it will work (this is a trick due to Dana Scott:[2] it works because the collection of objects with any given rank is a set).

Formally, the order among cardinal numbers is defined as follows: $|X| \leq |Y|$ means that there exists an injective function from X to Y. The Cantor–Bernstein–Schroeder theorem states that if $|X| \leq |Y|$ and $|Y| \leq |X|$ then $|X| = |Y|$. The axiom of choice is equivalent to the statement that given two sets X and Y, either $|X| \leq |Y|$ or $|Y| \leq |X|$.[3][4]

A set X is Dedekind-infinite if there exists a proper subset Y of X with $|X| = |Y|$, and Dedekind-finite if such a subset doesn't exist. The finite cardinals are just the natural numbers, i.e., a set X is finite if and only if $|X| = |n| = n$ for some natural number n. Any other set is infinite. Assuming the axiom of choice, it can be proved that the Dedekind notions correspond to the standard ones. It can also be proved that the cardinal $\aleph_0$ (aleph null or aleph-0, where aleph is the first letter in the Hebrew alphabet, represented $\aleph$) of the set of natural numbers is the smallest infinite cardinal, i.e. that any infinite set has a subset of cardinality $\aleph_0$. The next larger cardinal is denoted by $\aleph_1$ and so on. For every ordinal α there is a cardinal number $\aleph_\alpha$, and this list exhausts all infinite cardinal numbers.

33.4 Cardinal arithmetic

We can define arithmetic operations on cardinal numbers that generalize the ordinary operations for natural numbers. It can be shown that for finite cardinals these operations coincide with the usual operations for natural numbers. Furthermore, these operations share many properties with ordinary arithmetic.

33.4.1 Successor cardinal

For more details on this topic, see Successor cardinal.

If the axiom of choice holds, every cardinal κ has a successor $\kappa^+ > \kappa$, and there are no cardinals between κ and its successor. (Without the axiom of choice, using Hartogs' theorem, it can be shown that, for any cardinal number κ, there is a minimal cardinal κ^+, so that $\kappa^+ \nleq \kappa$.) For finite cardinals, the successor is simply $\kappa + 1$. For infinite cardinals, the successor cardinal differs from the successor ordinal.

33.4.2   Cardinal addition

If X and Y are disjoint, addition is given by the union of X and Y. If the two sets are not already disjoint, then they can be replaced by disjoint sets of the same cardinality, e.g., replace X by $X \times \{0\}$ and Y by $Y \times \{1\}$.

$|X| + |Y| = |X \cup Y|.$

Zero is an additive identity $\kappa + 0 = 0 + \kappa = \kappa$.

Addition is associative $(\kappa + \mu) + \nu = \kappa + (\mu + \nu)$.

Addition is commutative $\kappa + \mu = \mu + \kappa$.

Addition is non-decreasing in both arguments:

$(\kappa \leq \mu) \to ((\kappa + \nu \leq \mu + \nu) \text{ and } (\nu + \kappa \leq \nu + \mu)).$

Assuming the axiom of choice, addition of infinite cardinal numbers is easy. If either κ or μ is infinite, then

$\kappa + \mu = \max\{\kappa, \mu\}.$

Subtraction

Assuming the axiom of choice and, given an infinite cardinal σ and a cardinal μ, there exists a cardinal κ such that $\mu + \kappa = \sigma$ if and only if $\mu \leq \sigma$. It will be unique (and equal to σ) if and only if $\mu < \sigma$.

33.4.3   Cardinal multiplication

The product of cardinals comes from the cartesian product.

$|X| \cdot |Y| = |X \times Y|$

$\kappa \cdot 0 = 0 \cdot \kappa = 0.$

$\kappa \cdot \mu = 0 \to (\kappa = 0 \text{ or } \mu = 0).$

One is a multiplicative identity $\kappa \cdot 1 = 1 \cdot \kappa = \kappa$.

Multiplication is associative $(\kappa \cdot \mu) \cdot \nu = \kappa \cdot (\mu \cdot \nu)$.

Multiplication is commutative $\kappa \cdot \mu = \mu \cdot \kappa$.

Multiplication is non-decreasing in both arguments: $\kappa \leq \mu \to (\kappa \cdot \nu \leq \mu \cdot \nu \text{ and } \nu \cdot \kappa \leq \nu \cdot \mu)$.

Multiplication distributes over addition: $\kappa \cdot (\mu + \nu) = \kappa \cdot \mu + \kappa \cdot \nu$ and $(\mu + \nu) \cdot \kappa = \mu \cdot \kappa + \nu \cdot \kappa$.

Assuming the axiom of choice, multiplication of infinite cardinal numbers is also easy. If either κ or μ is infinite and both are non-zero, then

$\kappa \cdot \mu = \max\{\kappa, \mu\}.$

Division

Assuming the axiom of choice and, given an infinite cardinal π and a non-zero cardinal μ, there exists a cardinal κ such that $\mu \cdot \kappa = \pi$ if and only if $\mu \leq \pi$. It will be unique (and equal to π) if and only if $\mu < \pi$.

33.4.4 Cardinal exponentiation

Exponentiation is given by

$$|X|^{|Y|} = |X^Y|$$

where X^Y is the set of all functions from Y to X.

$\kappa^0 = 1$ (in particular $0^0 = 1$), see empty function.

If $1 \leq \mu$, then $0^\mu = 0$.

$1^\mu = 1$.

$\kappa^1 = \kappa$.

$\kappa^{\mu + \nu} = \kappa^\mu \cdot \kappa^\nu$.

$\kappa^{\mu \cdot \nu} = (\kappa^\mu)^\nu$.

$(\kappa \cdot \mu)^\nu = \kappa^\nu \cdot \mu^\nu$.

Exponentiation is non-decreasing in both arguments:

$(1 \leq \nu$ and $\kappa \leq \mu) \rightarrow (\nu^\kappa \leq \nu^\mu)$ and

$(\kappa \leq \mu) \rightarrow (\kappa^\nu \leq \mu^\nu)$.

Note that $2^{|X|}$ is the cardinality of the power set of the set X and Cantor's diagonal argument shows that $2^{|X|} > |X|$ for any set X. This proves that no largest cardinal exists (because for any cardinal κ, we can always find a larger cardinal 2^κ). In fact, the class of cardinals is a proper class. (This proof fails in some set theories, notably New Foundations.)

All the remaining propositions in this section assume the axiom of choice:

If κ and μ are both finite and greater than 1, and ν is infinite, then $\kappa^\nu = \mu^\nu$.

If κ is infinite and μ is finite and non-zero, then $\kappa^\mu = \kappa$.

If $2 \leq \kappa$ and $1 \leq \mu$ and at least one of them is infinite, then:

Max $(\kappa, 2^\mu) \leq \kappa^\mu \leq$ Max $(2^\kappa, 2^\mu)$.

Using König's theorem, one can prove $\kappa < \kappa^{\mathrm{cf}(\kappa)}$ and $\kappa < \mathrm{cf}(2^\kappa)$ for any infinite cardinal κ, where $\mathrm{cf}(\kappa)$ is the cofinality of κ.

Roots

Assuming the axiom of choice and, given an infinite cardinal κ and a finite cardinal μ greater than 0, the cardinal ν satisfying $\nu^\mu = \kappa$ will be κ.

Logarithms

Assuming the axiom of choice and, given an infinite cardinal κ and a finite cardinal μ greater than 1, there may or may not be a cardinal λ satisfying $\mu^\lambda = \kappa$. However, if such a cardinal exists, it is infinite and less than κ, and any finite cardinality ν greater than 1 will also satisfy $\nu^\lambda = \kappa$.

The logarithm of an infinite cardinal number κ is defined as the least cardinal number μ such that $\kappa \leq 2^\mu$. Logarithms of infinite cardinals are useful in some fields of mathematics, for example in the study of cardinal invariants of topological spaces, though they lack some of the properties that logarithms of positive real numbers possess.[5][6][7]

33.5 The continuum hypothesis

The continuum hypothesis (CH) states that there are no cardinals strictly between $\aleph_0$ and $2^{\aleph_0}$. The latter cardinal number is also often denoted by $\mathfrak{c}$; it is the cardinality of the continuum (the set of real numbers). In this case $2^{\aleph_0} = \aleph_1$. The generalized continuum hypothesis (GCH) states that for every infinite set X, there are no cardinals strictly between $|X|$ and $2^{|X|}$. The continuum hypothesis is independent of the usual axioms of set theory, the Zermelo-Fraenkel axioms together with the axiom of choice (ZFC).

33.6 See also

33.7 Notes

33.8 References

Notes

[1] Dauben 1990, pg. 54

[2] Deiser, Oliver (May 2010). "On the Development of the Notion of a Cardinal Number". *History and Philosophy of Logic* **31** (2): 123–143. doi:10.1080/01445340903545904.

[3] Enderton, Herbert. "Elements of Set Theory", Academic Press Inc., 1977. ISBN 0-12-238440-7

[4] Friedrich M. Hartogs (1915), Felix Klein, Walther von Dyck, David Hilbert, Otto Blumenthal, ed., "Über das Problem der Wohlordnung", *Math. Ann* (Leipzig: B. G. Teubner), Bd. 76 (4): 438–443, ISSN 0025-5831

[5] Robert A. McCoy and Ibula Ntantu, Topological Properties of Spaces of Continuous Functions, Lecture Notes in Mathematics 1315, Springer-Verlag.

[6] Eduard Čech, Topological Spaces, revised by Zdenek Frolík and Miroslav Katetov, John Wiley & Sons, 1966.

[7] D.A. Vladimirov, Boolean Algebras in Analysis, Mathematics and Its Applications, Kluwer Academic Publishers.

Bibliography

- Dauben, Joseph Warren (1990), *Georg Cantor: His Mathematics and Philosophy of the Infinite*, Princeton: Princeton University Press, ISBN 0691-02447-2

- Hahn, Hans, *Infinity*, Part IX, Chapter 2, Volume 3 of *The World of Mathematics*. New York: Simon and Schuster, 1956.

- Halmos, Paul, *Naive set theory*. Princeton, NJ: D. Van Nostrand Company, 1960. Reprinted by Springer-Verlag, New York, 1974. ISBN 0-387-90092-6 (Springer-Verlag edition).

33.9 External links

- Hazewinkel, Michiel, ed. (2001), "Cardinal number", *Encyclopedia of Mathematics*, Springer, ISBN 978-1-55608-010-4

- Weisstein, Eric W., "Cardinal Number", *MathWorld*.

- Cardinality at ProvenMath proofs of the basic theorems on cardinality.

Chapter 34

Second-order logic

In logic and mathematics **second-order logic** is an extension of first-order logic, which itself is an extension of propositional logic.[1] Second-order logic is in turn extended by higher-order logic and type theory.

First-order logic quantifies only variables that range over individuals (elements of the domain of discourse); second-order logic, in addition, also quantifies over relations. For example, the second-order sentence $\forall P \, \forall x (x \in P \lor x \notin P)$ says that for every unary relation (or set) P of individuals and every individual x, either x is in P or it is not (this is the principle of bivalence). Second-order logic also includes quantification over functions, and other variables as explained in the section *Syntax and fragments* below. Both first-order and second-order logic use the idea of a domain of discourse (often called simply the "domain" or the "universe"). The domain is a set of individual elements which can be quantified over.

34.1 Syntax and fragments

The syntax of second-order logic tells which expressions are well formed formulas. In addition to the syntax of first-order logic, second-order logic includes many new **sorts** (sometimes called **types**) of variables. These are:

- A sort of variables that range over sets of individuals. If S is a variable of this sort and t is a first-order term then the expression $t \in S$ (also written $S(t)$, or St to save parentheses) is an atomic formula. Sets of individuals can also be viewed as unary relations on the domain.

- For each natural number k there is a sort of variables that ranges over all k-ary relations on the individuals. If R is such a k-ary relation variable and $t_1,...,$ tk are first-order terms then the expression $R(t_1,...,tk)$ is an atomic formula.

- For each natural number k there is a sort of variables that ranges over all functions taking k elements of the domain and returning a single element of the domain. If f is such a k-ary function variable and $t_1,...,tk$ are first-order terms then the expression $f(t_1,...,tk)$ is a first-order term.

Each of the variables just defined may be universally and/or existentially quantified over, to build up formulas. Thus there are many kinds of quantifiers, two for each sort of variables. A **sentence** in second-order logic, as in first-order logic, is a well-formed formula with no free variables (of any sort).

It's possible to forgo the introduction of function variables in the definition given above (and some authors do this) because an n-ary function variable can be represented by a relation variable of arity $n+1$ and an appropriate formula for the uniqueness of the "result" in the $n+1$ argument of the relation. (Shapiro 2000, p. 63)

Monadic second-order logic (MSOL) is a restriction of second-order logic in which only quantification over unary relations (i.e.: sets) are allowed. Quantification over functions, owing to the equivalence to relations as described above, is thus also not allowed. The second-order logic without these restrictions is sometimes called **full second-order logic** to distinguish it from the monadic version. Monadic second-order logic is particularly used in the context of Courcelle's theorem, an algorithmic meta-theorem in graph theory.

Just as in first-order logic, second-order logic may include non-logical symbols in a particular second-order language. These are restricted, however, in that all terms that they form must be either first-order terms (which can be substituted for a first-order variable) or second-order terms (which can be substituted for a second-order variable of an appropriate sort).

A formula in second-order logic is said to be of first-order (and sometimes denoted Σ^1_0 or Π^1_0) if its quantifiers (which may be of either type) range only over variables of first order, although it may have free variables of second order. A Σ^1_1 (existential second-order) formula is one additionally having some existential quantifiers over second order variables, i.e. $\exists R_0 \ldots \exists R_m \phi$, where ϕ is a first-order formula. The fragment of second order logic consisting only of existential second-order formulas is called **existential second-order logic** and abbreviated as ESO, as Σ^1_1, or even as $\exists$SO. The fragment of Π^1_1 formulas is defined dually, it is called universal second-order logic. More expressive fragments are defined for any $k > 0$ by mutual recursion: Σ^1_{k+1} has the form $\exists R_0 \ldots \exists R_m \phi$, where ϕ is a Π^1_k formula, and similar, Π^1_{k+1} has the form $\forall R_0 \ldots \forall R_m \phi$, where ϕ is a Σ^1_k formula. (See analytical hierarchy for the analogous construction of second-order arithmetic.)

34.2   Semantics

The semantics of second-order logic establish the meaning of each sentence. Unlike first-order logic, which has only one standard semantics, there are two different semantics that are commonly used for second-order logic: **standard semantics** and **Henkin semantics**. In each of these semantics, the interpretations of the first-order quantifiers and the logical connectives are the same as in first-order logic. Only the ranges of quantifiers over second-order variables differ in the two types of semantics (Väänänen 2001).

In standard semantics, also called full semantics, the quantifiers range over *all* sets or functions of the appropriate sort. Thus once the domain of the first-order variables is established, the meaning of the remaining quantifiers is fixed. It is these semantics that give second-order logic its expressive power, and they will be assumed for the remainder of this article.

In Henkin semantics, each sort of second-order variable has a particular domain of its own to range over, which may be a proper subset of all sets or functions of that sort. Leon Henkin (1950) defined these semantics and proved that Gödel's completeness theorem and compactness theorem, which hold for first-order logic, carry over to second-order logic with Henkin semantics. This is because Henkin semantics are almost identical to many-sorted first-order semantics, where additional sorts of variables are added to simulate the new variables of second-order logic. Second-order logic with Henkin semantics is not more expressive than first-order logic. Henkin semantics are commonly used in the study of second-order arithmetic.

Väänänen (2001) argued that the choice between Henkin models and full models for second-order logic is analogous to the choice between ZFC and *V* as a basis for set theory: "As with second-order logic, we cannot really choose whether we axiomatize mathematics using *V* or ZFC. The result is the same in both cases, as ZFC *is* the best attempt so far to use *V* as an axiomatization of mathematics."

34.3   Expressive power

Second-order logic is more expressive than first-order logic. For example, if the domain is the set of all real numbers, one can assert in first-order logic the existence of an additive inverse of each real number by writing $\forall x \, \exists y \, (x + y = 0)$ but one needs second-order logic to assert the least-upper-bound property for sets of real numbers, which states that every bounded, nonempty set of real numbers has a supremum. If the domain is the set of all real numbers, the following second-order sentence (split over two lines) expresses the least upper bound property:

$(\forall A) \, ([(\exists w) \, (w \in A) \land (\exists z)(\forall u)(u \in A \rightarrow u \leq z)]$

$\quad \rightarrow (\exists x)(\forall y)[(\forall w)(w \in A \rightarrow w \leq y) \leftrightarrow (x \leq y)])$

This formula is a direct formalization of "every nonempty, bounded set A has a least upper bound." It can be shown that any ordered field that satisfies this property is isomorphic to the real number field. On the other hand, the set of first-order sentences valid in the reals has arbitrarily large models due to the compactness theorem. Thus the least-upper-bound property cannot be expressed by any set of sentences in first-order logic. (In fact, every real-closed field satisfies the same first-order sentences in the signature $\langle +, \cdot, \leq \rangle$ as the real numbers.)

In second-order logic, it is possible to write formal sentences which say "the domain is finite" or "the domain is of countable cardinality." To say that the domain is finite, use the sentence that says that every surjective function from the domain to itself is injective. To say that the domain has countable cardinality, use the sentence that says that there is a bijection between every two infinite subsets of the domain. It follows from the compactness theorem and the upward Löwenheim–Skolem theorem that it is not possible to characterize finiteness or countability, respectively, in first-order logic.

Certain fragments of second order logic like ESO are also more expressive than first-order logic even though they are strictly less expressive than the full second-order logic. ESO also enjoys translation equivalence with some extensions of first-order logic which allow non-linear ordering of quantifier dependencies, like first-order logic extended with Henkin quantifiers, Hintikka and Sandu's independence-friendly logic, and Väänänen's dependence logic.

34.4 Deductive systems

A deductive system for a logic is a set of inference rules and logical axioms that determine which sequences of formulas constitute valid proofs. Several deductive systems can be used for second-order logic, although none can be complete for the standard semantics (see below). Each of these systems is sound, which means any sentence they can be used to prove is logically valid in the appropriate semantics.

The weakest deductive system that can be used consists of a standard deductive system for first-order logic (such as natural deduction) augmented with substitution rules for second-order terms.[2] This deductive system is commonly used in the study of second-order arithmetic.

The deductive systems considered by Shapiro (1991) and Henkin (1950) add to the augmented first-order deductive scheme both comprehension axioms and choice axioms. These axioms are sound for standard second-order semantics. They are sound for Henkin semantics if only Henkin models that satisfy the comprehension and choice axioms are considered.[3]

34.5 Non-reducibility to first-order logic

One might attempt to reduce the second-order theory of the real numbers, with full second-order semantics, to the first-order theory in the following way. First expand the domain from the set of all real numbers to a two-sorted domain, with the second sort containing all *sets of* real numbers. Add a new binary predicate to the language: the membership relation. Then sentences that were second-order become first-order, with the formerly second-order quantifiers ranging over the second sort instead. This reduction can be attempted in a one-sorted theory by adding unary predicates that tell whether an element is a number or a set, and taking the domain to be the union of the set of real numbers and the power set of the real numbers.

But notice that the domain was asserted to include *all* sets of real numbers. That requirement cannot be reduced to a first-order sentence, as the Löwenheim–Skolem theorem shows. That theorem implies that there is some countably infinite subset of the real numbers, whose members we will call *internal numbers*, and some countably infinite collection of sets of internal numbers, whose members we will call "internal sets", such that the domain consisting of internal numbers and internal sets satisfies exactly the same first-order sentences as are satisfied by the domain of real numbers and sets of real numbers. In particular, it satisfies a sort of least-upper-bound axiom that says, in effect:

> Every nonempty *internal* set that has an *internal* upper bound has a least *internal* upper bound.

Countability of the set of all internal numbers (in conjunction with the fact that those form a densely ordered set) implies that that set does not satisfy the full least-upper-bound axiom. Countability of the set of all *internal* sets implies that it is not the set of *all* subsets of the set of all *internal* numbers (since Cantor's theorem implies that the set of all subsets of a countably infinite set is an uncountably infinite set). This construction is closely related to Skolem's paradox.

Thus the first-order theory of real numbers and sets of real numbers has many models, some of which are countable. The second-order theory of the real numbers has only one model, however. This follows from the classical theorem that there is only one Archimedean complete ordered field, along with the fact that all the axioms of an Archimedean complete ordered field are expressible in second-order logic. This shows that the second-order theory of the real numbers cannot be reduced to a first-order theory, in the sense that the second-order theory of the real numbers has only one model but the corresponding first-order theory has many models.

There are more extreme examples showing that second-order logic with standard semantics is more expressive than first-order logic. There is a finite second-order theory whose only model is the real numbers if the continuum hypothesis holds and which has no model if the continuum hypothesis does not hold (cf. Shapiro 2000, p. 105). This theory consists of a finite theory characterizing the real numbers as a complete Archimedean ordered field plus an axiom saying that the domain is of the first uncountable cardinality. This example illustrates that the question of whether a sentence in second-order logic is consistent is extremely subtle.

Additional limitations of second order logic are described in the next section.

34.6   Metalogical results

It is a corollary of Gödel's incompleteness theorem that there is no deductive system (that is, no notion of *provability*) for second-order formulas that simultaneously satisfies these three desired attributes:[4]

- (Soundness) Every provable second-order sentence is universally valid, i.e., true in all domains under standard semantics.

- (Completeness) Every universally valid second-order formula, under standard semantics, is provable.

- (Effectiveness) There is a proof-checking algorithm that can correctly decide whether a given sequence of symbols is a proof or not.

This corollary is sometimes expressed by saying that second-order logic does not admit a complete proof theory. In this respect second-order logic with standard semantics differs from first-order logic; Quine (1970, pp. 90–91) pointed to the lack of a complete proof system as a reason for thinking of second-order logic as not *logic*, properly speaking.

As mentioned above, Henkin proved that the standard deductive system for first-order logic is sound, complete, and effective for second-order logic with Henkin semantics, and the deductive system with comprehension and choice principles is sound, complete, and effective for Henkin semantics using only models that satisfy these principles.

The compactness theorem and the Löwenheim-Skolem theorem do not hold for full models of second-order logic. They do hold however for Henkin models. (Väänänen 2001)

34.7   History and disputed value

Predicate logic was primarily introduced to the mathematical community by C. S. Peirce, who coined the term *second-order logic* and whose notation is most similar to the modern form (Putnam 1982). However, today most students of logic are more familiar with the works of Frege, who actually published his work several years prior to Peirce but whose works remained in obscurity until Bertrand Russell and Alfred North Whitehead made them famous. Frege used different variables to distinguish quantification over objects from quantification over properties and sets; but he did not see himself as doing two different kinds of logic. After the discovery of Russell's paradox it was realized that something was wrong with his system. Eventually logicians found that restricting Frege's logic in various ways—to what is now called first-order logic—eliminated this problem: sets and properties cannot be quantified over in first-order-logic alone. The now-standard hierarchy of orders of logics dates from this time.

It was found that set theory could be formulated as an axiomatized system within the apparatus of first-order logic (at the cost of several kinds of completeness, but nothing so bad as Russell's paradox), and this was done (see Zermelo-Fraenkel set theory), as sets are vital for mathematics. Arithmetic, mereology, and a variety of other powerful logical theories could be formulated axiomatically without appeal to any more logical apparatus than first-order quantification, and this, along with Gödel and Skolem's adherence to first-order logic, led to a general decline in work in second (or any higher) order logic.

This rejection was actively advanced by some logicians, most notably W. V. Quine. Quine advanced the view that in predicate-language sentences like *Fx* the "*x*" is to be thought of as a variable or name denoting an object and hence can be quantified over, as in "For all things, it is the case that . . ." but the "*F*" is to be thought of as an *abbreviation* for an incomplete sentence, not the name of an object (not even of an abstract object like a property). For example, it might mean " . . . is a dog." But it makes no sense to think we can quantify over something like this. (Such a position is quite consistent with Frege's own arguments on the concept-object distinction). So to use a predicate as a variable is to have it occupy the place of a name which only individual variables should occupy. This reasoning has been rejected by Boolos.

In recent years second-order logic has made something of a recovery, buoyed by George Boolos' interpretation of second-order quantification as plural quantification over the same domain of objects as first-order quantification (Boolos 1984). Boolos furthermore points to the claimed nonfirstorderizability of sentences such as "Some critics admire only each other" and "Some of Fianchetto's men went into the warehouse unaccompanied by anyone else" which he argues can only be expressed by the full force of second-order quantification. However, generalized quantification and partially ordered, or branching, quantification may suffice to express a certain class of purportedly nonfirstorderizable sentences as well and it does not appeal to second-order quantification.

34.8 Relation to computational complexity

Main article: SO (complexity)

The expressive power of various forms of second-order logic on finite structures is intimately tied to computational complexity theory. The field of descriptive complexity studies which computational complexity classes can be characterized by the power of the logic needed to express languages (sets of finite strings) in them. A string $w = w_1 \cdots wn$ in a finite alphabet A can be represented by a finite structure with domain $D = \{1,...,n\}$, unary predicates Pa for each $a \in A$, satisfied by those indices i such that $wi = a$, and additional predicates which serve to uniquely identify which index is which (typically, one takes the graph of the successor function on D or the order relation $<$, possibly with other arithmetic predicates). Conversely, the table of any finite structure can be encoded by a finite string.

With this identification, we have the following characterizations of variants of second-order logic over finite structures:

- REG (the set of regular languages) is definable by monadic, second-order formulas (Büchi's theorem, 1960)

- NP is the set of languages definable by existential, second-order formulas (Fagin's theorem, 1974).

- co-NP is the set of languages definable by universal, second-order formulas.

- PH is the set of languages definable by second-order formulas.

- PSPACE is the set of languages definable by second-order formulas with an added transitive closure operator.

- EXPTIME is the set of languages definable by second-order formulas with an added least fixed point operator.

Relationships among these classes directly impact the relative expressiveness of the logics over finite structures; for example, if **PH = PSPACE**, then adding a transitive closure operator to second-order logic would not make it any more expressive over finite structures.

34.9   See also

- First-order logic

- Higher-order logic

- Hanf number

- Löwenheim number

- Second-order propositional logic

34.10   Notes

[1] Shapiro (1991) and Hinman (2005) give complete introductions to the subject, with full definitions.

[2] Such a system is used without comment by Hinman (2005).

[3] These are the models originally studied by Henkin (1950).

[4] The proof of this corollary is that a sound, complete, and effective deduction system for standard semantics could be used to produce a recursively enumerable completion of Peano arithmetic, which Gödel's theorem shows cannot exist.

34.11   References

- Andrews, Peter (2002). *An Introduction to Mathematical Logic and Type Theory: To Truth Through Proof* (2nd ed.). Kluwer Academic Publishers.

- Boolos, George (1984). "To Be Is To Be a Value of a Variable (or to Be Some Values of Some Variables)". *Journal of Philosophy* **81** (8): 430–50. doi:10.2307/2026308. JSTOR 2026308.. Reprinted in Boolos, *Logic, Logic and Logic*, 1998.

- Henkin, L. (1950). "Completeness in the theory of types". *Journal of Symbolic Logic* **15**(2): 81–91.doi:10.2307/2 JSTOR 2266967.

- Hinman, P. (2005). *Fundamentals of Mathematical Logic*. A K Peters. ISBN 1-56881-262-0.

- Putnam, Hilary (1982). "Peirce the Logician". *Historia Mathematica* **9**(3): 290–301.doi:10.1016/0315-0860 9.. Reprinted in Putnam, Hilary (1990), *Realism with a Human Face*, Harvard University Press, pp. 252–260.

- W.V. Quine (1970). *Philosophy of Logic*. Prentice-Hall.

- Rossberg, M. (2004). "First-Order Logic, Second-Order Logic, and Completeness" (PDF). In V. Hendricks *et al.*, eds. *First-order logic revisited*. Berlin: Logos-Verlag.

- Shapiro, S. (2000). *Foundations without Foundationalism: A Case for Second-order Logic*. Oxford University Press. ISBN 0-19-825029-0.

- Väänänen, J. (2001). "Second-Order Logic and Foundations of Mathematics". *Bulletin of Symbolic Logic* **7** (4): 504–520. doi:10.2307/2687796. JSTOR 2687796.

34.12   Further reading

- Grädel, Erich; Kolaitis, Phokion G.; Libkin, Leonid; Maarten, Marx; Spencer, Joel; Vardi, Moshe Y.; Venema, Yde; Weinstein, Scott (2007). *Finite model theory and its applications*. Texts in Theoretical Computer Science. An EATCS Series. Berlin: Springer-Verlag. ISBN 978-3-540-00428-8. Zbl 1133.03001.

Chapter 35

Skolem's paradox

In mathematical logic and philosophy, **Skolem's paradox** is a seeming contradiction that arises from the downward Löwenheim–Skolem theorem. Thoralf Skolem (1922) was the first to discuss the seemingly contradictory aspects of the theorem, and to discover the relativity of set-theoretic notions now known as non-absoluteness. Although it is not an actual antinomy like Russell's paradox, the result is typically called a paradox, and was described as a "paradoxical state of affairs" by Skolem (1922: p. 295).

Skolem's paradox is that every countable axiomatisation of set theory in first-order logic, if it is consistent, has a model that is countable. This appears contradictory because it is possible to prove, from those same axioms, a sentence that intuitively says (or that precisely says in the standard model of the theory) that there exist sets that are not countable. Thus the seeming contradiction is that a model that is itself countable, and which therefore contains only countable sets, satisfies the first order sentence that intuitively states "there are uncountable sets".

A mathematical explanation of the paradox, showing that it is not a contradiction in mathematics, was given by Skolem (1922). Skolem's work was harshly received by Ernst Zermelo, who argued against the limitations of first-order logic, but the result quickly came to be accepted by the mathematical community.

The philosophical implications of Skolem's paradox have received much study. One line of inquiry questions whether it is accurate to claim that any first-order sentence actually states "there are uncountable sets". This line of thought can be extended to question whether any set is uncountable in an absolute sense. More recently, the paper "Models and Reality" by Hilary Putnam, and responses to it, led to renewed interest in the philosophical aspects of Skolem's result.

35.1 Background

One of the earliest results in set theory, published by Georg Cantor in 1874, was the existence of uncountable sets, such as the powerset of the natural numbers, the set of real numbers, and the Cantor set. An infinite set X is countable if there is a function that gives a one-to-one correspondence between X and the natural numbers, and is uncountable if there is no such correspondence function. When Zermelo proposed his axioms for set theory in 1908, he proved Cantor's theorem from them to demonstrate their strength.

Löwenheim (1915) and Skolem (1920, 1923) proved the Löwenheim–Skolem theorem. The downward form of this theorem shows that if a countable first-order axiomatisation is satisfied by any infinite structure, then the same axioms are satisfied by some countable structure. In particular, this implies that if the first order versions of Zermelo's axioms of set theory are satisfiable, they are satisfiable in some countable model. The same is true of any consistent first order axiomatisation of set theory.

35.2   The paradoxical result and its mathematical implications

Skolem (1922) pointed out the seeming contradiction between the Löwenheim–Skolem theorem on the one hand, which implies that there is a countable model of Zermelo's axioms, and Cantor's theorem on the other hand, which states that uncountable sets exist, and which is provable from Zermelo's axioms. "So far as I know," Skolem writes, "no one has called attention to this peculiar and apparently paradoxical state of affairs. By virtue of the axioms we can prove the existence of higher cardinalities... How can it be, then, that the entire domain B [a countable model of Zermelo's axioms] can already be enumerated by means of the finite positive integers?" (Skolem 1922, p. 295, translation by Bauer-Mengelberg)

More specifically, let B be a countable model of Zermelo's axioms. Then there is some set u in B such that B satisfies the first-order formula saying that u is uncountable. For example, u could be taken as the set of real numbers in B. Now, because B is countable, there are only countably many elements c such that $c \in u$ according to B, because there are only countably many elements c in B to begin with. Thus it appears that u should be countable. This is Skolem's paradox.

Skolem went on to explain why there was no contradiction. In the context of a specific model of set theory, the term "set" does not refer to an arbitrary set, but only to a set that is actually included in the model. The definition of countability requires that a certain one-to-one correspondence, which is itself a set, must exist. Thus it is possible to recognize that a particular set u is countable, but not countable in a particular model of set theory, because there is no set in the model that gives a one-to-one correspondence between u and the natural numbers in that model.

Skolem used the term "relative" to describe this state of affairs, where the same set is included in two models of set theory, is countable in one model, and is not countable in the other model. He described this as the "most important" result in his paper. Contemporary set theorists describe concepts that do not depend on the choice of a transitive model as absolute. From their point of view, Skolem's paradox simply shows that countability is not an absolute property in first order logic. (Kunen 1980 p. 141; Enderton 2001 p. 152; Burgess 1977 p. 406).

Skolem described his work as a critique of (first-order) set theory, intended to illustrate its weakness as a foundational system:

> "I believed that it was so clear that axiomatization in terms of sets was not a satisfactory ultimate foundation of mathematics that mathematicians would, for the most part, not be very much concerned with it. But in recent times I have seen to my surprise that so many mathematicians think that these axioms of set theory provide the ideal foundation for mathematics; therefore it seemed to me that the time had come for a critique." (Ebbinghaus and van Dalen, 2000, p. 147)

35.3   Reception by the mathematical community

A central goal of early research into set theory was to find a first order axiomatisation for set theory which was categorical, meaning that the axioms would have exactly one model, consisting of all sets. Skolem's result showed this is not possible, creating doubts about the use of set theory as a foundation of mathematics. It took some time for the theory of first-order logic to be developed enough for mathematicians to understand the cause of Skolem's result; no resolution of the paradox was widely accepted during the 1920s. Fraenkel (1928) still described the result as an antinomy:

> "Neither have the books yet been closed on the antinomy, nor has agreement on its significance and possible solution yet been reached." (van Dalen and Ebbinghaus, 2000, p. 147).

In 1925, von Neumann presented a novel axiomatization of set theory, which developed into NBG set theory. Very much aware of Skolem's 1922 paper, von Neumann investigated countable models of his axioms in detail. In his concluding remarks, Von Neumann comments that there is no categorical axiomatization of set theory, or any other theory with an infinite model. Speaking of the impact of Skolem's paradox, he wrote,

> "At present we can do no more than note that we have one more reason here to entertain reservations about set theory and that for the time being no way of rehabilitating this theory is known."(Ebbinghaus and van Dalen, 2000, p. 148)

Zermelo at first considered the Skolem paradox a hoax (van Dalen and Ebbinghaus, 2000, p. 148 ff.), and spoke against it starting in 1929. Skolem's result applies only to what is now called first-order logic, but Zermelo argued against the finitary metamathematics that underlie first-order logic (Kanamori 2004, p. 519 ff.). Zermelo argued that his axioms should instead be studied in second-order logic, a setting in which Skolem's result does not apply. Zermelo published a second-order axiomatization in 1930 and proved several categoricity results in that context. Zermelo's further work on the foundations of set theory after Skolem's paper led to his discovery of the cumulative hierarchy and formalization of infinitary logic (van Dalen and Ebbinghaus, 2000, note 11).

Fraenkel *et al.* (1973, pp. 303–304) explain why Skolem's result was so surprising to set theorists in the 1920s. Gödel's completeness theorem and the compactness theorem were not proved until 1929. These theorems illuminated the way that first-order logic behaves and established its finitary nature, although Gödel's original proof of the completeness theorem was complicated. Leon Henkin's alternative proof of the completeness theorem, which is now a standard technique for constructing countable models of a consistent first-order theory, was not presented until 1947. Thus, in 1922, the particular properties of first-order logic that permit Skolem's paradox to go through were not yet understood. It is now known that Skolem's paradox is unique to first-order logic; if set theory is studied using higher-order logic with full semantics then it does not have any countable models, due to the semantics being used.

35.4 Current mathematical opinion

Current mathematical logicians do not view Skolem's paradox as any sort of fatal flaw in set theory. Kleene (1967, p. 324) describes the result as "not a paradox in the sense of outright contradiction, but rather a kind of anomaly". After surveying Skolem's argument that the result is not contradictory, Kleene concludes "there is no absolute notion of countability." Hunter (1971, p. 208) describes the contradiction as "hardly even a paradox". Fraenkel *et al.* (1973, p. 304) explain that contemporary mathematicians are no more bothered by the lack of categoricity of first-order theories than they are bothered by the conclusion of Gödel's incompleteness theorem that no consistent, effective, and sufficiently strong set of first-order axioms is complete.

Countable models of ZF have become common tools in the study of set theory. Forcing, for example, is often explained in terms of countable models. The fact that these countable models of ZF still satisfy the theorem that there are uncountable sets is not considered a pathology; van Heijenoort (1967) describes it as "a novel and unexpected feature of formal systems." (van Heijenoort 1967, p. 290)

Although mathematicians no longer consider Skolem's result paradoxical, the result is often discussed by philosophers. In the setting of philosophy, a merely mathematical resolution of the paradox may be less than satisfactory.

35.5 References

- Barwise, Jon (1977), "An introduction to first-order logic", in Barwise, Jon, ed. (1982), *Handbook of Mathematical Logic*, Studies in Logic and the Foundations of Mathematics, Amsterdam: North-Holland, ISBN 978-0-444-86388-1

- Timothy Bays (2000). *Reflections on Skolem's Paradox* (PDF) (Ph.D. thesis). UCLA Philosophy Department.

- Crossley, J.N.; Ash, C.J.; Brickhill, C.J.; Stillwell, J.C.; Williams, N.H. (1972), *What is mathematical logic?*, London-Oxford-New York: Oxford University Press, ISBN 0-19-888087-1, Zbl 0251.02001

- Dirk Van Dalen; Heinz-Dieter Ebbinghaus (Jun 2000). "Zermelo and the Skolem Paradox". *The Bulletin of Symbolic Logic* **6** (2): 145—161.

- Dragalin, A.G. (2001), "S/s085750", in Hazewinkel, Michiel, *Encyclopedia of Mathematics*, Springer, ISBN 978-1-55608-010-4

- Abraham Fraenkel, Yehoshua Bar-Hillel, Azriel Levy, Dirk van Dalen (1973), *Foundations of Set Theory*, North-Holland.

- Henkin, L. (1950), "Completeness in the theory of types", *The Journal of Symbolic Logic* **15**(2): 81–91, doi:10. JSTOR 2266967.

- Kanamori, Akihiro (2004), "Zermelo and set theory", *The Bulletin of Symbolic Logic* **10**(4): 487–553, doi:10. ISSN 1079-8986, MR 2136635

- Stephen Cole Kleene, (1952, 1971 with emendations, 1991 10th printing), *Introduction to Metamathematics*, North-Holland Publishing Company, Amsterdam NY. ISBN 0-444-10088-1. cf pages 420-432: § 75. Axiom systems, Skolem's paradox, the natural number sequence.

- Stephen Cole Kleene, (1967). *Mathematical Logic.*

- Kunen, Kenneth (1980), *Set Theory: An Introduction to Independence Proofs*, Amsterdam: North-Holland, ISBN 978-0-444-85401-8

- Löwenheim, Leopold (1915), "Über Möglichkeiten im Relativkalkül" (PDF), *Mathematische Annalen* **76** (4): 447–470, doi:10.1007/BF01458217, ISSN 0025-5831

- Moore, A.W., "Set Theory, Skolem's Paradox and the Tractatus", *Analysis* **1985**: 45, doi:10.2307/3327397.

- Hilary Putnam (Sep 1980). "Models and Reality" (PDF). *The Journal of Symbolic Logic* **45** (3): 464—482.

- Rautenberg, Wolfgang (2010), *A Concise Introduction to Mathematical Logic* (3rd ed.), New York: Springer Science+Business Media, doi:10.1007/978-1-4419-1221-3, ISBN 978-1-4419-1220-6

- Skolem, Thoralf (1922). "Axiomatized set theory". Reprinted in *From Frege to Gödel*, van Heijenoort, 1967, in English translation by Stefan Bauer-Mengelberg, pp. 291–301.

35.6   External links

- Vaughan Pratt's celebration of his academic ancestor Skolem's 120th birthday

- Extract from Moore's discussion of the paradox (broken link)

Chapter 36

Lindström's theorem

In mathematical logic, **Lindström's theorem** (named after Swedish logician Per Lindström, who published it in 1969) states that first-order logic is the *strongest logic* [1] (satisfying certain conditions, e.g. closure under classical negation) having both the (countable) compactness property and the (downward) Löwenheim–Skolem property.[2]

Lindström's theorem is perhaps the best known result of what later became known as abstract model theory,[3] the basic notion of which is an abstract logic;[4] the more general notion of an institution was later introduced, which advances from a set-theoretical notion of model to a category theoretical one.[5] Lindström had previously obtained a similar result in studying first-order logics extended with Lindström quantifiers.[6]

Lindström's theorem has been extended to various other systems of logic in particular modal logics by Johan van Benthem and Sebastian Enqvist.

36.1 Notes

[1] In the sense of Heinz-Dieter Ebbinghaus *Extended logics: the general framework* in K. J. Barwise and S. Feferman, editors, *Model-theoretic logics*, 1985 ISBN 0-387-90936-2 page 43

[2] *A companion to philosophical logic* by Dale Jacquette 2005 ISBN 1-4051-4575-7 page 329

[3] Chen Chung Chang; H. Jerome Keisler (1990). *Model theory*. Elsevier. p. 127. ISBN 978-0-444-88054-3.

[4] Jean-Yves Béziau (2005). *Logica universalis: towards a general theory of logic*. Birkhäuser. p. 20. ISBN 978-3-7643-7259-0.

[5] Dov M. Gabbay, ed. (1994). *What is a logical system?*. Clarendon Press. p. 380. ISBN 978-0-19-853859-2.

[6] Jouko Väänänen, Lindström's Theorem

36.2 References

• Per Lindström, "On Extensions of Elementary Logic", *Theoria* 35, 1969, 1–11. doi:10.1111/j.1755-2567.1969.t

• Johan van Benthem, "A New Modal Lindström Theorem", *Logica Universalis* 1, 2007, 125–128. doi:10.1007/s11787-006-0006-3

• Ebbinghaus, Heinz-Dieter; Flum, Jörg; Thomas, Wolfgang (1994), *Mathematical Logic* (2nd ed.), Berlin, New York: Springer-Verlag, ISBN 978-0-387-94258-2

• Sebastian Enqvist, "A General Lindström Theorem for Some Normal Modal Logics", *Logica Universalis* 7, 2013, 233–264. doi:10.1007/s11787-013-0078-9

- Monk, J. Donald (1976), *Mathematical Logic*, Graduate Texts in Mathematics, Berlin, New York: Springer-Verlag, ISBN 978-0-387-90170-1

- Shawn Hedman, *A first course in logic: an introduction to model theory, proof theory, computability, and complexity*, Oxford University Press, 2004, ISBN 0-19-852981-3, section 9.4

Chapter 37

Compactness theorem

In mathematical logic, the **compactness theorem** states that a set of first-order sentences has a model if and only if every finite subset of it has a model. This theorem is an important tool in model theory, as it provides a useful method for constructing models of any set of sentences that is finitely consistent.

The compactness theorem for the propositional calculus is a consequence of Tychonoff's theorem (which says that the product of compact spaces is compact) applied to compact Stone spaces;[1] hence, the theorem's name. Likewise, it is analogous to the finite intersection property characterization of compactness in topological spaces: a collection of closed sets in a compact space has a non-empty intersection if every finite subcollection has a non-empty intersection.

The compactness theorem is one of the two key properties, along with the downward Löwenheim–Skolem theorem, that is used in Lindström's theorem to characterize first-order logic. Although there are some generalizations of the compactness theorem to non-first-order logics, the compactness theorem itself does not hold in them.

37.1 History

Kurt Gödel proved the countable compactness theorem in 1930. Anatoly Maltsev proved the uncountable case in 1936.[2] [3]

37.2 Applications

The compactness theorem has many applications in model theory; a few typical results are sketched here.

The compactness theorem implies Robinson's principle: If a first-order sentence holds in every field of characteristic zero, then there exists a constant p such that the sentence holds for every field of characteristic larger than p. This can be seen as follows: suppose φ is a sentence that holds in every field of characteristic zero. Then its negation $\neg\varphi$, together with the field axioms and the infinite sequence of sentences $1+1 \neq 0$, $1+1+1 \neq 0$, …, is not satisfiable (because there is no field of characteristic 0 in which $\neg\varphi$ holds, and the infinite sequence of sentences ensures any model would be a field of characteristic 0). Therefore, there is a finite subset A of these sentences that is not satisfiable. We can assume that A contains $\neg\varphi$, the field axioms, and, for some k, the first k sentences of the form $1+1+...+1 \neq 0$ (because adding more sentences doesn't change unsatisfiability). Let B contain all the sentences of A except $\neg\varphi$. Then any model of B is a field of characteristic greater than k, and $\neg\varphi$ together with B is not satisfiable. This means that φ must hold in every model of B, which means precisely that φ holds in every field of characteristic greater than k.

A second application of the compactness theorem shows that any theory that has arbitrarily large finite models, or a single infinite model, has models of arbitrary large cardinality (this is the Upward Löwenheim–Skolem theorem). So, for instance, there are nonstandard models of Peano arithmetic with uncountably many 'natural numbers'. To achieve this, let T be the initial theory and let κ be any cardinal number. Add to the language of T one constant symbol for every element

269

of κ. Then add to T a collection of sentences that say that the objects denoted by any two distinct constant symbols from the new collection are distinct (this is a collection of κ^2 sentences). Since every *finite* subset of this new theory is satisfiable by a sufficiently large finite model of T, or by any infinite model, the entire extended theory is satisfiable. But any model of the extended theory has cardinality at least κ

A third application of the compactness theorem is the construction of nonstandard models of the real numbers, that is, consistent extensions of the theory of the real numbers that contain "infinitesimal" numbers. To see this, let Σ be a first-order axiomatization of the theory of the real numbers. Consider the theory obtained by adding a new constant symbol ε to the language and adjoining to Σ the axiom $\varepsilon > 0$ and the axioms $\varepsilon < 1/n$ for all positive integers n. Clearly, the standard real numbers $\mathbf{R}$ are a model for every finite subset of these axioms, because the real numbers satisfy everything in Σ and, by suitable choice of ε, can be made to satisfy any finite subset of the axioms about ε. By the compactness theorem, there is a model $*\mathbf{R}$ that satisfies Σ and also contains an infinitesimal element ε. A similar argument, adjoining axioms $\omega > 0$, $\omega > 1$, etc., shows that the existence of infinitely large integers cannot be ruled out by any axiomatization Σ of the reals.[4]

37.3 Proofs

One can prove the compactness theorem using Gödel's completeness theorem, which establishes that a set of sentences is satisfiable if and only if no contradiction can be proven from it. Since proofs are always finite and therefore involve only finitely many of the given sentences, the compactness theorem follows. In fact, the compactness theorem is equivalent to Gödel's completeness theorem, and both are equivalent to the Boolean prime ideal theorem, a weak form of the axiom of choice.[5]

Gödel originally proved the compactness theorem in just this way, but later some "purely semantic" proofs of the compactness theorem were found, i.e., proofs that refer to *truth* but not to *provability*. One of those proofs relies on ultraproducts hinging on the axiom of choice as follows:

Proof: Fix a first-order language L, and let Σ be a collection of L-sentences such that every finite subcollection of L-sentences, $i \subseteq \Sigma$ of it has a model $\mathcal{M}_i$. Also let $\prod_{i \subseteq \Sigma} \mathcal{M}_i$ be the direct product of the structures and I be the collection of finite subsets of Σ. For each i in I let $\mathrm{A}i := \{\, j \in I : j \supseteq i\}$. The family of all of these sets $\mathrm{A}i$ generates a proper filter, so there is an ultrafilter U containing all sets of the form $\mathrm{A}i$.

Now for any formula φ in Σ we have:

- the set $\mathrm{A}\{_\varphi\}$ is in U

- whenever $j \in \mathrm{A}\{_\varphi\}$, then $\varphi \in j$, hence φ holds in $\mathcal{M}_j$

- the set of all j with the property that φ holds in $\mathcal{M}_j$ is a superset of $\mathrm{A}\{_\varphi\}$, hence also in U

Using Łoś's theorem we see that φ holds in the ultraproduct $\prod_{i \subseteq \Sigma} \mathcal{M}_i/U$. So this ultraproduct satisfies all formulas in Σ.

37.4 See also

- List of Boolean algebra topics

- Löwenheim-Skolem theorem

- Herbrand's theorem

- Barwise compactness theorem

37.5 Notes

[1] See Truss (1997).

[2] Vaught, Robert L.: Alfred Tarski's work in model theory. J. Symbolic Logic 51 (1986), no. 4, 869–882

[3] Robinson, A.: Non-standard analysis. North-Holland Publishing Co., Amsterdam 1966. page 48.

[4] Goldblatt, Robert (1998). *Lectures on the Hyperreals.* New York: Springer. pp. 10–11. ISBN 0-387-98464-X.

[5] See Hodges (1993).

37.6 References

- Boolos, George; Jeffrey, Richard; Burgess, John (2004). Computability and Logic (fourth ed.). "Cambridge University Press.

- Chang, C.C.; Keisler, H. Jerome (1989). *Model Theory* (third ed.). Elsevier. ISBN 0-7204-0692-7.

- Dawson, John W. junior (1993). "The compactness of first-order logic: From Gödel to Lindström". *History and Philosophy of Logic* **14**: 15–37. doi:10.1080/01445349308837208.

- Hodges, Wilfrid (1993). *Model theory.* Cambridge University Press. ISBN 0-521-30442-3.

- Marker, David (2002). *Model Theory: An Introduction.* Graduate Texts in Mathematics 217. Springer. ISBN 0-387-98760-6.

- Truss, John K. (1997). *Foundations of Mathematical Analysis.* Oxford University Press. ISBN 0-19-853375-6.

37.7 Further reading

- Hummel, Christoph (1997). *Gromov's compactness theorem for pseudo-holomorphic curves.* Basel, Switzerland: Birkhäuser. ISBN 3-7643-5735-5.

Chapter 38

Ultraproduct

The **ultraproduct** is a mathematical construction that appears mainly in abstract algebra and in model theory, a branch of mathematical logic. An ultraproduct is a quotient of the direct product of a family of structures. All factors need to have the same signature. The **ultrapower** is the special case of this construction in which all factors are equal.

For example, ultrapowers can be used to construct new fields from given ones. The hyperreal numbers, an ultrapower of the real numbers, are a special case of this.

Some striking applications of ultraproducts include very elegant proofs of the compactness theorem and the completeness theorem, Keisler's ultrapower theorem, which gives an algebraic characterization of the semantic notion of elementary equivalence, and the Robinson-Zakon presentation of the use of superstructures and their monomorphisms to construct nonstandard models of analysis, leading to the growth of the area of non-standard analysis, which was pioneered (as an application of the compactness theorem) by Abraham Robinson.

38.1 Definition

The general method for getting ultraproducts uses an index set I, a structure M_i for each element i of I (all of the same signature), and an ultrafilter U on I. The usual choice is for I to be infinite and U to contain all cofinite subsets of I. Otherwise the ultrafilter is principal, and the ultraproduct is isomorphic to one of the factors.

Algebraic operations on the Cartesian product

$$\prod_{i \in I} M_i$$

are defined in the usual way (for example, for a binary function +, $(a + b)\, i = a i + b i$), and an equivalence relation is defined by $a \sim b$ if

$$\{i \in I : a_i = b_i\} \in U,$$

and the **ultraproduct** is the quotient set with respect to ~. The ultraproduct is therefore sometimes denoted by

$$\prod_{i \in I} M_i / U.$$

One may define a finitely additive measure m on the index set I by saying $m(A) = 1$ if $A \in U$ and $= 0$ otherwise. Then two members of the Cartesian product are equivalent precisely if they are equal almost everywhere on the index set. The ultraproduct is the set of equivalence classes thus generated.

Other relations can be extended the same way:

$$R([a^1], \ldots, [a^n]) \iff \{i \in I : R^{M_i}(a_i^1, \ldots, a_i^n)\} \in U,$$

where $[a]$ denotes the equivalence class of a with respect to ~.

In particular, if every Mi is an ordered field, then so is the ultraproduct.

An **ultrapower** is an ultraproduct for which all the factors Mi are equal:

$$M^\kappa / U = \prod_{\alpha < \kappa} M / U.$$

More generally, the construction above can be carried out whenever U is a filter on I; the resulting model $\prod_{i \in I} M_i / U$ is then called a **reduced product**.

38.2 Examples

The hyperreal numbers are the ultraproduct of one copy of the real numbers for every natural number, with regard to an ultrafilter over the natural numbers containing all cofinite sets. Their order is the extension of the order of the real numbers. For example, the sequence ω given by $\omega i = i$ defines an equivalence class representing a hyperreal number that is greater than any real number.

Analogously, one can define nonstandard integers, nonstandard complex numbers, etc., by taking the ultraproduct of copies of the corresponding structures.

As an example of the carrying over of relations into the ultraproduct, consider the sequence ψ defined by $\psi i = 2i$. Because $\psi i > \omega i = i$ for all i, it follows that the equivalence class of $\psi i = 2i$ is greater than the equivalence class of $\omega i = i$, so that it can be interpreted as an infinite number which is greater than the one originally constructed. However, let $\chi i = i$ for i not equal to 7, but $\chi_7 = 8$. The set of indices on which ω and χ agree is a member of any ultrafilter (because ω and χ agree almost everywhere), so ω and χ belong to the same equivalence class.

In the theory of large cardinals, a standard construction is to take the ultraproduct of the whole set-theoretic universe with respect to some carefully chosen ultrafilter U. Properties of this ultrafilter U have a strong influence on (higher order) properties of the ultraproduct; for example, if U is σ-complete, then the ultraproduct will again be well-founded. (See measurable cardinal for the prototypical example.)

38.3 Łoś's theorem

Łoś's theorem, also called *the fundamental theorem of ultraproducts*, is due to Jerzy Łoś (the surname is pronounced ['wɔɕ], approximately "wash"). It states that any first-order formula is true in the ultraproduct if and only if the set of indices i such that the formula is true in Mi is a member of U. More precisely:

Let σ be a signature, U an ultrafilter over a set I, and for each $i \in I$ let M_i be a σ-structure. Let M be the ultraproduct of the M_i with respect to U, that is, $M = \prod_{i \in I} M_i / U$. Then, for each $a^1, \ldots, a^n \in \prod M_i$, where $a^k = (a_i^k)_{i \in I}$, and for every σ-formula ϕ,

$$M \models \phi[[a^1], \ldots, [a^n]] \iff \{i \in I : M_i \models \phi[a_i^1, \ldots, a_i^n]\} \in U.$$

The theorem is proved by induction on the complexity of the formula ϕ. The fact that U is an ultrafilter (and not just a filter) is used in the negation clause, and the axiom of choice is needed at the existential quantifier step. As an application, one obtains the transfer theorem for hyperreal fields.

38.3.1   Examples

Let R be a unary relation in the structure M, and form the ultrapower of M. Then the set $S = \{x \in M | Rx\}$ has an analog $^*S$ in the ultrapower, and first-order formulas involving S are also valid for $^*S$. For example, let M be the reals, and let Rx hold if x is a rational number. Then in M we can say that for any pair of rationals x and y, there exists another number z such that z is not rational, and $x < z < y$. Since this can be translated into a first-order logical formula in the relevant formal language, Łoś's theorem implies that $^*S$ has the same property. That is, we can define a notion of the hyperrational numbers, which are a subset of the hyperreals, and they have the same first-order properties as the rationals.

Consider, however, the Archimedean property of the reals, which states that there is no real number x such that $x > 1$, $x > 1 + 1$, $x > 1 + 1 + 1$, ... for every inequality in the infinite list. Łoś's theorem does not apply to the Archimedean property, because the Archimedean property cannot be stated in first-order logic. In fact, the Archimedean property is false for the hyperreals, as shown by the construction of the hyperreal number ω above.

38.4   Ultralimit

> *For the ultraproduct of a sequence of metric spaces, see Ultralimit.*

In model theory and set theory, an **ultralimit** or **limiting ultrapower** is a direct limit of a sequence of ultrapowers.

Beginning with a structure, A_0, and an ultrafilter, D_0, form an ultrapower, A_1. Then repeat the process to form A_2, and so forth. For each n there is a canonical diagonal embedding $A_n \to A_{n+1}$. At limit stages, such as $A\omega$, form the direct limit of earlier stages. One may continue into the transfinite.

38.5   References

- Bell, John Lane; Slomson, Alan B. (2006) [1969]. *Models and Ultraproducts: An Introduction* (reprint of 1974 ed.). Dover Publications. ISBN 0-486-44979-3.

- Burris, Stanley N.; Sankappanavar, H.P. (2000) [1981]. *A Course in Universal Algebra* (Millennium ed.).

Chapter 39

Quantifier elimination

Quantifier elimination is a concept of simplification used in mathematical logic, model theory, and theoretical computer science. One way of classifying formulas is by the amount of quantification. Formulae with less depth of quantifier alternation are thought of as being simpler, with the quantifier-free formulae as the simplest. A theory has quantifier elimination if for every formula α, there exists another formula α_{QF} without quantifiers that is equivalent to it (modulo the theory).

39.1 Examples

Examples of theories that have been shown decidable using quantifier elimination are Presburger arithmetic,[1] algebraically closed fields, real closed fields,[1][2] atomless Boolean algebras, term algebras, dense linear orders,[1] random graphs, Feature trees, as well as many of their combinations such as Boolean Algebra with Presburger arithmetic, and Term Algebras with Queues.

Quantifier eliminator for the theory of the real numbers as an ordered additive group is *Fourier–Motzkin elimination*; for the theory of the field of real numbers it is the *Tarski–Seidenberg theorem*.[1]

Quantifier elimination can also be used to show that "combining" decidable theories leads to new decidable theories. Such constructions include the Feferman-Vaught theorem and Term Powers.

39.2 Algorithms and decidability

If a theory has quantifier elimination, then a specific question can be addressed: Is there a method of determining α_{QF} for each α? If there is such a method we call it a quantifier elimination algorithm. If there is such an algorithm, then decidability for the theory reduces to deciding the truth of the quantifier-free sentences. Quantifier-free sentences have no variables, so their validity in a given theory can often be computed, which enables the use of quantifier elimination algorithms to decide validity of sentences.

39.3 Related concepts

Various model theoretic ideas are related to quantifier elimination, and there are various equivalent conditions.

Every theory with quantifier elimination is model complete.

A first-order theory T has quantifier elimination if and only if for any two models B and C of T and for any common substructure A of B and C, B and C are elementarily equivalent in the language of T augmented with constants from A. In fact, it is sufficient here to show that any sentence with only existential quantifiers have the same truth value in B and C.

39.4   Basic ideas

To show constructively that a theory has quantifier elimination, it suffices to show that we can eliminate an existential quantifier applied to a conjunction of literals, that is, show that each formula of the form:

$$\exists x. \bigwedge_{i=1}^{n} L_i$$

where each L_i is a literal, is equivalent to a quantifier-free formula. Indeed, suppose we know how to eliminate quantifiers from conjunctions of formulae, then if F is a quantifier-free formula, we can write it in disjunctive normal form

$$\bigvee_{j=1}^{m} \bigwedge_{i=1}^{n} L_{ij},$$

and use the fact that

$$\exists x. \bigvee_{j=1}^{m} \bigwedge_{i=1}^{n} L_{ij}$$

is equivalent to

$$\bigvee_{j=1}^{m} \exists x. \bigwedge_{i=1}^{n} L_{ij}.$$

Finally, to eliminate a universal quantifier

$$\forall x.F$$

where F is quantifier-free, we transform $\neg F$ into disjunctive normal form, and use the fact that $\forall x.F$ is equivalent to $\neg \exists x. \neg F$.

39.5   History

In early model theory, quantifier elimination was used to demonstrate that various theories possess certain model-theoretic properties like decidability and completeness. A common technique was to show first that a theory admits elimination of quantifiers and thereafter prove decidability or completeness by considering only the quantifier-free formulas. This technique is used to show that Presburger arithmetic, i.e. the theory of the additive natural numbers, is decidable.

Theories could be decidable yet not admit quantifier elimination. Strictly speaking, the theory of the additive natural numbers did not admit quantifier elimination, but it was an expansion of the additive natural numbers that was shown to be decidable. Whenever a theory in a countable language is decidable, it is possible to extend its language with countably many relations to ensure that it admits quantifier elimination (for example, one can introduce a relation symbol for each formula).

Example: Nullstellensatz in ACF and DCF.

39.6 See also

- elimination theory

- conjunction elimination

39.7 References

[1] Grädel, Erich; Kolaitis, Phokion G.; Libkin, Leonid; Maarten, Marx; Spencer, Joel; Vardi, Moshe Y.; Venema, Yde; Weinstein, Scott (2007). *Finite model theory and its applications*. Texts in Theoretical Computer Science. An EATCS Series. Berlin: Springer-Verlag. ISBN 978-3-540-00428-8. Zbl 1133.03001.

[2] Fried, Michael D.; Jarden, Moshe (2008). *Field arithmetic*. Ergebnisse der Mathematik und ihrer Grenzgebiete. 3. Folge **11** (3rd revised ed.). Springer-Verlag. p. 171. ISBN 978-3-540-77269-9. Zbl 1145.12001.

- Wilfrid Hodges. "Model Theory". Cambridge University Press. 1993.

- Viktor Kuncak and Martin Rinard. "Structural Subtyping of Non-Recursive Types is Decidable". In *Eighteenth Annual IEEE Symposium on Logic in Computer Science,* 2003.

Chapter 40

Model complete theory

In model theory, a first-order theory is called **model complete** if every embedding of models is an elementary embedding. Equivalently, every first-order formula is equivalent to a universal formula. This notion was introduced by Abraham Robinson.

40.1 Model companion and model completion

A **companion** of a theory T is a theory T^* such that every model of T can be embedded in a model of T^* and vice versa.

A **model companion** of a theory T is a companion of T that is model complete. Robinson proved that a theory has at most one model companion.

A **model completion** for a theory T is a model companion T^* such that for any model M of T, the theory of T^* together with the diagram of M is complete. Roughly speaking, this means every model of T is embeddable in a model of T^* in a unique way.

If T^* is a model companion of T then the following conditions are equivalent:

- T^* is a model completion of T

- T has the amalgamation property.

If T also has universal axiomatization, both of the above are also equivalent to:

- T^* has elimination of quantifiers

40.2 Examples

- The theory of dense linear orders with a first and last element is complete but not model complete.

- The theory of dense linear orders with two constant symbols is model complete but not complete.

- The theory of algebraically closed fields is the model completion of the theory of fields. It is model complete but not complete.

- The theory of real closed fields, in the language of ordered rings, is a model completion of the theory of ordered fields (or even ordered domains). The theory of real closed fields, in the language of rings, is the model companion for the theory of formally real fields, but is not a model completion.

278

- Any theory with elimination of quantifiers is model complete.

- The model completion of the theory of equivalence relations is the theory of equivalence relations with infinitely many equivalence classes.

- The theory of groups (in a language with symbols for the identity, product, and inverses) has the amalgamation property but does not have a model companion.

40.3 References

- Chang, Chen Chung; Keisler, H. Jerome (1990) [1973], *Model Theory*, Studies in Logic and the Foundations of Mathematics (3rd ed.), Elsevier, ISBN 978-0-444-88054-3

- Hirschfeld, Joram; Wheeler, William H. (1975), "Model-completions and model-companions", *Forcing, Arithmetic, Division Rings*, Lecture Notes in Mathematics **454**, Springer, pp. 44–54, doi:10.1007/BFb0064085, ISBN 978-3-540-07157-0, MR 0389581

Chapter 41

Elementary equivalence

In model theory, a branch of mathematical logic, two structures M and N of the same signature σ are called **elementarily equivalent** if they satisfy the same first-order σ-sentences.

If N is a substructure of M, one often needs a stronger condition. In this case N is called an **elementary substructure** of M if every first-order σ-formula $\varphi(a_1, ..., an)$ with parameters $a_1, ..., an$ from N is true in N if and only if it is true in M. If N is an elementary substructure of M, M is called an **elementary extension** of N. An embedding $h: N \to M$ is called an **elementary embedding** of N into M if $h(N)$ is an elementary substructure of M.

A substructure N of M is elementary if and only if it passes the **Tarski–Vaught test**: every first-order formula $\varphi(x, b_1, ..., bn)$ with parameters in N that has a solution in M also has a solution in N when evaluated in M. One can prove that two structures are elementary equivalent with the Ehrenfeucht–Fraïssé games.

41.1 Elementarily equivalent structures

Two structures M and N of the same signature σ are **elementarily equivalent** if every first-order sentence (formula without free variables) over σ is true in M if and only if it is true in N, i.e. if M and N have the same complete first-order theory. If M and N are elementarily equivalent, one writes $M \equiv N$.

A first-order theory is complete if and only if any two of its models are elementarily equivalent.

For example, consider the language with one binary relation symbol '<'. The model **R** of real numbers with its usual order and the model **Q** of rational numbers with its usual order are elementarily equivalent, since they both interpret '<' as an unbounded dense linear ordering. This is sufficient to ensure elementary equivalence, because the theory of unbounded dense linear orderings is complete, as can be shown by Vaught's test.

More generally, any first-order theory has non-isomorphic, elementary equivalent models, which can be obtained via the Löwenheim–Skolem theorem. Thus, for example, there are non-standard models of Peano arithmetic, which contain other objects than just the numbers 0, 1, 2, etc., and yet are elementarily equivalent to the standard model.

41.2 Elementary substructures and elementary extensions

N is an **elementary substructure** of M if N and M are structures of the same signature σ such that for all first-order σ-formulas $\varphi(x_1, ..., xn)$ with free variables $x_1, ..., xn$, and all elements $a_1, ..., a_n$ of N, $\varphi(a_1, ..., a_n)$ holds in N if and only if it holds in M:

$$N \models \varphi(a_1, ..., an) \text{ iff } M \models \varphi(a_1, ..., an).$$

It follows that N is a substructure of M.

If N is a substructure of M, then both N and M can be interpreted as structures in the signature σN consisting of σ together with a new constant symbol for every element of N. N is an elementary substructure of M if and only if N is a substructure of M and N and M are elementarily equivalent as σN-structures.

If N is an elementary substructure of M, one writes $N \preceq M$ and says that M is an **elementary extension** of N: $M \succeq N$.

The downward Löwenheim–Skolem theorem gives a countable elementary substructure for any infinite first-order structure; the upward Löwenheim–Skolem theorem gives elementary extensions of any infinite first-order structure of arbitrarily large cardinality.

41.3 Tarski–Vaught test

The **Tarski–Vaught test** (or **Tarski–Vaught criterion**) is a necessary and sufficient condition for a substructure N of a structure M to be an elementary substructure. It can be useful for constructing an elementary substructure of a large structure.

Let M be a structure of signature σ and N a substructure of M. N is an elementary substructure of M if and only if for every first-order formula $\varphi(x, y_1, \ldots, yn)$ over σ and all elements $b_1, \ldots, bn$ from N, if $M \models x\, \varphi(x, b_1, \ldots, bn)$, then there is an element a in N such that $M \models \varphi(a, b_1, \ldots, bn)$.

41.4 Elementary embeddings

An **elementary embedding** of a structure N into a structure M of the same signature σ is a map $h\colon N \to M$ such that for every first-order σ-formula $\varphi(x_1, \ldots, xn)$ and all elements $a_1, \ldots, a_n$ of N,

$$N \models \varphi(a_1, \ldots, an) \text{ if and only if } M \models \varphi(h(a_1), \ldots, h(an)).$$

Every elementary embedding is a strong homomorphism, and its image is an elementary substructure.

Elementary embeddings are the most important maps in model theory. In set theory, elementary embeddings whose domain is V (the universe of set theory) play an important role in the theory of large cardinals (see also critical point).

41.5 References

- Chang, Chen Chung; Keisler, H. Jerome (1990) [1973], *Model Theory*, Studies in Logic and the Foundations of Mathematics (3rd ed.), Elsevier, ISBN 978-0-444-88054-3.

- Hodges, Wilfrid (1997), *A shorter model theory*, Cambridge: Cambridge University Press, ISBN 978-0-521-58713-6.

- Monk, J. Donald (1976), *Mathematical Logic*, Graduate Texts in Mathematics, New York • Heidelberg • Berlin: Springer Verlag, ISBN 0-387-90170-1

Chapter 42

Omega-categorical theory

In mathematical logic, an **omega-categorical theory** is a theory that has only one countable model up to isomorphism. Omega-categoricity is the special case $\kappa = \aleph_0 = \omega$ of κ-categoricity, and omega-categorical theories are also referred to as $\boldsymbol{\omega}$**-categorical**. The notion is most important for countable first-order theories.

42.1 Equivalent conditions for omega-categoricity

Many conditions on a theory are equivalent to the property of omega-categoricity. In 1959 Erwin Engeler, Czesław Ryll-Nardzewski and Lars Svenonius, proved several independently.[1] Despite this, the literature still widely refers to the Ryll-Nardzewski theorem as a name for these conditions. The conditions included with the theorem vary between authors.[2][3]

Given a countable complete first-order theory T with infinite models, the following are equivalent:

- The theory T is omega-categorical.

- Every countable model of T has an oligomorphic automorphism group.

- Some countable model of T has an oligomorphic automorphism group.[4]

- The theory T has a model which, for every natural number n, realizes only finitely many n-types, that is, the Stone space $Sn(T)$ is finite.

- For every natural number n, T has only finitely many n-types.

- For every natural number n, every n-type is isolated.

- For every natural number n, up to equivalence modulo T there are only finitely many formulas with n free variables, in other words, every nth Lindenbaum-Tarski algebra of T is finite.

- Every model of T is atomic.

- Every countable model of T is atomic.

- The theory T has a countable atomic and saturated model.

- The theory T has a saturated prime model.

42.2 Notes

[1] Rami Grossberg, José Iovino and Olivier Lessmann, *A primer of simple theories*

[2] Hodges, Model Theory, p. 341.

[3] Rothmaler, p. 200.

[4] Cameron (1990) p.30

42.3 References

- Cameron, Peter J. (1990), *Oligomorphic permutation groups*, London Mathematical Society Lecture Note Series **152**, Cambridge: Cambridge University Press, ISBN 0-521-38836-8, Zbl 0813.20002

- Chang, Chen Chung; Keisler, H. Jerome (1989) [1973], *Model Theory*, Elsevier, ISBN 978-0-7204-0692-4

- Hodges, Wilfrid (1993), *Model theory*, Cambridge: Cambridge University Press, ISBN 978-0-521-30442-9

- Hodges, Wilfrid (1997), *A shorter model theory*, Cambridge: Cambridge University Press, ISBN 978-0-521-58713-6

- Poizat, Bruno (2000), *A Course in Model Theory: An Introduction to Contemporary Mathematical Logic*, Berlin, New York: Springer-Verlag, ISBN 978-0-387-98655-5

- Rothmaler, Philipp (2000), *Introduction to Model Theory*, New York: Taylor & Francis Group, ISBN 978-90-5699-313-9

Chapter 43

Type (model theory)

In model theory and related areas of mathematics, a **type** is an object that, loosely speaking, describes how a (real or possible) element or elements in a mathematical structure might behave. More precisely, it is a set of first-order formulas in a language L with free variables $x_1, x_2, \ldots, xn$ which are true of a sequence of elements of an L-structure $\mathcal{M}$. Depending on the context, types can be **complete** or **partial** and they may use a fixed set of constants, A, from the structure $\mathcal{M}$. The question of which types represent actual elements of $\mathcal{M}$ leads to the ideas of saturated models and **omitting types**.

43.1 Formal definition

Consider a structure $\mathcal{M}$ for a language L. Let M be the universe of the structure. For every $A \subseteq M$, let $L(A)$ be the language which is obtained from L by adding a constant ca for every $a \in A$. In other words,

$$L(A) = L \cup \{c_a : a \in A\}.$$

A **1-type (of** $\mathcal{M}$ **) over** A is a set $p(x)$ of formulas in $L(A)$ with at most one free variable x (therefore 1-type) such that for every finite subset $p_0(x) \subseteq p(x)$ there is some $b \in M$, depending on $p_0(x)$, with $\mathcal{M} \models p_0(b)$ (i.e. all formulas in $p_0(x)$ are true in $\mathcal{M}$ when x is replaced by b).

Similarly an **n-type (of** $\mathcal{M}$ **) over** A is defined to be a set $p(x_1, \ldots, xn) = p(x)$ of formulas in $L(A)$, each having its free variables occurring only among the given n free variables $x_1, \ldots, xn$, such that for every finite subset $p_0(x) \subseteq p(x)$ there are some elements $b_1, \ldots, bn \in M$ with $\mathcal{M} \models p_0(b_1, \ldots, b_n)$.

Complete type refers to those types which are maximal with respect to inclusion, i.e. if $p(x)$ is a complete type, then for every $\phi(x) \in L(A, x)$ either $\phi(x) \in p(x)$ or $\neg\phi(x) \in p(x)$. Any non-complete type is called a **partial type**. So, the word **type** in general refers to any n-type, partial or complete, over any chosen set of parameters (possibly the empty set).

An n-type $p(x)$ is said to be **realized in** $\mathcal{M}$ if there is an element $b \in M^n$ such that $\mathcal{M} \models p(b)$. The existence of such a realization is guaranteed for any type by the Compactness theorem, although the realization might take place in some elementary extension of $\mathcal{M}$, rather than in $\mathcal{M}$ itself. If a complete type is realized by b in $\mathcal{M}$, then the type is typically denoted $tp_n^{\mathcal{M}}(b/A)$ and referred to as **the complete type of b over A**.

A type $p(x)$ is said to be **isolated by** φ if there is a formula $\varphi(x)$ with the property that $\forall \psi(x) \in p(x), \varphi(x) \to \psi(x)$. Since finite subsets of a type are always realized in $\mathcal{M}$, there is always an element $b \in M^n$ such that $\varphi(b)$ is true in $\mathcal{M}$; i.e. $\mathcal{M} \models \varphi(b)$, thus b realizes the entire isolated type. So isolated types will be realized in every elementary substructure or extension. Because of this, isolated types can never be omitted (see below).

A model that realizes the maximum possible variety of types is called a saturated model, and the ultrapower construction provides one way of producing saturated models.

43.2 Examples of types

Consider the language with one binary connective, which we denote as $\in$. Let $\mathcal{M}$ be the model $\langle \omega, \in_\omega \rangle$, which is the ordinal ω with its standard well-ordering. Let $\mathcal{T}$ denote the theory of this model.

Consider the set of formulas $p(x) := \{ n \in x \mid n \in_\omega \omega \}$. First, we claim this is a type. Let $p_0(x) \subseteq p(x)$ be a finite subset of $p(x)$. We need to find an $n \in \omega$ that satisfies all the formulas in p_0 . Well, we can just take the successor of the largest ordinal mentioned in the set of formulas $p_0(x)$. Then this will clearly contain all the ordinals mentioned in $p_0(x)$. Thus we have that $p(x)$ is a type. Next, note that $p(x)$ is not realized in $\mathcal{M}$. For, if it were there would be some $n \in \omega$ that contains every element of ω . If we wanted to realize the type, we might be tempted to consider the model $\langle \omega + 1, \in_{\omega+1} \rangle$, which is indeed a supermodel of $\mathcal{M}$ which realizes the type. Unfortunately, this extension is not elementary, that is this model does not have to satisfy $\mathcal{T}$. In particular, the sentence $\exists x \forall y (y \in x)$ is satisfied by this model and not by $\mathcal{M}$.

So, we wish to realize the type in an elementary extension. We can do this by defining a new structure in this language, which we will denote $\mathcal{M}'$. The domain of the structure will be $\omega \cup \mathbb{Z}'$ where $\mathbb{Z}'$ is the set of integers adorned in such a way that $\mathbb{Z}' \cap \omega = \emptyset$. Let $<$ denote the usual order of $\mathbb{Z}'$. We interpret the symbol $\in$ in our new structure by $\in_{\mathcal{M}'} = \in_\omega \cup < \cup (\omega \times \mathbb{Z}')$. The idea being that we are adding a "Z-chain", or copy of the integers, above all the finite ordinals. Clearly any element of $\mathbb{Z}'$ realizes the type $p(x)$. Moreover, one can verify that this extension is elementary.

Another example: the complete type of the number 2 over the emptyset, considered as a member of the natural numbers, would be the set of all first-order statements describing a variable x that are true for $x = 2$. This set would include formulas such as $x \neq 1 + 1 + 1$, $x \leq 1 + 1 + 1 + 1$, and $\exists y (y < x)$. This is an example of an isolated type, since the formula $x = 1 + 1$ implies all other formulas that are true about the number 2.

For example, the statements

$$\forall y (y^2 < 2 \implies y < x)$$

and

$$\forall y ((y > 0 \wedge y^2 > 2) \implies y > x)$$

describing the square root of 2 are consistent with the axioms of ordered fields, and can be extended to a complete type. This type is not realized in the ordered field of rational numbers, but is realized in the ordered field of reals. Similarly, the infinite set of formulas (over the emptyset) {x>1, x>1+1, x>1+1+1, ...} is not realized in the ordered field of real numbers, but is realized in the ordered field of hyperreals. If we allow more parameters, for instance all of the reals, we can specify a type $\{0 < x < r : r \in \mathbb{R}\}$ that is realized by an infinitesimal hyperreal that violates the Archimedean property.

The reason it is useful to restrict the parameters to a certain subset of the model is that it helps to distinguish the types that can be satisfied from those that cannot. For example, using the entire set of real numbers as parameters one could generate an uncountably infinite set of formulas like $x \neq 1$, $x \neq \pi$, ... that would explicitly rule out every possible real value for x, and therefore could never be realized within the real numbers.

43.3 Stone spaces

It is useful to consider the set of complete n-types over A as a topological space. Consider the following equivalence relation on formulae in the free variables $x_1,\dots, x_n$ with parameters in M:

$$\psi \equiv \phi \Leftrightarrow \mathcal{M} \models \forall x_1, \dots, x_n (\psi(x_1, \dots, x_n) \leftrightarrow \phi(x_1, \dots, x_n)).$$

One can show that $\psi \equiv \phi$ iff they are contained in exactly the same complete types.

The set of formulae in free variables $x_1,\ldots,xn$ over A up to this equivalence relation is a Boolean algebra (and is canonically isomorphic to the set of A-definable subsets of M^n). The complete n-types correspond to ultrafilters of this boolean algebra. The set of complete n-types can be made into a topological space by taking the sets of types containing a given formula as basic open sets. This constructs the Stone space which is compact, Hausdorff, and totally disconnected.

Example. The complete theory of algebraically closed fields of characteristic 0 has quantifier elimination which allows one to show that the possible complete 1-types correspond to:

- Roots of a given irreducible non-constant polynomial over the rationals with leading coefficient 1. For example, the type of square roots of 2. Each of these types is an open point of the Stone space.

- Transcendental elements, that are not roots of any non-zero polynomial. This type is a point in the Stone space that is closed but not open.

In other words, the 1-types correspond exactly to the prime ideals of the polynomial ring $\mathbf{Q}[x]$ over the rationals $\mathbf{Q}$: if r is an element of the model of type p, then the ideal corresponding to p is the set of polynomials with r as a root. More generally, the complete n-types correspond to the prime ideals of the polynomial ring $\mathbf{Q}[x_1,\ldots,x_n]$, in other words to the points of the prime spectrum of this ring. (The Stone space topology can in fact be viewed as the Zariski topology of a Boolean ring induced in a natural way from the lattice structure of the Boolean Algebra; while the Zariski topology is not in general Hausdorff, it is in the case of Boolean rings.) For example, if $q(x,y)$ is an irreducible polynomial in 2 variables, there is a 2-type whose realizations are (informally) pairs (x,y) of transcendental elements with $q(x,y)=0$.

43.4   The omitting types theorem

Given a complete n-type p one can ask if there is a model of the theory that **omits** p, in other words there is no n-tuple in the model which realizes p. If p is an isolated point in the Stone space, i.e. if $\{p\}$ is an open set, it is easy to see that every model realizes p (at least if the theory is complete). The **omitting types theorem** says that conversely if p is not isolated then there is a countable model omitting p (provided that the language is countable).

Example: In the theory of algebraically closed fields of characteristic 0, there is a 1-type represented by elements that are transcendental over the prime field. This is a non-isolated point of the Stone space (in fact, the only non-isolated point). The field of algebraic numbers is a model omitting this type, and the algebraic closure of any transcendental extension of the rationals is a model realizing this type.

All the other types are "algebraic numbers" (more precisely, they are the sets of first order statements satisfied by some given algebraic number), and all such types are realized in all algebraically closed fields of characteristic 0.

43.5   References

- Hodges, Wilfrid (1997). *A shorter model theory*. Cambridge University Press. ISBN 0-521-58713-1.

- Chang, C.C.; Keisler, H. Jerome (1989). *Model Theory* (third ed.). Elsevier. ISBN 0-7204-0692-7.

- Marker, David (2002). *Model Theory: An Introduction*. Graduate Texts in Mathematics 217. Springer. ISBN 0-387-98760-6.

Chapter 44

Oligomorphic group

In group theory, a branch of mathematics, an **oligomorphic group** is a particular kind of permutation group. If a group G acts on a set S (usually infinite), then G is said to be oligomorphic if this action has only finitely many orbits on every Cartesian product S^n of S (n-tuples of elements of S for every natural number n). The interest in oligomorphic groups is partly based on their application to model theory, e.g. automorphisms in countably categorical theories.[1]

44.1 References

[1] Bhattacharjee, Meenaxi; Macpherson, Dugald; Möller, Rögnvaldur G.; Neumann, Peter M. (1998). *Notes on infinite permutation groups*. Lecture Notes in Mathematics **1698**. Berlin: Springer-Verlag. p. 83. ISBN 3-540-64965-4. Zbl 0916.20002.

- Cameron, Peter J. (1990). *Oligomorphic permutation groups*. London Mathematical Society Lecture Note Series **152**. Cambridge: Cambridge University Press. ISBN 0-521-38836-8. Zbl 0813.20002.

44.2 External links

- Oligomorphic permutation groups - Isaac Newton Institute preprint, Peter J. Cameron

Chapter 45

Morley's categoricity theorem

"Vaught's test" redirects here. Not to be confused with the Tarski–Vaught test.
"Categorical theory" redirects here. Not to be confused with Category Theory.

In model theory, a branch of mathematical logic, a theory is κ-**categorical** (or **categorical in** κ) if it has exactly one model of cardinality κ up to isomorphism. **Morley's categoricity theorem** is a theorem of Michael D. Morley (1965) which states that if a first-order theory in a countable language is categorical in some uncountable cardinality, then it is categorical in all uncountable cardinalities.

Saharon Shelah (1974) extended Morley's theorem to uncountable languages: if the language has cardinality κ and a theory is categorical in some uncountable cardinal greater than or equal to κ then it is categorical in all cardinalities greater than κ.

45.1   History and motivation

Oswald Veblen in 1904 defined a theory to be **categorical** if all of its models are isomorphic. It follows from the definition above and the Löwenheim–Skolem theorem that any first-order theory with a model of infinite cardinality cannot be categorical. One is then immediately led to the more subtle notion of κ-categoricity, which asks: for which cardinals κ is there exactly one model of cardinality κ of the given theory T up to isomorphism? This is a deep question and significant progress was only made in 1954 when Jerzy Łoś noticed that, at least for complete theories T over countable languages with at least one infinite model, he could only find three ways for T to be κ-categorical at some κ:

- T is **totally categorical**, *i.e.* T is κ-categorical for all infinite cardinals κ.
- T is **uncountably categorical**, *i.e.* T is κ-categorical if and only if κ is an uncountable cardinal.
- T is **countably categorical**, *i.e.* T is κ-categorical if and only if κ is a countable cardinal.

In other words, he observed that, in all the cases he could think of, κ-categoricity at any one uncountable cardinal implied κ-categoricity at all other uncountable cardinals. This observation spurred a great amount of research into the 1960s, eventually culminating in Michael Morley's famous result that these are in fact the only possibilities. The theory was subsequently extended and refined by Saharon Shelah in the 1970s and beyond, leading to stability theory and Shelah's more general programme of classification theory.

45.2   Examples

There are not many natural examples of theories that are categorical in some uncountable cardinal. The known examples include:

- Pure identity theory (with no functions, constants, predicates other than "=", or axioms).

- The classic example is the theory of algebraically closed fields of a given characteristic. Categoricity does *not* say that all algebraically closed fields of characteristic 0 as large as the complex numbers **C** are the same as **C**; it only asserts that they are isomorphic *as fields* to **C**. It follows that although the completed p-adic closures **C***p* are all isomorphic as fields to **C**, they may (and in fact do) have completely different topological and analytic properties. The theory of algebraically closed fields of given characteristic is **not** categorical in ω (the countable infinite cardinal); there are models of transcendence degree 0, 1, 2, ..., ω.

- Vector spaces over a given countable field. This includes abelian groups of given prime exponent (essentially the same as vector spaces over a finite field) and divisible torsion-free abelian groups (essentially the same as vector spaces over the rationals).

- The theory of the set of natural numbers with a successor function.

There are also examples of theories that are categorical in ω but not categorical in uncountable cardinals. The simplest example is the theory of an equivalence relation with exactly two equivalence classes both of which are infinite. Another example is the theory of dense linear orders with no endpoints; Cantor proved that any such countable linear order is isomorphic to the rational numbers.

Any theory T categorical in some infinite cardinal κ is very close to being complete. More precisely, the **Łoś–Vaught test** states that if a satisfiable theory has no finite models and is categorical in some infinite cardinal κ at least equal to the cardinality of its language, then the theory is complete. The reason is that all infinite models are equivalent to some model of cardinal κ by the Löwenheim–Skolem theorem, and so are all equivalent as the theory is categorical in κ. Therefore the theory is complete as all models are equivalent. The assumption that the theory have no finite models is necessary.[1]

45.3   See also

- Spectrum of a theory

45.4   References

[1] Marker (2002) p.42

- Chang, Chen Chung; Keisler, H. Jerome (1990) [1973], *Model Theory*, Studies in Logic and the Foundations of Mathematics, Elsevier, ISBN 978-0-444-88054-3

- Corcoran, John (1980), "Categoricity", *History and Philosophy of Logic* **1** (1): 187–207

- Hodges, Wilfrid, "First-order Model Theory", The Stanford Encyclopedia of Philosophy (Summer 2005 Edition), Edward N. Zalta (ed.).

- Marker, David (2002), *Model theory: An introduction*, Graduate Texts in Mathematics **217**, New York, NY: Springer-Verlag, ISBN 0-387-98760-6, Zbl 1003.03034

- Morley, Michael (1965), "Categoricity in Power", *Transactions of the American Mathematical Society* (Transactions of the American Mathematical Society, Vol. 114, No. 2) **114** (2): 514–538, doi:10.2307/1994188, ISSN 0002-9947, JSTOR 1994188

- Palyutin, E.A. (2001), "Categoricity in cardinality", in Hazewinkel, Michiel, *Encyclopedia of Mathematics*, Springer, ISBN 978-1-55608-010-4

- Shelah, Saharon (1974), "Categoricity of uncountable theories", *Proceedings of the Tarski Symposium (Proc. Sympos. Pure Math., Vol. XXV, Univ. of California, Berkeley, Calif., 1971)*, Providence, R.I.: American Mathematical Society, pp. 187–203, MR 0373874

- Shelah, Saharon (1990) [1978], *Classification theory and the number of nonisomorphic models*, Studies in Logic and the Foundations of Mathematics (2nd ed.), Elsevier, ISBN 978-0-444-70260-9 (IX, 1.19, pg.49)

- Veblen, Oswald (1904), "A System of Axioms for Geometry", *Transactions of the American Mathematical Society* (Transactions of the American Mathematical Society, Vol. 5, No. 3) **5** (3): 343–384, doi:10.2307/1986462, ISSN 0002-9947, JSTOR 1986462

Chapter 46

Reduct

This article is about a relation on algebraic structures. For reducts in abstract rewriting, see Confluence (abstract rewriting).

In universal algebra and in model theory, a **reduct** of an algebraic structure is obtained by omitting some of the operations and relations of that structure. The converse of "reduct" is "expansion."

46.1 Definition

Let A be an algebraic structure (in the sense of universal algebra) or equivalently a structure in the sense of model theory, organized as a set X together with an indexed family of operations and relations φ_i on that set, with index set I. Then the **reduct** of A defined by a subset J of I is the structure consisting of the set X and J-indexed family of operations and relations whose j-th operation or relation for $j \in J$ is the j-th operation or relation of A. That is, this reduct is the structure A with the omission of those operations and relations φi for which i is not in J.

A structure A is an **expansion** of B just when B is a reduct of A. That is, reduct and expansion are mutual converses.

46.2 Examples

The monoid $(\mathbf{Z}, +, 0)$ of integers under addition is a reduct of the group $(\mathbf{Z}, +, -, 0)$ of integers under addition and negation, obtained by omitting negation. By contrast, the monoid $(\mathbf{N}, +, 0)$ of natural numbers under addition is not the reduct of any group.

Conversely the group $(\mathbf{Z}, +, -, 0)$ is the expansion of the monoid $(\mathbf{Z}, +, 0)$, expanding it with the operation of negation.

46.3 References

- Burris, Stanley N.; H. P. Sankappanavar (1981). *A Course in Universal Algebra*. Springer. ISBN 3-540-90578-2.

- Hodges, Wilfrid (1993). *Model theory*. Cambridge University Press. ISBN 0-521-30442-3.

Chapter 47

Interpretation (model theory)

For other uses, see Interpretation (disambiguation).

In model theory, **interpretation** of a structure M in another structure N (typically of a different signature) is a technical notion that approximates the idea of representing M inside N. For example every reduct or definitional expansion of a structure N has an interpretation in N.

Many model-theoretic properties are preserved under interpretability. For example if the theory of N is stable and M is interpretable in N, then the theory of M is also stable.

47.1 Definition

An **interpretation** of M in N **with parameters** (or **without parameters**, respectively) is a pair (n, f) where n is a natural number and f is a surjective map from a subset of N^n onto M such that the f-preimage (more precisely the f^k-preimage) of every set $X \subseteq M^k$ definable in M by a first-order formula without parameters is definable (in N) by a first-order formula with parameters (or without parameters, respectively). Since the value of n for an interpretation (n, f) is often clear from context, the map f itself is also called an interpretation.

To verify that the preimage of every definable (without parameters) set in M is definable in N (with or without parameters), it is sufficient to check the preimages of the following definable sets:

- the domain of M;

- the diagonal of M;

- every relation in the signature of M;

- the graph of every function in the signature of M.

In model theory the term *definable* often refers to definability with parameters; if this convention is used, definability without parameters is expressed by the term *0-definable*. Similarly, an interpretation with parameters may be referred to as simply an interpretation, and an interpretation without parameters as a **0-interpretation**.

47.2 Bi-interpretability

If L, M and N are three structures, L is interpreted in M, and M is interpreted in N, then one can naturally construct a composite interpretation of L in N. If two structures M and N are interpreted in each other, then by combining the interpretations in two possible ways, one obtains an interpretation of each of the two structures in itself. This observation permits one to define an equivalence relation among structures, reminiscent of the homotopy equivalence among topological spaces.

Two structures M and N are **bi-interpretable** if there exists an interpretation of M in N and an interpretation of N in M such that the composite interpretations of M in itself and of N in itself are definable in M and in N, respectively (the composite interpretations being viewed as operations on M and on N).

47.3 Example

The partial map f from $\mathbf{Z} \times \mathbf{Z}$ onto $\mathbf{Q}$ which maps (x, y) to x/y provides an interpretation of the field $\mathbf{Q}$ of rational numbers in the ring $\mathbf{Z}$ of integers (to be precise, the interpretation is $(2, f)$). In fact, this particular interpretation is often used to *define* the rational numbers. To see that it is an interpretation (without parameters), one needs to check the following preimages of definable sets in $\mathbf{Q}$:

- the preimage of $\mathbf{Q}$ is defined by the formula $\varphi(x, y)$ given by $\neg\,(y = 0)$;

- the preimage of the diagonal of $\mathbf{Q}$ is defined by the formula $\varphi(x_1, y_1, x_2, y_2)$ given by $x_1 \times y_2 = x_2 \times y_1$;

- the preimages of 0 and 1 are defined by the formulas $\varphi(x, y)$ given by $x = 0$ and $x = y$;

- the preimage of the graph of addition is defined by the formula $\varphi(x_1, y_1, x_2, y_2, x_3, y_3)$ given by $x_1 \times y_2 \times y_3 + x_2 \times y_1 \times y_3 = x_3 \times y_1 \times y_2$;

- the preimage of the graph of multiplication is defined by the formula $\varphi(x_1, y_1, x_2, y_2, x_3, y_3)$ given by $x_1 \times x_2 \times y_3 = x_3 \times y_1 \times y_2$.

47.4 References

- Ahlbrandt, Gisela; Ziegler, Martin (1986), "Quasi finitely axiomatizable totally categorical theories", *Annals of Pure and Applied Logic* **30**: 63–82, doi:10.1016/0168-0072(86)90037-0

- Hodges, Wilfrid (1997), *A shorter model theory*, Cambridge: Cambridge University Press, ISBN 978-0-521-58713-6 (Section 4.3)

- Poizat, Bruno (2000), *A Course in Model Theory*, Springer, ISBN 0-387-98655-3 (Section 9.4)

Chapter 48

Decidability (logic)

In logic, the term **decidable** refers to the decision problem, the question of the existence of an effective method for determining membership in a set of formulas, or, more precisely, an algorithm that can and will return a Boolean true or false value (instead of looping indefinitely). Logical systems such as propositional logic are decidable if membership in their set of logically valid formulas (or theorems) can be effectively determined. A theory (set of sentences closed under logical consequence) in a fixed logical system is decidable if there is an effective method for determining whether arbitrary formulas are included in the theory. Many important problems are undecidable, that is, it has been proven that no effective method can exist for them.

48.1   Relationship to computability

As with the concept of a decidable set, the definition of a decidable theory or logical system can be given either in terms of *effective methods* or in terms of *computable functions*. These are generally considered equivalent per Church's thesis. Indeed, the proof that a logical system or theory is undecidable will use the formal definition of computability to show that an appropriate set is not a decidable set, and then invoke Church's thesis to show that the theory or logical system is not decidable by any effective method (Enderton 2001, pp. 206*ff.*).

48.2   Decidability of a logical system

Each logical system comes with both a syntactic component, which among other things determines the notion of provability, and a semantic component, which determines the notion of logical validity. The logically valid formulas of a system are sometimes called the **theorems** of the system, especially in the context of first-order logic where Gödel's completeness theorem establishes the equivalence of semantic and syntactic consequence. In other settings, such as linear logic, the syntactic consequence (provability) relation may be used to define the theorems of a system.

A logical system is decidable if there is an effective method for determining whether arbitrary formulas are theorems of the logical system. For example, propositional logic is decidable, because the truth-table method can be used to determine whether an arbitrary propositional formula is logically valid.

First-order logic is not decidable in general; in particular, the set of logical validities in any signature that includes equality and at least one other predicate with two or more arguments is not decidable.[1] Logical systems extending first-order logic, such as second-order logic and type theory, are also undecidable.

The validities of monadic predicate calculus with identity are decidable, however. This system is first-order logic restricted to signatures that have no function symbols and whose relation symbols other than equality never take more than one argument.

Some logical systems are not adequately represented by the set of theorems alone. (For example, Kleene's logic has no

294

theorems at all.) In such cases, alternative definitions of decidability of a logical system are often used, which ask for an effective method for determining something more general than just validity of formulas; for instance, validity of sequents, or the consequence relation $\{(\Gamma, A) \mid \Gamma \vDash A\}$ of the logic.

48.3 Decidability of a theory

A theory is a set of formulas, which here is assumed to be closed under logical consequence. The question of decidability for a theory is whether there is an effective procedure that, given an arbitrary formula in the signature of the theory, decides whether the formula is a member of the theory or not. This problem arises naturally when a theory is defined as the set of logical consequences of a fixed set of axioms. Examples of decidable first-order theories include the theory of real closed fields, and Presburger arithmetic, while the theory of groups and Robinson arithmetic are examples of undecidable theories.

There are several basic results about decidability of theories. Every inconsistent theory is decidable, as every formula in the signature of the theory will be a logical consequence of, and thus member of, the theory. Every complete recursively enumerable first-order theory is decidable. An extension of a decidable theory may not be decidable. For example, there are undecidable theories in propositional logic, although the set of validities (the smallest theory) is decidable.

A consistent theory that has the property that every consistent extension is undecidable is said to be **essentially undecidable**. In fact, every consistent extension will be essentially undecidable. The theory of fields is undecidable but not essentially undecidable. Robinson arithmetic is known to be essentially undecidable, and thus every consistent theory that includes or interprets Robinson arithmetic is also (essentially) undecidable.

48.4 Some decidable theories

Some decidable theories include (Monk 1976, p. 234):[2]

- The set of first-order logical validities in the signature with only equality, established by Leopold Löwenheim in 1915.

- The set of first-order logical validities in a signature with equality and one unary function, established by Ehrenfeucht in 1959.

- The first-order theory of the integers in the signature with equality and addition, also called Presburger arithmetic. The completeness was established by Mojżesz Presburger in 1929.

- The first-order theory of Boolean algebras, established by Alfred Tarski in 1949.

- The first-order theory of algebraically closed fields of a given characteristic, established by Tarski in 1949.

- The first-order theory of real-closed ordered fields, established by Tarski in 1949 (see also Tarski's exponential function problem).

- The first-order theory of Euclidean geometry, established by Tarski in 1949.

- The first-order theory of hyperbolic geometry, established by Schwabhäuser in 1959.

- Specific decidable sublanguages of set theory investigated in the 1980s through today.(Cantone *et al.*, 2001)

Methods used to establish decidability include quantifier elimination, model completeness, and Vaught's test.

48.5   Some undecidable theories

Some undecidable theories include (Monk 1976, p. 279):[2]

- The set of logical validities in any first-order signature with equality and either: a relation symbol of arity no less than 2, or two unary function symbols, or one function symbol of arity no less than 2, established by Trakhtenbrot in 1953.

- The first-order theory of the natural numbers with addition, multiplication, and equality, established by Tarski and Andrzej Mostowski in 1949.

- The first-order theory of the rational numbers with addition, multiplication, and equality, established by Julia Robinson in 1949.

- The first-order theory of groups, established by Alfred Tarski in 1953.[3] Remarkably, not only the general theory of groups is undecidable, but also several more specific theories, for example (as established by Mal'cev 1961) the theory of finite groups. Mal'cev also established that the theory of semigroups and the theory of rings are undecidable. Robinson established in 1949 that the theory of fields is undecidable.

- Robinson arithmetic (and therefore any consistent extension, such as Peano arithmetic) is essentially undecidable, as established by Raphael Robinson in 1950.

- The first-order theory with equality and two function symbols[4]

The interpretability method is often used to establish undecidability of theories. If an essentially undecidable theory T is interpretable in a consistent theory S, then S is also essentially undecidable. This is closely related to the concept of a many-one reduction in computability theory.

48.6   Semidecidability

A property of a theory or logical system weaker than decidability is **semidecidability**. A theory is semidecidable if there is an effective method which, given an arbitrary formula, will always tell correctly when the formula is in the theory, but may give either a negative answer or no answer at all when the formula is not in the theory. A logical system is semidecidable if there is an effective method for generating theorems (and only theorems) such that every theorem will eventually be generated. This is different from decidability because in a semidecidable system there may be no effective procedure for checking that a formula is *not* a theorem.

Every decidable theory or logical system is semidecidable, but in general the converse is not true; a theory is decidable if and only if both it and its complement are semi-decidable. For example, the set of logical validities V of first-order logic is semi-decidable, but not decidable. In this case, it is because there is no effective method for determining for an arbitrary formula A whether A is not in V. Similarly, the set of logical consequences of any recursively enumerable set of first-order axioms is semidecidable. Many of the examples of undecidable first-order theories given above are of this form.

48.7   Relationship with completeness

Decidability should not be confused with completeness. For example, the theory of algebraically closed fields is decidable but incomplete, whereas the set of all true first-order statements about nonnegative integers in the language with + and × is complete but undecidable. Unfortunately, as a terminological ambiguity, the term "undecidable statement" is sometimes used as a synonym for independent statement.

48.8   See also

- László Kalmár (1936)

- Alonzo Church (1956)

- W.V.O. Quine (1953)

- Meyer and Lambert (1967)

48.9   References

48.9.1   Notes

[1] Trakhtenbrot, 1953

[2] Donald Monk (1976). *Mathematical Logic*. Springer-Verlag. ISBN 9780387901701.

[3] Tarski, A.; Mostovski, A.; Robinson, R. (1953), *Undecidable Theories*, Studies in Logic and the Foundation of Mathematics, North-Holland, Amsterdam

[4] Gurevich, Yuri (1976). "The Decision Problem for Standard Classes". *J. Symb. Log.* **41** (2): 460—464. Retrieved 5 August 2014.

48.9.2   Bibliography

- Barwise, Jon (1982), "Introduction to first-order logic", in Barwise, Jon, *Handbook of Mathematical Logic*, Studies in Logic and the Foundations of Mathematics, Amsterdam: North-Holland, ISBN 978-0-444-86388-1

- Cantone, D., E. G. Omodeo and A. Policriti, "Set Theory for Computing. From Decision Procedures to Logic Programming with Sets," Monographs in Computer Science, Springer, 2001.

- Chagrov, Alexander; Zakharyaschev, Michael (1997), *Modal logic*, Oxford Logic Guides **35**, The Clarendon Press Oxford University Press, ISBN 978-0-19-853779-3, MR 1464942

- Davis, Martin (1958), *Computability and Unsolvability*, McGraw-Hill Book Company, Inc, New York

- Enderton, Herbert (2001), *A mathematical introduction to logic* (2nd ed.), Boston, MA: Academic Press, ISBN 978-0-12-238452-3

- Keisler, H. J. (1982), "Fundamentals of model theory", in Barwise, Jon, *Handbook of Mathematical Logic*, Studies in Logic and the Foundations of Mathematics, Amsterdam: North-Holland, ISBN 978-0-444-86388-1

- Monk, J. Donald (1976), *Mathematical Logic*, Berlin, New York: Springer-Verlag

Chapter 49

Boolean algebra (structure)

For an introduction to the subject, see Boolean algebra. For an alternative presentation, see Boolean algebras canonically defined.

In abstract algebra, a **Boolean algebra** or **Boolean lattice** is a complemented distributive lattice. This type of algebraic structure captures essential properties of both set operations and logic operations. A Boolean algebra can be seen as a generalization of a power set algebra or a field of sets, or its elements can be viewed as generalized truth values. It is also a special case of a De Morgan algebra and a Kleene algebra (with involution).

Every Boolean algebra gives rise to a Boolean ring, and vice versa, with ring multiplication corresponding to conjunction or meet ∧, and ring addition to exclusive disjunction or symmetric difference (not disjunction ∨). However, the theory of Boolean rings has an inherent asymmetry between the two operators, while the axioms and theorems of Boolean algebra express the symmetry of the theory described by the duality principle.[1]

49.1 History

The term "Boolean algebra" honors George Boole (1815–1864), a self-educated English mathematician. He introduced the algebraic system initially in a small pamphlet, *The Mathematical Analysis of Logic*, published in 1847 in response to an ongoing public controversy between Augustus De Morgan and William Hamilton, and later as a more substantial book, *The Laws of Thought*, published in 1854. Boole's formulation differs from that described above in some important respects. For example, conjunction and disjunction in Boole were not a dual pair of operations. Boolean algebra emerged in the 1860s, in papers written by William Jevons and Charles Sanders Peirce. The first systematic presentation of Boolean algebra and distributive lattices is owed to the 1890 *Vorlesungen* of Ernst Schröder. The first extensive treatment of Boolean algebra in English is A. N. Whitehead's 1898 *Universal Algebra*. Boolean algebra as an axiomatic algebraic structure in the modern axiomatic sense begins with a 1904 paper by Edward V. Huntington. Boolean algebra came of age as serious mathematics with the work of Marshall Stone in the 1930s, and with Garrett Birkhoff's 1940 *Lattice Theory*. In the 1960s, Paul Cohen, Dana Scott, and others found deep new results in mathematical logic and axiomatic set theory using offshoots of Boolean algebra, namely forcing and Boolean-valued models.

49.2 Definition

A **Boolean algebra** is a six-tuple consisting of a set A, equipped with two binary operations ∧ (called "meet" or "and"), ∨ (called "join" or "or"), a unary operation ¬ (called "complement" or "not") and two elements 0 and 1 (called "bottom" and "top", or "least" and "greatest" element, also denoted by the symbols ⊥ and ⊤, respectively), such that for all elements a, b and c of A, the following axioms hold:[2]

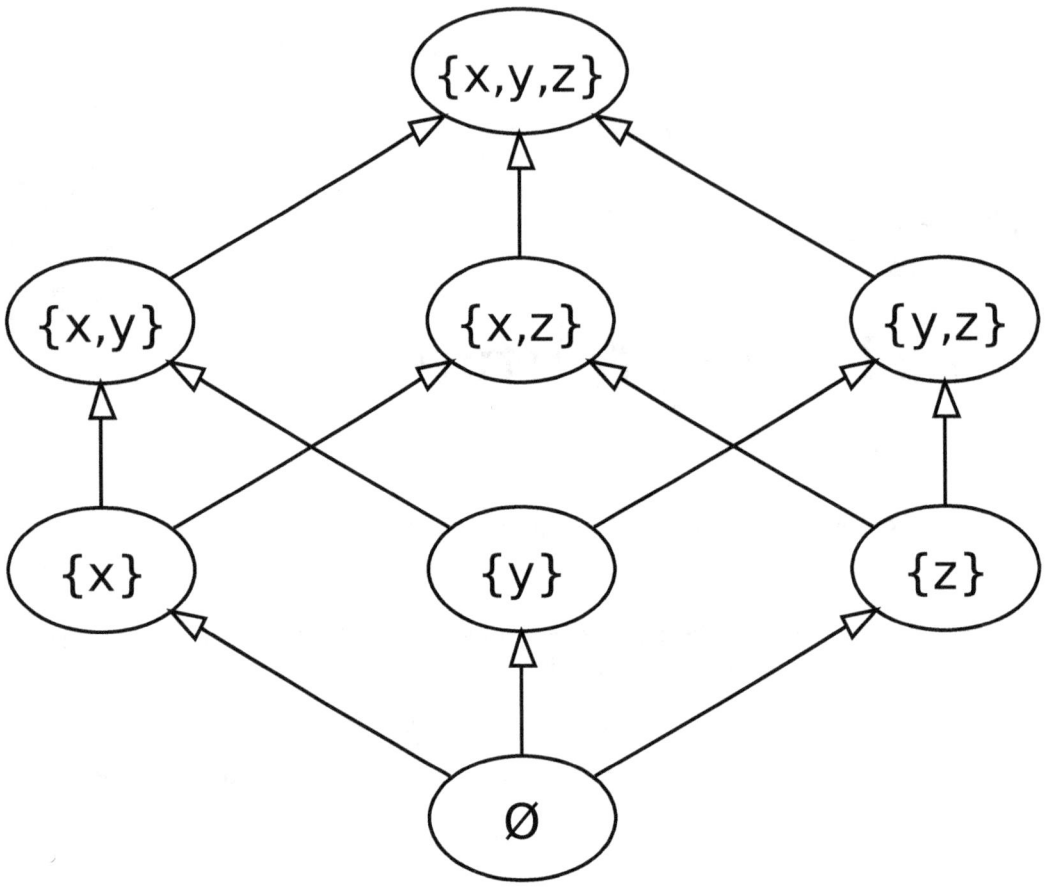

Boolean lattice of subsets

Note, however, that the absorption law can be excluded from the set of axioms as it can be derived by the other axioms.

A Boolean algebra with only one element is called a **trivial Boolean algebra** or a **degenerate Boolean algebra**. (Some authors require 0 and 1 to be *distinct* elements in order to exclude this case.)

It follows from the last three pairs of axioms above (identity, distributivity and complements), or from the absorption axiom, that

$$a = b \wedge a \text{ if and only if } a \vee b = b.$$

The relation $\leq$ defined by $a \leq b$ if these equivalent conditions hold, is a partial order with least element 0 and greatest element 1. The meet $a \wedge b$ and the join $a \vee b$ of two elements coincide with their infimum and supremum, respectively, with respect to $\leq$.

The first four pairs of axioms constitute a definition of a bounded lattice.

It follows from the first five pairs of axioms that any complement is unique.

The set of axioms is self-dual in the sense that if one exchanges $\vee$ with $\wedge$ and 0 with 1 in an axiom, the result is again an axiom. Therefore by applying this operation to a Boolean algebra (or Boolean lattice), one obtains another Boolean algebra with the same elements; it is called its **dual**.[3]

49.3 Examples

- The simplest non-trivial Boolean algebra, the two-element Boolean algebra, has only two elements, 0 and 1, and is defined by the rules:

  - It has applications in logic, interpreting 0 as *false*, 1 as *true*, ∧ as *and*, ∨ as *or*, and ¬ as *not*. Expressions involving variables and the Boolean operations represent statement forms, and two such expressions can be shown to be equal using the above axioms if and only if the corresponding statement forms are logically equivalent.

  - The two-element Boolean algebra is also used for circuit design in electrical engineering; here 0 and 1 represent the two different states of one bit in a digital circuit, typically high and low voltage. Circuits are described by expressions containing variables, and two such expressions are equal for all values of the variables if and only if the corresponding circuits have the same input-output behavior. Furthermore, every possible input-output behavior can be modeled by a suitable Boolean expression.

  - The two-element Boolean algebra is also important in the general theory of Boolean algebras, because an equation involving several variables is generally true in all Boolean algebras if and only if it is true in the two-element Boolean algebra (which can be checked by a trivial brute force algorithm for small numbers of variables). This can for example be used to show that the following laws (*Consensus theorems*) are generally valid in all Boolean algebras:

    - $(a \lor b) \land (\neg a \lor c) \land (b \lor c) \equiv (a \lor b) \land (\neg a \lor c)$
    - $(a \land b) \lor (\neg a \land c) \lor (b \land c) \equiv (a \land b) \lor (\neg a \land c)$

- The power set (set of all subsets) of any given nonempty set S forms a Boolean algebra, an algebra of sets, with the two operations $\lor := \cup$ (union) and $\land := \cap$ (intersection). The smallest element 0 is the empty set and the largest element 1 is the set S itself.

  - After the two-element Boolean algebra, the simplest Boolean algebra is that defined by the power set of two atoms:

- The set of all subsets of S that are either finite or cofinite is a Boolean algebra, an algebra of sets.

- Starting with the propositional calculus with κ sentence symbols, form the Lindenbaum algebra (that is, the set of sentences in the propositional calculus modulo tautology). This construction yields a Boolean algebra. It is in fact the free Boolean algebra on κ generators. A truth assignment in propositional calculus is then a Boolean algebra homomorphism from this algebra to the two-element Boolean algebra.

- Given any linearly ordered set L with a least element, the interval algebra is the smallest algebra of subsets of L containing all of the half-open intervals $[a, b)$ such that a is in L and b is either in L or equal to ∞. Interval algebras are useful in the study of Lindenbaum-Tarski algebras; every countable Boolean algebra is isomorphic to an interval algebra.

- For any natural number n, the set of all positive divisors of n, defining $a \leq b$ if a divides b, forms a distributive lattice. This lattice is a Boolean algebra if and only if n is square-free. The bottom and the top element of this Boolean algebra is the natural number 1 and n, respectively. The complement of a is given by n/a. The meet and the join of a and b is given by the greatest common divisor (gcd) and the least common multiple (lcm) of a and b, respectively. The ring addition $a+b$ is given by $\mathrm{lcm}(a,b)/\gcd(a,b)$. The picture shows an example for $n = 30$. As a counter-example, considering the non-square-free $n=60$, the greatest common divisor of 30 and its complement 2 would be 2, while it should be the bottom element 1.

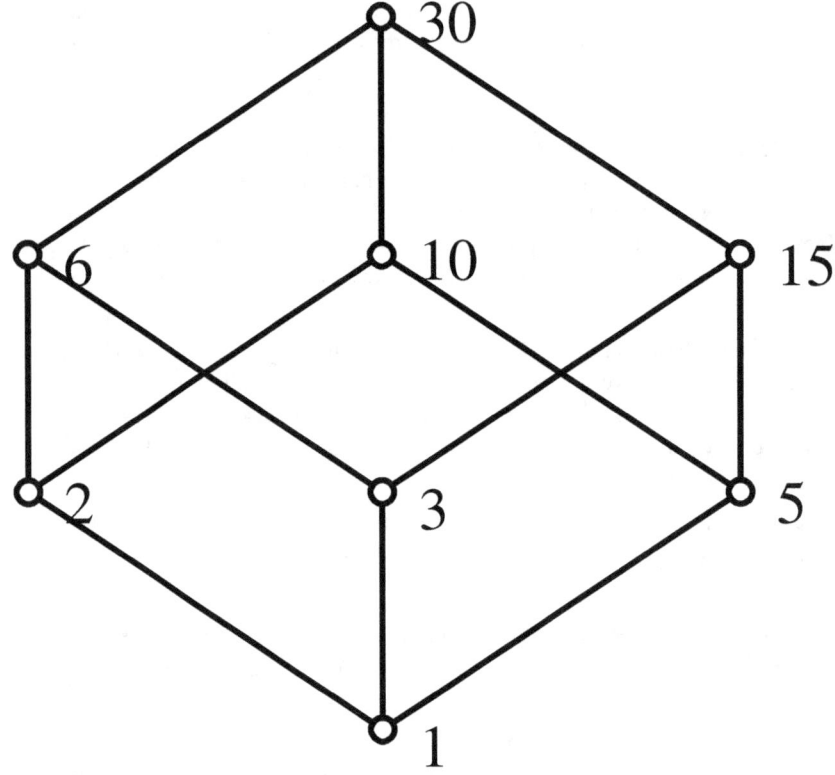

Hasse diagram of the Boolean algebra of divisors of 30.

- Other examples of Boolean algebras arise from topological spaces: if X is a topological space, then the collection of all subsets of X which are both open and closed forms a Boolean algebra with the operations $\vee := \cup$ (union) and $\wedge := \cap$ (intersection).

- If R is an arbitrary ring and we define the set of *central idempotents* by
  $A = \{\, e \in R : e^2 = e,\ ex = xe,\ \forall x \in R \,\}$
  then the set A becomes a Boolean algebra with the operations $e \vee f := e + f - ef$ and $e \wedge f := ef$.

49.4   Homomorphisms and isomorphisms

A *homomorphism* between two Boolean algebras A and B is a function $f : A \to B$ such that for all a, b in A:

$f(a \vee b) = f(a) \vee f(b),$

$f(a \wedge b) = f(a) \wedge f(b),$

$f(0) = 0,$

$f(1) = 1.$

It then follows that $f(\neg a) = \neg f(a)$ for all a in A. The class of all Boolean algebras, together with this notion of morphism, forms a full subcategory of the category of lattices.

49.5   Boolean rings

Main article: Boolean ring

Every Boolean algebra $(A, \wedge, \vee)$ gives rise to a ring $(A, +, \cdot)$ by defining $a + b := (a \wedge \neg b) \vee (b \wedge \neg a) = (a \vee b) \wedge \neg(a \wedge b)$ (this operation is called symmetric difference in the case of sets and XOR in the case of logic) and $a \cdot b := a \wedge b$. The zero element of this ring coincides with the 0 of the Boolean algebra; the multiplicative identity element of the ring is the 1 of the Boolean algebra. This ring has the property that $a \cdot a = a$ for all a in A; rings with this property are called Boolean rings.

Conversely, if a Boolean ring A is given, we can turn it into a Boolean algebra by defining $x \vee y := x + y + (x \cdot y)$ and $x \wedge y := x \cdot y$. [4][5] Since these two constructions are inverses of each other, we can say that every Boolean ring arises from a Boolean algebra, and vice versa. Furthermore, a map $f : A \rightarrow B$ is a homomorphism of Boolean algebras if and only if it is a homomorphism of Boolean rings. The categories of Boolean rings and Boolean algebras are equivalent.[6]

Hsiang (1985) gave a rule-based algorithm to check whether two arbitrary expressions denote the same value in every Boolean ring. More generally, Boudet, Jouannaud, and Schmidt-Schauß (1989) gave an algorithm to solve equations between arbitrary Boolean-ring expressions. Employing the similarity of Boolean rings and Boolean algebras, both algorithms have applications in automated theorem proving.

49.6   Ideals and filters

Main articles: Ideal (order theory) and Filter (mathematics)

An *ideal* of the Boolean algebra A is a subset I such that for all x, y in I we have x $\vee$ y in I and for all a in A we have $a \wedge x$ in I. This notion of ideal coincides with the notion of ring ideal in the Boolean ring A. An ideal I of A is called *prime* if $I \neq A$ and if $a \wedge b$ in I always implies a in I or b in I. Furthermore, for every $a \in A$ we have that $a \wedge -a = 0 \in I$ and then $a \in I$ or $-a \in I$ for every $a \in A$, if I is prime. An ideal I of A is called *maximal* if $I \neq A$ and if the only ideal properly containing I is A itself. For an ideal I, if $a \notin I$ and $-a \notin I$, then $I \cup \{a\}$ or $I \cup \{-a\}$ is properly contained in another ideal J. Hence, that an I is not maximal and therefore the notions of prime ideal and maximal ideal are equivalent in Boolean algebras. Moreover, these notions coincide with ring theoretic ones of prime ideal and maximal ideal in the Boolean ring A.

The dual of an *ideal* is a *filter*. A *filter* of the Boolean algebra A is a subset p such that for all x, y in p we have $x \wedge y$ in p and for all a in A we have $a \vee x$ in p. The dual of a *maximal* (or *prime*) *ideal* in a Boolean algebra is *ultrafilter*. Ultrafilters can alternatively be described as 2-valued morphisms from A to the two-element Boolean algebra. The statement *every filter in a Boolean algebra can be extended to an ultrafilter* is called the *Ultrafilter Theorem* and can not be proved in ZF, if ZF is consistent. Within ZF, it is strictly weaker than the axiom of choice. The Ultrafilter Theorem has many equivalent formulations: *every Boolean algebra has an ultrafilter*, *every ideal in a Boolean algebra can be extended to a prime ideal*, etc.

49.7 Representations

It can be shown that every *finite* Boolean algebra is isomorphic to the Boolean algebra of all subsets of a finite set. Therefore, the number of elements of every finite Boolean algebra is a power of two.

Stone's celebrated *representation theorem for Boolean algebras* states that *every* Boolean algebra A is isomorphic to the Boolean algebra of all clopen sets in some (compact totally disconnected Hausdorff) topological space.

49.8 Axiomatics

The first axiomatization of Boolean lattices/algebras in general was given by Alfred North Whitehead in 1898.[7][8] It included the above axioms and additionally $x \vee 1 = 1$ and $x \wedge 0 = 0$. In 1904, the American mathematician Edward V. Huntington (1874–1952) gave probably the most parsimonious axiomatization based on $\wedge$, $\vee$, $\neg$, even proving the associativity laws (see box).[9] He also proved that these axioms are independent of each other.[10] In 1933, Huntington set out the following elegant axiomatization for Boolean algebra. It requires just one binary operation + and a unary functional symbol n, to be read as 'complement', which satisfy the following laws:

1. *Commutativity*: $x + y = y + x$.

2. *Associativity*: $(x + y) + z = x + (y + z)$.

3. *Huntington equation*: $n(n(x) + y) + n(n(x) + n(y)) = x$.

Herbert Robbins immediately asked: If the Huntington equation is replaced with its dual, to wit:

4. *Robbins Equation*: $n(n(x + y) + n(x + n(y))) = x$,

do (1), (2), and (4) form a basis for Boolean algebra? Calling (1), (2), and (4) a *Robbins algebra*, the question then becomes: Is every Robbins algebra a Boolean algebra? This question (which came to be known as the Robbins conjecture) remained open for decades, and became a favorite question of Alfred Tarski and his students. In 1996, William McCune at Argonne National Laboratory, building on earlier work by Larry Wos, Steve Winker, and Bob Veroff, answered Robbins's question in the affirmative: Every Robbins algebra is a Boolean algebra. Crucial to McCune's proof was the automated reasoning program EQP he designed. For a simplification of McCune's proof, see Dahn (1998).

49.9 Generalizations

Removing the requirement of existence of a unit from the axioms of Boolean algebra yields "generalized Boolean algebras". Formally, a distributive lattice B is a generalized Boolean lattice, if it has a smallest element 0 and for any elements a and b in B such that $a \leq b$, there exists an element x such that $a \wedge x = 0$ and $a \vee x = b$. Defining $a \setminus b$ as the unique x such that $(a \wedge b) \vee x = a$ and $(a \wedge b) \wedge x = 0$, we say that the structure $(B, \wedge, \vee, \setminus, 0)$ is a *generalized Boolean algebra*, while $(B, \vee, 0)$ is a *generalized Boolean semilattice*. Generalized Boolean lattices are exactly the ideals of Boolean lattices.

A structure that satisfies all axioms for Boolean algebras except the two distributivity axioms is called an orthocomplemented lattice. Orthocomplemented lattices arise naturally in quantum logic as lattices of closed subspaces for separable Hilbert spaces.

49.10 See also

49.11 Notes

[1] Givant and Paul Halmos, 2009, p. 20

[2] Davey, Priestley, 1990, p.109, 131, 144

[3] Goodstein, R. L. (2012), "Chapter 2: The self-dual system of axioms", *Boolean Algebra*, Courier Dover Publications, pp. 21ff, ISBN 9780486154978.

[4] Stone, 1936

[5] Hsiang, 1985, p.260

[6] Cohn (2003), p. 81.

[7] Padmanabhan, p. 73

[8] Whitehead, 1898, p.37

[9] Huntington, 1904, p.292-293, (first of several axiomatizations by Huntington)

[10] Huntington, 1904, p.296

49.12   References

- Brown, Stephen; Vranesic, Zvonko (2002), *Fundamentals of Digital Logic with VHDL Design* (2nd ed.), McGraw–Hill, ISBN 978-0-07-249938-4. See Section 2.5.

- A. Boudet, J.P. Jouannaud, M. Schmidt-Schauß (1989). "Unification in Boolean Rings and Abelian Groups" (PDF). *Journal of Symbolic Computation* **8**: 449–477. doi:10.1016/s0747-7171(89)80054-9.

- Cohn, Paul M. (2003), *Basic Algebra: Groups, Rings, and Fields*, Springer, pp. 51, 70–81, ISBN 9781852335878

- Cori, Rene; Lascar, Daniel (2000), *Mathematical Logic: A Course with Exercises*, Oxford University Press, ISBN 978-0-19-850048-3. See Chapter 2.

- Dahn, B. I. (1998), "Robbins Algebras are Boolean: A Revision of McCune's Computer-Generated Solution of the Robbins Problem", *Journal of Algebra* **208** (2): 526–532, doi:10.1006/jabr.1998.7467.

- B.A. Davey, H.A. Priestley (1990). *Introduction to Lattices and Order*. Cambridge Mathematical Textbooks. Cambridge University Press.

- Givant, Steven; Halmos, Paul (2009), *Introduction to Boolean Algebras*, Undergraduate Texts in Mathematics, Springer, ISBN 978-0-387-40293-2.

- Halmos, Paul (1963), *Lectures on Boolean Algebras*, Van Nostrand, ISBN 978-0-387-90094-0.

- Halmos, Paul; Givant, Steven (1998), *Logic as Algebra*, Dolciani Mathematical Expositions **21**, Mathematical Association of America, ISBN 978-0-88385-327-6.

- Hsiang, Jieh (1985). "Refutational Theorem Proving Using Term Rewriting Systems" (PDF). *AI* **25**: 255–300. doi:10.1016/0004-3702(85)90074-8.

- Edward V. Huntington (1904). "Sets of Independent Postulates for the Algebra of Logic" (PDF). *These Transactions* **5**: 288–309. doi:10.1090/s0002-9947-1904-1500675-4.

- Huntington, E. V. (1933), "New sets of independent postulates for the algebra of logic" (PDF), *Transactions of the American Mathematical Society* (American Mathematical Society) **35** (1): 274–304, doi:10.2307/1989325, JSTOR 1989325.

- Huntington, E. V. (1933), "Boolean algebra: A correction", *Transactions of the American Mathematical Society* (American Mathematical Society) **35** (2): 557–558, doi:10.2307/1989783, JSTOR 1989783.

- Mendelson, Elliott (1970), *Boolean Algebra and Switching Circuits*, Schaum's Outline Series in Mathematics, McGraw–Hill, ISBN 978-0-07-041460-0.

- Monk, J. Donald; Bonnet, R., eds. (1989), *Handbook of Boolean Algebras*, North-Holland, ISBN 978-0-444-87291-3. In 3 volumes. (Vol.1:ISBN 978-0-444-70261-6, Vol.2:ISBN 978-0-444-87152-7, Vol.3:ISBN 978-0-444-87153-4)

- Padmanabhan, Ranganathan; Rudeanu, Sergiu (2008), *Axioms for lattices and boolean algebras*, World Scientific, ISBN 978-981-283-454-6.

- Sikorski, Roman (1966), *Boolean Algebras*, Ergebnisse der Mathematik und ihrer Grenzgebiete, Springer Verlag.

- Stoll, R. R. (1963), *Set Theory and Logic*, W. H. Freeman, ISBN 978-0-486-63829-4. Reprinted by Dover Publications, 1979.

- Marshall H. Stone (1936). "The Theory of Representations for Boolean Algebra". *Trans. AMS* **40**: 37–111.

- A.N. Whitehead (1898). *A Treatise on Universal Algebra*. Cambridge University Press. ISBN 1-4297-0032-7.

49.13 External links

- Hazewinkel, Michiel, ed. (2001), "Boolean algebra", *Encyclopedia of Mathematics*, Springer, ISBN 978-1-55608-010-4

- Boolean Algebra from AllAboutCircuits

- Stanford Encyclopedia of Philosophy: "The Mathematics of Boolean Algebra," by J. Donald Monk.

- McCune W., 1997. *Robbins Algebras Are Boolean* JAR 19(3), 263—276

- "Boolean Algebra" by Eric W. Weisstein, Wolfram Demonstrations Project, 2007.

A monograph available free online:

- Burris, Stanley N.; Sankappanavar, H. P., 1981. *A Course in Universal Algebra*. Springer-Verlag. ISBN 3-540-90578-2.

- Weisstein, Eric W., "Boolean Algebra", *MathWorld*.

Chapter 50

Algebraically closed field

In abstract algebra, an **algebraically closed field** F contains a root for every non-constant polynomial in $F[x]$, the ring of polynomials in the variable x with coefficients in F.

50.1 Examples

As an example, the field of real numbers is not algebraically closed, because the polynomial equation $x^2 + 1 = 0$ has no solution in real numbers, even though all its coefficients (1 and 0) are real. The same argument proves that no subfield of the real field is algebraically closed; in particular, the field of rational numbers is not algebraically closed. Also, no finite field F is algebraically closed, because if $a_1, a_2, \ldots, an$ are the elements of F, then the polynomial $(x - a_1)(x - a_2) \cdots (x - an) + 1$ has no zero in F. By contrast, the fundamental theorem of algebra states that the field of complex numbers is algebraically closed. Another example of an algebraically closed field is the field of (complex) algebraic numbers.

50.2 Equivalent properties

Given a field F, the assertion "F is algebraically closed" is equivalent to other assertions:

50.2.1 The only irreducible polynomials are those of degree one

The field F is algebraically closed if and only if the only irreducible polynomials in the polynomial ring $F[x]$ are those of degree one.

The assertion "the polynomials of degree one are irreducible" is trivially true for any field. If F is algebraically closed and $p(x)$ is an irreducible polynomial of $F[x]$, then it has some root a and therefore $p(x)$ is a multiple of $x - a$. Since $p(x)$ is irreducible, this means that $p(x) = k(x - a)$, for some $k \in F \setminus \{0\}$. On the other hand, if F is not algebraically closed, then there is some non-constant polynomial $p(x)$ in $F[x]$ without roots in F. Let $q(x)$ be some irreducible factor of $p(x)$. Since $p(x)$ has no roots in F, $q(x)$ also has no roots in F. Therefore, $q(x)$ has degree greater than one, since every first degree polynomial has one root in F.

50.2.2 Every polynomial is a product of first degree polynomials

The field F is algebraically closed if and only if every polynomial $p(x)$ of degree $n \geq 1$, with coefficients in F, splits into linear factors. In other words, there are elements $k, x_1, x_2, \ldots, xn$ of the field F such that $p(x) = k(x - x_1)(x - x_2) \cdots (x - xn)$.

If F has this property, then clearly every non-constant polynomial in $F[x]$ has some root in F; in other words, F is algebraically closed. On the other hand, that the property stated here holds for F if F is algebraically closed follows from the previous property together with the fact that, for any field K, any polynomial in $K[x]$ can be written as a product of irreducible polynomials.

50.2.3 Polynomials of prime degree have roots

J. Shipman showed in 2007 that if every polynomial over F of prime degree has a root in F, then every non-constant polynomial has a root in F, thus F is algebraically closed.

50.2.4 The field has no proper algebraic extension

The field F is algebraically closed if and only if it has no proper algebraic extension.

If F has no proper algebraic extension, let $p(x)$ be some irreducible polynomial in $F[x]$. Then the quotient of $F[x]$ modulo the ideal generated by $p(x)$ is an algebraic extension of F whose degree is equal to the degree of $p(x)$. Since it is not a proper extension, its degree is 1 and therefore the degree of $p(x)$ is 1.

On the other hand, if F has some proper algebraic extension K, then the minimal polynomial of an element in $K \setminus F$ is irreducible and its degree is greater than 1.

50.2.5 The field has no proper finite extension

The field F is algebraically closed if and only if it has no finite algebraic extension because if, within the previous proof, the word "algebraic" is replaced by the word "finite", then the proof is still valid.

50.2.6 Every endomorphism of F^n has some eigenvector

The field F is algebraically closed if and only if, for each natural number n, every linear map from F^n into itself has some eigenvector.

An endomorphism of F^n has an eigenvector if and only if its characteristic polynomial has some root. Therefore, when F is algebraically closed, every endomorphism of F^n has some eigenvector. On the other hand, if every endomorphism of F^n has an eigenvector, let $p(x)$ be an element of $F[x]$. Dividing by its leading coefficient, we get another polynomial $q(x)$ which has roots if and only if $p(x)$ has roots. But if $q(x) = x^n + a_{n-1}x^{n-1} + \cdots + a_0$, then $q(x)$ is the characteristic polynomial of the companion matrix

$$\begin{pmatrix} 0 & 0 & \cdots & 0 & -a_0 \\ 1 & 0 & \cdots & 0 & -a_1 \\ 0 & 1 & \cdots & 0 & -a_2 \\ \vdots & \vdots & \ddots & \vdots & \vdots \\ 0 & 0 & \cdots & 1 & -a_{n-1} \end{pmatrix}.$$

50.2.7 Decomposition of rational expressions

The field F is algebraically closed if and only if every rational function in one variable x, with coefficients in F, can be written as the sum of a polynomial function with rational functions of the form $a/(x - b)^n$, where n is a natural number, and a and b are elements of F.

If F is algebraically closed then, since the irreducible polynomials in $F[x]$ are all of degree 1, the property stated above holds by the theorem on partial fraction decomposition.

On the other hand, suppose that the property stated above holds for the field F. Let $p(x)$ be an irreducible element in $F[x]$. Then the rational function $1/p$ can be written as the sum of a polynomial function q with rational functions of the form $a/(x-b)^n$. Therefore, the rational expression

$$\frac{1}{p(x)} - q(x) = \frac{1 - p(x)q(x)}{p(x)}$$

can be written as a quotient of two polynomials in which the denominator is a product of first degree polynomials. Since $p(x)$ is irreducible, it must divide this product and, therefore, it must also be a first degree polynomial.

50.2.8   Relatively prime polynomials and roots

For any field F, if two polynomials $p(x),q(x) \in F[x]$ are relatively prime then they do not have a common root, for if $a \in F$ was a common root, then $p(x)$ and $q(x)$ would both be multiples of $x - a$ and therefore they would not be relatively prime. The fields for which the reverse implication holds (that is, the fields such that whenever two polynomials have no common root then they are relatively prime) are precisely the algebraically closed fields.

If the field F is algebraically closed, let $p(x)$ and $q(x)$ be two polynomials which are not relatively prime and let $r(x)$ be their greatest common divisor. Then, since $r(x)$ is not constant, it will have some root a, which will be then a common root of $p(x)$ and $q(x)$.

If F is not algebraically closed, let $p(x)$ be a polynomial whose degree is at least 1 without roots. Then $p(x)$ and $p(x)$ are not relatively prime, but they have no common roots (since none of them has roots).

50.3   Other properties

If F is an algebraically closed field and n is a natural number, then F contains all nth roots of unity, because these are (by definition) the n (not necessarily distinct) zeroes of the polynomial $x^n - 1$. A field extension that is contained in an extension generated by the roots of unity is a *cyclotomic extension*, and the extension of a field generated by all roots of unity is sometimes called its *cyclotomic closure*. Thus algebraically closed fields are cyclotomically closed. The converse is not true. Even assuming that every polynomial of the form $x^n - a$ splits into linear factors is not enough to assure that the field is algebraically closed.

If a proposition which can be expressed in the language of first-order logic is true for an algebraically closed field, then it is true for every algebraically closed field with the same characteristic. Furthermore, if such a proposition is valid for an algebraically closed field with characteristic 0, then not only is it valid for all other algebraically closed fields with characteristic 0, but there is some natural number N such that the proposition is valid for every algebraically closed field with characteristic p when $p > N$.[1]

Every field F has some extension which is algebraically closed. Among all such extensions there is one and (up to isomorphism, but not unique isomorphism) only one which is an algebraic extension of F;[2] it is called the algebraic closure of F.

The theory of algebraically closed fields has quantifier elimination.

50.4   Notes

[1] See subsections *Rings and fields* and *Properties of mathematical theories* in §2 of J. Barwise's "An introduction to first-order logic".

[2] See Lang's *Algebra*, §VII.2 or van der Waerden's *Algebra I*, §10.1.

50.5 References

- Barwise, Jon (1978), "An introduction to first-order logic", in Barwise, Jon, *Handbook of Mathematical Logic*, Studies in Logic and the Foundations of Mathematics, North Holland, ISBN 0-7204-2285-X

- Lang, Serge (2002), *Algebra*, Graduate Texts in Mathematics **211** (Revised third ed.), New York: Springer-Verlag, ISBN 978-0-387-95385-4, MR 1878556

- Shipman, Joseph (2007), "Improving the Fundamental Theorem of Algebra", *Mathematical Intelligencer* **29** (4): 9–14, doi:10.1007/BF02986170, ISSN 0343-6993

- van der Waerden, Bartel Leendert (2003), *Algebra* **I** (7th ed.), Springer-Verlag, ISBN 0-387-40624-7

Chapter 51

Bijection

In mathematics, a **bijection**, **bijective function** or **one-to-one correspondence** is a function between the elements of two sets, where every element of one set is paired with exactly one element of the other set, and every element of the other set is paired with exactly one element of the first set. There are no unpaired elements. In mathematical terms, a bijective function $f: X \rightarrow Y$ is a one-to-one (injective) and onto (surjective) mapping of a set X to a set Y.

A bijection from the set X to the set Y has an inverse function from Y to X. If X and Y are finite sets, then the existence of a bijection means they have the same number of elements. For infinite sets the picture is more complicated, leading to the concept of cardinal number, a way to distinguish the various sizes of infinite sets.

A bijective function from a set to itself is also called a *permutation*.

Bijective functions are essential to many areas of mathematics including the definitions of isomorphism, homeomorphism, diffeomorphism, permutation group, and projective map.

51.1 Definition

For more details on notation, see Function (mathematics) § Notation.

For a pairing between X and Y (where Y need not be different from X) to be a bijection, four properties must hold:

1. each element of X must be paired with at least one element of Y,

2. no element of X may be paired with more than one element of Y,

3. each element of Y must be paired with at least one element of X, and

4. no element of Y may be paired with more than one element of X.

Satisfying properties (1) and (2) means that a bijection is a function with domain X. It is more common to see properties (1) and (2) written as a single statement: Every element of X is paired with exactly one element of Y. Functions which satisfy property (3) are said to be "onto Y " and are called surjections (or **surjective functions**). Functions which satisfy property (4) are said to be "one-to-one functions" and are called injections (or **injective functions**).[1] With this terminology, a bijection is a function which is both a surjection and an injection, or using other words, a bijection is a function which is both "one-to-one" and "onto".

51.2 Examples

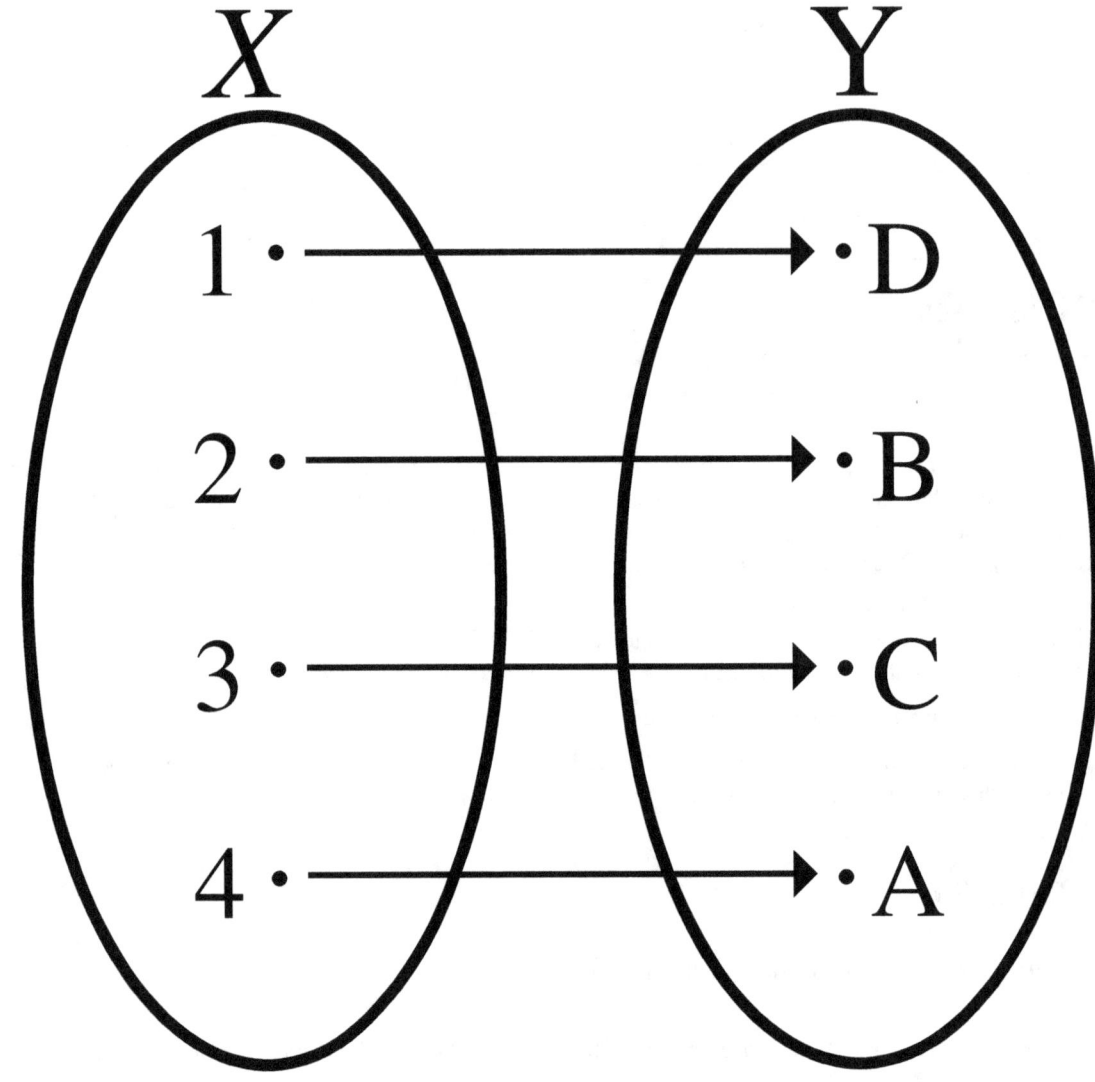

A bijective function, f: X → Y, where set X is {1, 2, 3, 4} and set Y is {A, B, C, D}. For example, f(1) = D.

51.2.1 Batting line-up of a baseball team

Consider the batting line-up of a baseball team (or any list of all the players of any sports team). The set X will be the nine players on the team and the set Y will be the nine positions in the batting order (1st, 2nd, 3rd, etc.) The "pairing" is given by which player is in what position in this order. Property (1) is satisfied since each player is somewhere in the list. Property (2) is satisfied since no player bats in two (or more) positions in the order. Property (3) says that for each position in the order, there is some player batting in that position and property (4) states that two or more players are never batting in the same position in the list.

51.2.2 Seats and students of a classroom

In a classroom there are a certain number of seats. A bunch of students enter the room and the instructor asks them all to be seated. After a quick look around the room, the instructor declares that there is a bijection between the set of students and the set of seats, where each student is paired with the seat they are sitting in. What the instructor observed in order to reach this conclusion was that:

1. Every student was in a seat (there was no one standing),

2. No student was in more than one seat,

3. Every seat had someone sitting there (there were no empty seats), and

4. No seat had more than one student in it.

The instructor was able to conclude that there were just as many seats as there were students, without having to count either set.

51.3 More mathematical examples and some non-examples

- For any set X, the identity function $1X: X \to X$, $1X(x) = x$, is bijective.

- The function $f: \mathbf{R} \to \mathbf{R}$, $f(x) = 2x + 1$ is bijective, since for each y there is a unique $x = (y - 1)/2$ such that $f(x) = y$. In more generality, any linear function over the reals, $f: \mathbf{R} \to \mathbf{R}$, $f(x) = ax + b$ (where a is non-zero) is a bijection. Each real number y is obtained from (paired with) the real number $x = (y - b)/a$.

- The function $f: R \to (-\pi/2, \pi/2)$, given by $f(x) = \arctan(x)$ is bijective since each real number x is paired with exactly one angle y in the interval $(-\pi/2, \pi/2)$ so that $\tan(y) = x$ (that is, $y = \arctan(x)$). If the codomain $(-\pi/2, \pi/2)$ was made larger to include an integer multiple of $\pi/2$ then this function would no longer be onto (surjective) since there is no real number which could be paired with the multiple of $\pi/2$ by this arctan function.

- The exponential function, $g: \mathbf{R} \to \mathbf{R}$, $g(x) = e^x$, is not bijective: for instance, there is no x in $\mathbf{R}$ such that $g(x) = -1$, showing that g is not onto (surjective). However if the codomain is restricted to the positive real numbers $\mathbb{R}^+ \equiv (0, +\infty)$, then g becomes bijective; its inverse (see below) is the natural logarithm function ln.

- The function $h: \mathbf{R} \to \mathbf{R}^+$, $h(x) = x^2$ is not bijective: for instance, $h(-1) = h(1) = 1$, showing that h is not one-to-one (injective). However, if the domain is restricted to $\mathbb{R}_0^+ \equiv [0, +\infty)$, then h becomes bijective; its inverse is the positive square root function.

51.4 Inverses

A bijection f with domain X ("functionally" indicated by $f: X \to Y$) also defines a relation starting in Y and going to X (by turning the arrows around). The process of "turning the arrows around" for an arbitrary function does not, *in general*, yield a function, but properties (3) and (4) of a bijection say that this inverse relation is a function with domain Y. Moreover, properties (1) and (2) then say that this inverse *function* is a surjection and an injection, that is, the inverse function exists and is also a bijection. Functions that have inverse functions are said to be invertible. A function is invertible if and only if it is a bijection.

Stated in concise mathematical notation, a function $f: X \to Y$ is bijective if and only if it satisfies the condition

for every y in Y there is a unique x in X with $y = f(x)$.

Continuing with the baseball batting line-up example, the function that is being defined takes as input the name of one of the players and outputs the position of that player in the batting order. Since this function is a bijection, it has an inverse function which takes as input a position in the batting order and outputs the player who will be batting in that position.

51.5 Composition

The composition $g \circ f$ of two bijections $f: X \to Y$ and $g: Y \to Z$ is a bijection. The inverse of $g \circ f$ is $(g \circ f)^{-1} = (f^{-1}) \circ (g^{-1})$

.

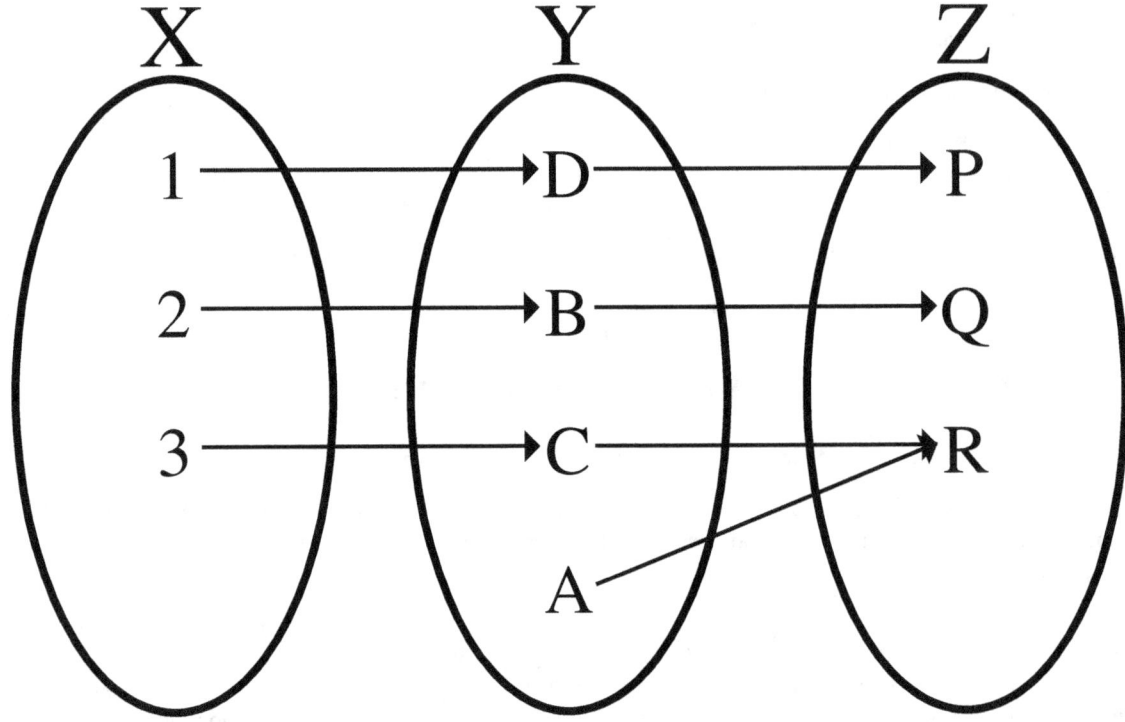

A bijection composed of an injection (left) and a surjection (right).

Conversely, if the composition $g \circ f$ of two functions is bijective, we can only say that f is injective and g is surjective.

51.6 Bijections and cardinality

If X and Y are finite sets, then there exists a bijection between the two sets X and Y if and only if X and Y have the same number of elements. Indeed, in axiomatic set theory, this is taken as the definition of "same number of elements" (equinumerosity), and generalising this definition to infinite sets leads to the concept of cardinal number, a way to distinguish the various sizes of infinite sets.

51.7 Properties

- A function $f: \mathbf{R} \to \mathbf{R}$ is bijective if and only if its graph meets every horizontal and vertical line exactly once.

- If X is a set, then the bijective functions from X to itself, together with the operation of functional composition ($\circ$), form a group, the symmetric group of X, which is denoted variously by S(X), SX, or $X!$ (X factorial).

- Bijections preserve cardinalities of sets: for a subset A of the domain with cardinality $|A|$ and subset B of the codomain with cardinality $|B|$, one has the following equalities:

$$|f(A)| = |A| \text{ and } |f^{-1}(B)| = |B|.$$

- If X and Y are finite sets with the same cardinality, and $f: X \to Y$, then the following are equivalent:

    1. f is a bijection.
    2. f is a surjection.
    3. f is an injection.

- For a finite set S, there is a bijection between the set of possible total orderings of the elements and the set of bijections from S to S. That is to say, the number of permutations of elements of S is the same as the number of total orderings of that set—namely, $n!$.

51.8   Bijections and category theory

Bijections are precisely the isomorphisms in the category **Set** of sets and set functions. However, the bijections are not always the isomorphisms for more complex categories. For example, in the category **Grp** of groups, the morphisms must be homomorphisms since they must preserve the group structure, so the isomorphisms are *group isomorphisms* which are bijective homomorphisms.

51.9   Generalization to partial functions

The notion of one-one correspondence generalizes to partial functions, where they are called **partial bijections**, although partial bijections are only required to be injective. The reason for this relaxation is that a (proper) partial function is already undefined for a portion of its domain; thus there is no compelling reason to constrain its inverse to be a total function, i.e. defined everywhere on its domain. The set of all partial bijections on a given base set is called the symmetric inverse semigroup.[2]

Another way of defining the same notion is to say that a partial bijection from A to B is any relation R (which turns out to be a partial function) with the property that R is the graph of a bijection $f:A' \to B'$, where A' is a subset of A and likewise $B' \subseteq B$.[3]

When the partial bijection is on the same set, it is sometimes called a **one-to-one partial transformation**.[4] An example is the Möbius transformation simply defined on the complex plane, rather than its completion to the extended complex plane.[5]

51.10   Contrast with

This list is incomplete; you can help by expanding it.

- Multivalued function

51.11   See also

- Injective function

- Surjective function

- Bijection, injection and surjection

- Symmetric group

- Bijective numeration

- Bijective proof

- Cardinality

- Category theory

- Ax–Grothendieck theorem

51.12   Notes

[1] There are names associated to properties (1) and (2) as well. A relation which satisfies property (1) is called a *total relation* and a relation satisfying (2) is a *single valued relation*.

[2] Christopher Hollings (16 July 2014). *Mathematics across the Iron Curtain: A History of the Algebraic Theory of Semigroups.* American Mathematical Society. p. 251. ISBN 978-1-4704-1493-1.

[3] Francis Borceux (1994). *Handbook of Categorical Algebra: Volume 2, Categories and Structures.* Cambridge University Press. p. 289. ISBN 978-0-521-44179-7.

[4] Pierre A. Grillet (1995). *Semigroups: An Introduction to the Structure Theory.* CRC Press. p. 228. ISBN 978-0-8247-9662-4.

[5] John Meakin (2007). "Groups and semigroups: connections and contrasts". In C.M. Campbell, M.R. Quick, E.F. Robertson, G.C. Smith. *Groups St Andrews 2005 Volume 2.* Cambridge University Press. p. 367. ISBN 978-0-521-69470-4. preprint citing Lawson, M. V. (1998). "The Möbius Inverse Monoid". *Journal of Algebra* **200** (2): 428. doi:10.1006/jabr.1997.7242.

51.13   References

This topic is a basic concept in set theory and can be found in any text which includes an introduction to set theory. Almost all texts that deal with an introduction to writing proofs will include a section on set theory, so the topic may be found in any of these:

- Wolf (1998). *Proof, Logic and Conjecture: A Mathematician's Toolbox.* Freeman.

- Sundstrom (2003). *Mathematical Reasoning: Writing and Proof.* Prentice-Hall.

- Smith; Eggen; St.Andre (2006). *A Transition to Advanced Mathematics (6th Ed.).* Thomson (Brooks/Cole).

- Schumacher (1996). *Chapter Zero: Fundamental Notions of Abstract Mathematics.* Addison-Wesley.

- O'Leary (2003). *The Structure of Proof: With Logic and Set Theory.* Prentice-Hall.

- Morash. *Bridge to Abstract Mathematics.* Random House.

- Maddox (2002). *Mathematical Thinking and Writing.* Harcourt/ Academic Press.

- Lay (2001). *Analysis with an introduction to proof.* Prentice Hall.

- Gilbert; Vanstone (2005). *An Introduction to Mathematical Thinking.* Pearson Prentice-Hall.

- Fletcher; Patty. *Foundations of Higher Mathematics.* PWS-Kent.

- Iglewicz; Stoyle. *An Introduction to Mathematical Reasoning.* MacMillan.

- Devlin, Keith (2004). *Sets, Functions, and Logic: An Introduction to Abstract Mathematics.* Chapman & Hall/ CRC Press.

- D'Angelo; West (2000). *Mathematical Thinking: Problem Solving and Proofs.* Prentice Hall.

- Cupillari. *The Nuts and Bolts of Proofs.* Wadsworth.

- Bond. *Introduction to Abstract Mathematics*. Brooks/Cole.

- Barnier; Feldman (2000). *Introduction to Advanced Mathematics*. Prentice Hall.

- Ash. *A Primer of Abstract Mathematics*. MAA.

51.14   External links

- Hazewinkel, Michiel, ed. (2001), "Bijection", *Encyclopedia of Mathematics*, Springer, ISBN 978-1-55608-010-4

- Weisstein, Eric W., "Bijection", *MathWorld*.

- Earliest Uses of Some of the Words of Mathematics: entry on Injection, Surjection and Bijection has the history of Injection and related terms.

Chapter 52

Constructible topology

In commutative algebra, the **constructible topology** on the spectrum $\mathrm{Spec}(A)$ of a commutative ring A is a topology where each closed set is the image of $\mathrm{Spec}(B)$ in $\mathrm{Spec}(A)$ for some algebra B over A. An important feature of this construction is that the map $\mathrm{Spec}(B) \to \mathrm{Spec}(A)$ is a closed map with respect to the constructible topology.

With respect to this topology, $\mathrm{Spec}(A)$ is a compact,[1] Hausdorff, and totally disconnected topological space. In general the constructible topology is a finer topology than the Zariski topology, but the two topologies will coincide if and only if $A/\mathrm{nil}(A)$ is a von Neumann regular ring, where $\mathrm{nil}(A)$ is the nilradical of A.

52.1 See also

- Constructible set (topology)

52.2 References

[1] Some authors prefer the term *quasicompact* here.

- Atiyah, Michael Francis; Macdonald, I.G. (1969), *Introduction to Commutative Algebra*, Westview Press, p. 87, ISBN 978-0-201-40751-8

- Knight, J. T. (1971), *Commutative Algebra*, Cambridge University Press, pp. 121–123, ISBN 0-521-08193-9

Chapter 53

Zariski topology

In algebraic geometry and commutative algebra, the **Zariski topology** is a topology which has been primarily by Oscar Zariski on algebraic varieties and has been later generalized for making the set of prime ideals of a commutative ring a topological space, called the spectrum of the ring.

The Zariski topology allows using tools of topology for the study of algebraic varieties, even when the underlying field is not a topological field. This is one of the basic ideas of scheme theory, which allows to build general algebraic varieties by gluing together affine varieties in a similar way as it is done in manifold theory, where manifolds are built by gluing together charts, which are open subset of real affine spaces.

The **Zariski topology** of an algebraic variety is the topology whose closed sets are the algebraic subsets of the variety. In the case of an algebraic variety over the complex numbers, the Zariski topology is thus coarser than the usual topology, as every algebraic set is closed for the usual topology.

The generalization of the Zariski topology to the set prime ideals of a commutative ring follows from Hilbert's Nullstellensatz that establishes a bijective correspondence between the points of an affine variety defined over an algebraically closed field and the maximal ideals of the ring of its regular functions. This suggests to define the Zariski topology on the set of the maximal ideals of a commutative ring as the topology such that a set of maximal ideals is closed if and only if it is the set of all maximal ideals that contain a given ideal. Another basic ideas of Grothendieck's scheme theory is to consider as *points*, not only the usual points corresponding to maximal ideals, but also all (irreducible) algebraic varieties, which correspond to prime ideals. Thus the **Zariski topology** on the set of prime ideals (spectrum) of a commutative ring is the topology such a set of prime ideals is closed if and only it is the set of all prime ideals that contain a fixed ideal.

53.1 Zariski topology of varieties

In classical algebraic geometry (that is the part of algebraic geometry in which one does not use schemes, that have been introduced by Grothendieck around 1960), the Zariski topology was defined on algebraic varieties.[1] The Zariski topology, defined on the points of the variety, is the topology such that the closed sets are the algebraic subsets of the variety. As the most elementary algebraic varieties are affine and projective varieties, it is useful to make this definition more eplicit in both cases. We assume that we are working over a fixed, algebraically closed field k (in classical geometry k is almost always the complex numbers).

53.1.1 Affine varieties

First we define the topology on affine spaces $\mathbb{A}^n$, which as sets are just n-dimensional vector spaces over k. The topology is defined by specifying its closed sets, rather than its open sets, and these are taken simply to be all the algebraic sets in $\mathbb{A}^n$. That is, the closed sets are those of the form

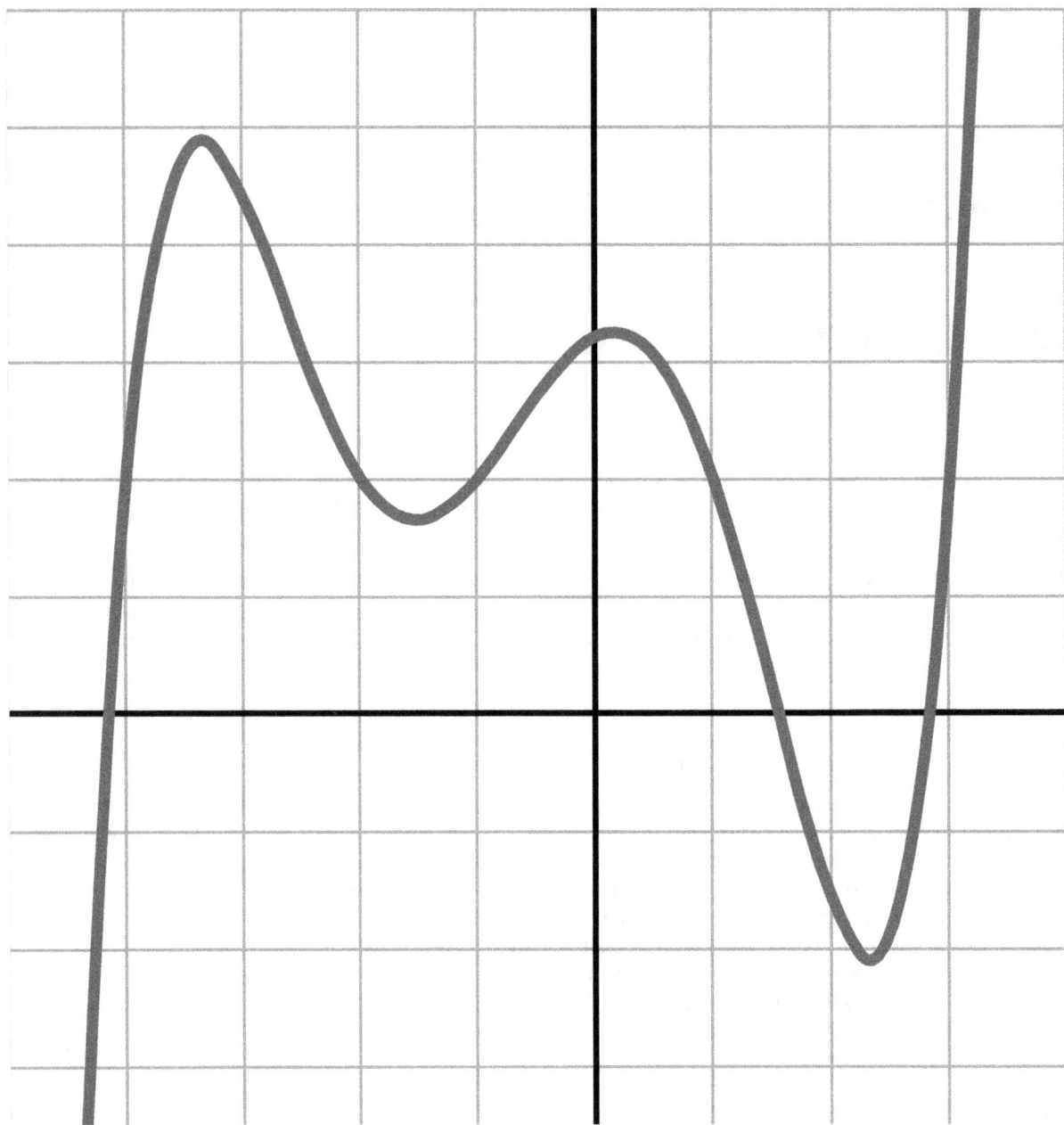

In the Zariski topology on the affine plane, this graph of a polynomial is closed.

$$V(S) = \{x \in \mathbb{A}^n \mid f(x) = 0, \forall f \in S\}$$

where S is any set of polynomials in n variables over k. It is a straightforward verification to show that:

- $V(S) = V((S))$, where (S) is the ideal generated by the elements of S;

- For any two ideals of polynomials I, J, we have

    1. $V(I) \cup V(J) = V(IJ)$;
    2. $V(I) \cap V(J) = V(I + J)$.

It follows that finite unions and arbitrary intersections of the sets $V(S)$ are also of this form, so that these sets form the closed sets of a topology (equivalently, their complements, denoted $D(S)$ and called *principal open sets*, form the topology itself). This is the Zariski topology on $\mathbb{A}^n$.

If X is an affine algebraic set (irreducible or not) then the Zariski topology on it is defined simply to be the subspace topology induced by its inclusion into some $\mathbb{A}^n$. Equivalently, it can be checked that:

- The elements of the affine coordinate ring

$$A(X) = k[x_1, \ldots, x_n]/I(X)$$

act as functions on X just as the elements of $k[x_1, \ldots, x_n]$ act as functions on $\mathbb{A}^n$;

- For any set of polynomials S, let T be the set of their images in $A(X)$. Then the subset of X

$$V'(T) = \{x \in X \mid f(x) = 0, \forall f \in T\}$$

(these notations are not standard) is equal to the intersection with X of $V(S)$.

This establishes that the above equation, clearly a generalization of the previous one, defines the Zariski topology on any affine variety.

53.1.2   Projective varieties

Recall that n-dimensional projective space $\mathbb{P}^n$ is defined to be the set of equivalence classes of non-zero points in $\mathbb{A}^{n+1}$ by identifying two points that differ by a scalar multiple in k. The elements of the polynomial ring $k[x_0, \ldots, x_n]$ are not functions on $\mathbb{P}^n$ because any point has many representatives that yield different values in a polynomial; however, for homogeneous polynomials the condition of having zero or nonzero value on any given projective point is well-defined since the scalar multiple factors out of the polynomial. Therefore if S is any set of homogeneous polynomials we may reasonably speak of

$$V(S) = \{x \in \mathbb{P}^n \mid f(x) = 0, \forall f \in S\}.$$

The same facts as above may be established for these sets, except that the word "ideal" must be replaced by the phrase "homogeneous ideal", so that the $V(S)$, for sets S of homogeneous polynomials, define a topology on $\mathbb{P}^n$. As above the complements of these sets are denoted $D(S)$, or, if confusion is likely to result, $D'(S)$.

The projective Zariski topology is defined for projective algebraic sets just as the affine one is defined for affine algebraic sets, by taking the subspace topology. Similarly, it may be shown that this topology is defined intrinsically by sets of elements of the projective coordinate ring, by the same formula as above.

53.1.3   Properties

A very useful fact about these topologies is that we may exhibit a basis for them consisting of particularly simple elements, namely the $D(f)$ for individual polynomials (or for projective varieties, homogeneous polynomials) f. Indeed, that these form a basis follows from the formula for the intersection of two Zariski-closed sets given above (apply it repeatedly to the principal ideals generated by the generators of (S)). These are called *distinguished* or *basic* open sets.

By Hilbert's basis theorem and some elementary properties of Noetherian rings, every affine or projective coordinate ring is Noetherian. As a consequence, affine or projective spaces with the Zariski topology are Noetherian topological spaces, which implies that any closed subset of these spaces is compact.

However, unless for finite algebraic sets, no algebraic set is ever a Hausdorff space. In the old topological literature "compact" was taken to include the Hausdorff property, and this convention is still honored in algebraic geometry; therefore compactness in the modern sense is called "quasicompactness" in algebraic geometry. However, since every point $(a_1, ..., an)$ is the zero set of the polynomials $x_1 - a_1, ..., xn - an$, points are closed and so every variety satisfies the T_1 axiom.

Every regular map of varieties is continuous in the Zariski topology. In fact, the Zariski topology is the weakest topology (with the fewest open sets) in which this is true and in which points are closed. This is easily verified by noting that the Zariski-closed sets are simply the intersections of the inverse images of 0 by the polynomial functions, considered as regular maps into $\mathbb{A}^1$.

53.2 Spectrum of a ring

In modern algebraic geometry, an algebraic variety is often represented by its associated scheme, which is topological spaces (equipped with additional structures) that is locally homeomorphic to the spectrum of a ring.[2] The *spectrum of a commutative ring A*, denoted Spec(A), is the set of the primes ideals of A, equipped with the **Zariski topology**, for which the closed sets are the sets

$$V(I) = \{P \in \text{Spec}\,(A) \mid I \subseteq P\}$$

where I is an ideal.

To see the connection with the classical picture, note that for any set S of polynomials (over an algebraically closed field), it follows from Hilbert's Nullstellensatz that the points of $V(S)$ (in the old sense) are exactly the tuples $(a_1, ..., an)$ such that $(x_1 - a_1, ..., xn - an)$ contains S; moreover, these are maximal ideals and by the "weak" Nullstellensatz, an ideal of any affine coordinate ring is maximal if and only if it is of this form. Thus, $V(S)$ is "the same as" the maximal ideals containing S. Grothendieck's innovation in defining Spec was to replace maximal ideals with all prime ideals; in this formulation it is natural to simply generalize this observation to the definition of a closed set in the spectrum of a ring.

Another way, perhaps more similar to the original, to interpret the modern definition is to realize that the elements of A can actually be thought of as functions on the prime ideals of A; namely, as functions on Spec A. Simply, any prime ideal P has a corresponding residue field, which is the field of fractions of the quotient A/P, and any element of A has a reflection in this residue field. Furthermore, the elements that are actually in P are precisely those whose reflection vanishes at P. So if we think of the map, associated to any element a of A:

$$e_a \colon \left(P \in \text{Spec}(A)\right) \mapsto \left(\frac{a \bmod P}{1} \in \text{Frac}(A/P)\right)$$

("evaluation of a"), which assigns to each point its reflection in the residue field there, as a function on Spec A (whose values, admittedly, lie in different fields at different points), then we have

$$e_a(P) = 0 \Leftrightarrow P \in V(a)$$

More generally, $V(I)$ for any ideal I is the common set on which all the "functions" in I vanish, which is formally similar to the classical definition. In fact, they agree in the sense that when A is the ring of polynomials over some algebraically closed field k, the maximal ideals of A are (as discussed in the previous paragraph) identified with n-tuples of elements of k, their residue fields are just k, and the "evaluation" maps are actually evaluation of polynomials at the corresponding n-tuples. Since as shown above, the classical definition is essentially the modern definition with only maximal ideals considered, this shows that the interpretation of the modern definition as "zero sets of functions" agrees with the classical definition where they both make sense.

Just as Spec replaces affine varieties, the Proj construction replaces projective varieties in modern algebraic geometry. Just as in the classical case, to move from the affine to the projective definition we need only replace "ideal" by "homogeneous ideal", though there is a complication involving the "irrelevant maximal ideal," which is discussed in the cited article.

53.2.1   Examples

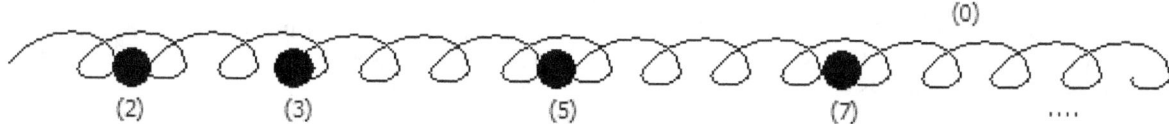

The spectrum of $\mathbb{Z}$

- Spec k, the spectrum of a field k is the topological space with one element.

- Spec $\mathbb{Z}$, the spectrum of the integers has a closed point for every prime number p corresponding to the maximal ideal $(p) \subset \mathbb{Z}$, and one non-closed generic point (i.e., whose closure is the whole space) corresponding to the zero ideal (0). So the closed subsets of Spec $\mathbb{Z}$ are precisely finite unions of closed points and the whole space.

- Spec $k[t]$, the spectrum of the polynomial ring over a field k: such a polynomial ring is known to be a principal ideal domain and the irreducible polynomials are the prime elements of $k[t]$. If k is algebraically closed, for example the field of complex numbers, a non-constant polynomial is irreducible if and only if it is linear, of the form $t - a$, for some element a of k. So, the spectrum consists of one closed point for every element a of k and a generic point, corresponding to the zero ideal, and the set of the closed points is homeomorphic with the affine line k equipped with its Zariski topology. Because of this homeomorphism, some authors call *affine line* the spectrum of $k[t]$. If k is not algebraically closed, for example the field of the real numbers, the picture becomes more complicated because of the existence of non-linear irreducible polynomials. For example, the spectrum of $\mathbb{R}[t]$ consists of the closed points $(x - a)$, for a in $\mathbb{R}$, the closed points $(x^2 + px + q)$ where p, q are in $\mathbb{R}$ and with negative discriminant $p^2 - 4q < 0$, and finally a generic point (0). For any field, the closed subsets of Spec $k[t]$ are finite unions of closed points, and the whole space. (This is clear from the above discussion for algebraically closed fields. The proof of the general case requires some commutative algebra, namely the fact, that the Krull dimension of $k[t]$ is one — see Krull's principal ideal theorem).

53.2.2   Properties

The most dramatic change in the topology from the classical picture to the new is that points are no longer necessarily closed; by expanding the definition, Grothendieck introduced generic points, which are the points with maximal closure, that is the minimal prime ideals. The closed points correspond to maximal ideals of A. Note, however, that the spectrum and projective spectrum are still T_0 spaces: given two points P, Q, which are prime ideals of A, at least one of them, say P, does not contain the other. Then $D(Q)$ contains P but, of course, not Q.

Just as in classical algebraic geometry, any spectrum or projective spectrum is compact, and if the ring in question is Noetherian then the space is a Noetherian space. However, these facts are counterintuitive: we do not normally expect open sets, other than connected components, to be compact, and for affine varieties (for example, Euclidean space) we do not even expect the space itself to be compact. This is one instance of the geometric unsuitability of the Zariski topology. Grothendieck solved this problem by defining the notion of properness of a scheme (actually, of a morphism of schemes), which recovers the intuitive idea of compactness: Proj is proper, but Spec is not.

53.3 See also

- Spectrum of a ring

- Spectral space

53.4 References

[1] Mumford, David (1999) [1967], *The red book of varieties and schemes*, Lecture Notes in Mathematics **1358** (expanded, Includes Michigan Lectures (1974) on Curves and their Jacobians ed.), Berlin, New York: Springer-Verlag, doi:10.1007/b62130, ISBN 978-3-540-63293-1, MR 1748380

[2] Dummit, D. S.; Foote, R. (2004). *Abstract Algebra* (3 ed.). Wiley. pp. 71–72. ISBN 9780471433347.

53.5 Further reading

- Hartshorne, Robin (1977), *Algebraic Geometry*, Berlin, New York: Springer-Verlag, ISBN 978-0-387-90244-9, MR 0463157, OCLC 13348052

- Todd Rowland, "Zariski Topology", *MathWorld*.

53.6   Text and image sources, contributors, and licenses

53.6.1   Text

- **Model theory** *Source:* https://en.wikipedia.org/wiki/Model_theory?oldid=687131708 *Contributors:* Damian Yerrick, Zundark, The Anome, Iwnbap, Toby Bartels, Youandme, Michael Hardy, David Martland, MartinHarper, Bcrowell, Chinju, Eric119, Snoyes, Rotem Dan, Revolver, Charles Matthews, Dcoetzee, Dysprosia, Hyacinth, Fred ulisses maranhão, Robbot, Josh Cherry, Sparky, MathMartin, Henrygb, Halibutt, Tobias Bergemann, Filemon, Ancheta Wis, Giftlite, Lupin, Karl-Henner, Sam Hocevar, Creidieki, Smimram, Rcog, Paul August, Elwikipedista~enwiki, Gauge, Floorsheim, Chalst, EmilJ, Peter M Gerdes, Jonsafari, Mdd, Msh210, EmmetCaulfield, Trylks, Jim O'Donnell, Oleg Alexandrov, Joriki, Linas, Ruud Koot, Payrard, RuM, Graham87, BD2412, Grammarbot, R.e.b., JonathanZ, SDaniel, Tillmo, Algebraist, YurikBot, Hairy Dude, NTBot~enwiki, Archelon, CarlHewitt, NawlinWiki, Trovatore, Bota47, Ott2, Arthur Rubin, Sardanaphalus, SmackBot, Rtc, Melchoir, SaxTeacher, Ppntori, Spellchecker, Charles Moss, Viebel, Tompsci, Jon Awbrey, Byelf2007, Lambiam, Loadmaster, WAREL, Jason.grossman, Lottamiata, Zero sharp, Cyrusc, James pic, CRGreathouse, CBM, Gregbard, Gryakj, Blaisorblade, Julian Mendez, Bernard the Varanid, Klausness, Escarbot, WinBot, Majorly, Avaya1, David Eppstein, Thehalfone, Joejunsun, Pavel Jelínek, Joshua Davis, Pomte, Tomaz.slivnik, Policron, Dessources, Thefrettinghand, VolkovBot, JohnBlackburne, Danadocus, Popopp, Synthebot, Sapphic, Logan, Barkeep, SieBot, Shellgirl, Pseudonomous, Mrw7, DesolateReality, Valeria.depaiva, WikiSBTR, Pi zero, Mild Bill Hiccup, Jusdafax, Bender2k14, Hans Adler, Qwfp, Little Mountain 5, Addbot, MagnusA.Bot, Loupeter, Wireless friend, Luckas-bot, Yobot, Ht686rg90, Pcap, AnomieBOT, Materialscientist, Citation bot, ArthurBot, Groovenstein, VladimirReshetnikov, SassoBot, Ringspectrum, FrescoBot, LucienBOT, Paine Ellsworth, Citation bot 1, Tkuvho, Foobarnix, Gf uip, John of Reading, Stephan Spahn, Masssly, Ultracoffee, Helpful Pixie Bot, Itzuvit, Freitagj, Brad7777, Dexbot, Thierry Le Provost, Monkbot, SolidPhase, SoSivr, SocraticOath, KasparBot and Anonymous: 100

- **Mathematical model** *Source:* https://en.wikipedia.org/wiki/Mathematical_model?oldid=687215977 *Contributors:* AxelBoldt, Derek Ross, Wesley, The Anome, Tbackstr, Jan Hidders, BenBaker, Arvindn, Jdpipe, Olivier, JohnOwens, Michael Hardy, Lexor, Dori, JWSchmidt, Kingturtle, Nikai, Emperorbma, Dysprosia, Tpbradbury, Altenmann, Mayooranathan, MathMartin, Henrygb, TittoAssini, Goodralph, Pdenapo, Giftlite, Neilc, Vina, Frau Holle, Abdull, Brianjd, Discospinster, Rich Farmbrough, Oliver Lineham, Bender235, Billlion, Rgdboer, RoyBoy, Obradovic Goran, Hesperian, Helix84, Mdd, Passw0rd, Msh210, Atlant, GiantSloth, Oleg Alexandrov, Tbsmith, Woohookitty, Karnesky, Pol098, Ruud Koot, Smmurphy, Palica, Graham87, Magister Mathematicae, BD2412, Kbdank71, Porcher, Sjö, John Deas, Winhunter, YurikBot, Wavelength, RobotE, Elapsed, Filippof, Hede2000, Hydrargyrum, Grubber, Manop, CarlHewitt, Ahills60, Trovatore, Eurosong, Arthur Rubin, Abune, Wsiegmund, MaNeMeBasat, Silverhelm, SmackBot, Maksim-e~enwiki, Eskimbot, Binarypower, Cazort, Commander Keane bot, Bluebot, Trebor, Ph7five, DHN-bot~enwiki, Audriusa, Cybercobra, Tilin, Spiritia, Lambiam, G.de.Lange, Spacemika, Dakart, JRSpriggs, Jonnat, CBM, Thomasmeeks, Penbat, Gregbard, Gogo Dodo, SeniorScribbler, Kwaku~enwiki, Omicronpersei8, MCrawford, AndrewDressel, Al Lemos, Marek69, Dvsphanindra, HolgerK~enwiki, Dawnseeker2000, AntiVandalBot, Seaphoto, Phanerozoic, Deadbeef, JAnDbot, Stephanhartmannde, EmersonLowry, Sangwinc, VoABot II, Squidonius, Gwern, Matqkks, EyeSerene, Sam Douglas, R'n'B, Tgeairn, Maurice Carbonaro, T.ogar, Tparameter, Rbakker99, The enemies of god, Josephfrodo, Camrn86, JohnBlackburne, Philip Trueman, Ranilb5, TXiKiBoT, Martin451, Abdullais4u, PDFbot, Millancad, Nshackles, Wavehunter, Carlodn6, SieBot, Flyer22 Reborn, Soy ivan, Paolo.dL, DancingPhilosopher, Vanished user qkqknjitkcse45u3, Martarius, ClueBot, Fioravante Patrone en, Tomas e, PMDrive1061, Djr32, Awickert, Skbkekas, Hans Adler, Agor153, DumZiBoT, GKantaris, XLinkBot, Good Olfactory, Addbot, Globalsolidarity, Spradlig, Cst17, MrOllie, Vlado0605, Zorrobot, Wireless friend, Legobot, Luckas-bot, Yobot, Julia W, Yotaloop, Piano non troppo, Materialscientist, HanPritcher, Farazgiki, Xqbot, Tasudrty, Cureden, Prazan, Paine Ellsworth, Hxxvxxy, Aliotra, Meemee878, Sîmbotin, Dterp, Pinethicket, HRoestBot, RedBot, Cnwilliams, TobeBot, Howard.noble323, Math.geek3.1415926, WikiTome, DARTH SIDIOUS 2, Onel5969, Dalba, HisashiKobayashi, Kai.velten, Emble64, EmausBot, JustinTime55, Azuris, Stricklandjs59, Donner60, Sameer143, ClueBot NG, Sina92, Aiwing, Gilderien, Fioravante Patrone, Widr, Helpful Pixie Bot, Brad7777, ChrisGualtieri, Dexbot, Stevebillings, Yourmomsdadsbrotherscousin, Csreseracher, Bluemix, ModalPeak, BillWhiten, Aubreybardo, Science.philosophy.arts, Pff1971, Hoogamaphone, TranquilHope, Flyb4ck88, Loraof, Marab05, KasparBot, Claudiobong and Anonymous: 187

- **Structure (mathematical logic)** *Source:* https://en.wikipedia.org/wiki/Structure_(mathematical_logic)?oldid=687105719 *Contributors:* MichaelHardy, Hyacinth, Rorro, Tobias Bergemann, Giftlite, Jason Quinn, Jabowery, D6, Spayrard, Oleg Alexandrov, Linas, Salix alba, John Baez, Jrtayloriv, Algebraist, Grubber, Archelon, Trovatore, Zwobot, Mhss, Byelf2007, Physis, Simon12, CBM, Gregbard, Salgueiro~enwiki, DWIII, Swpb, Thehalfone, R'n'B, Ps ttf, VolkovBot, Crisperdue, Jorgen W, Niceguyedc, Hans Adler, Addbot, DOI bot, Drpickem, Ht 686rg90, Pcap, KamikazeBot, Citation bot, Xqbot, FrescoBot, Kwiki, Citation bot 1, Adlerbot, RedBot, Nascar1996, Wojowu, RjwilmsiBot, Tijfo098, Jiri 1984, Helpful Pixie Bot and Anonymous: 28

- **Group (mathematics)** *Source:* https://en.wikipedia.org/wiki/Group_(mathematics)?oldid=686401714 *Contributors:* AxelBoldt, Brion VIBBER, Mav, Uriyan, Bryan Derksen, Zundark, Tarquin, XJaM, Darius Bacon, Toby~enwiki, Toby Bartels, Zippy, Olivier, Patrick, Chas zzz brown, Michael Hardy, Wshun, Kku, Bcrowell, Chinju, Tango, TakuyaMurata, Jimfbleak, LittleDan, Glenn, Marco Krohn, Andres, Jonik, Revolver, Charles Matthews, Dcoetzee, Dysprosia, Jitse Niesen, The Anomebot, Taxman, Fibonacci, Topbanana, Raul654, Daran, PuzzletChung, Donarreiskoffer, Robbot, Gandalf61, Puckly, Rursus, Ashwin, Fuelbottle, Tobias Bergemann, Pdenapo, Tosha, Giftlite, Lee J Haywood, Michael Devore, Python eggs, Stephen Ducret, DRE, APH, Pmanderson, ELApro, Shahab, Mormegil, Rich Farmbrough, Guanabot, ArnoldReinhold, Mani1, Wadewitz, Paul August, Goochelaar, Bender235, RJHall, Joanjoc~enwiki, Hayabusa future, Art LaPella, Wood Thrush, C S, Dungodung, Helix84, Jumbuck, Msh210, Guy Harris, Sl, Hippophaë~enwiki, PAR, Woodstone, HenryLi, Oleg Alexandrov, Arneth, SP-KP, OdedSchramm, Mpatel, Jdiemer, Ryan Reich, Graham87, Ilya, Qwertyus, Jshadias, Chenxlee, Josh Parris, Rjwilmsi, Jarretinha, OneWeirdDude, MarSch, Pako, Salix alba, R.e.b., DoubleBlue, Penumbra2000, VKokielov, Nihiltres, Jrtayloriv, Mongreilf, Chobot, MithrandirMage, Algebraist, Debivort, YurikBot, Wavelength, Hairy Dude, Grubber, Archelon, Gaius Cornelius, Canadaduane, Rick Norwood, Dtrebbien, Kinser, PAStheLoD, DYLAN LENNON~enwiki, Natkeeran, KarlHeg, David Underdown, LarryLACa, Zzuuzz, Arthur Rubin, Redgolpe, SmackBot, Melchoir, Stifle, Gilliam, Dan Hoey, Bh3u4m, Bluebot, Soru81, Oli Filth, Silly rabbit, Nbarth, Emurphy42, Kjetil1001, Mark Wolfe, Vanished User 0001, Lesnail, TKD, LkNsngth, Nibuod, Slawekk, DMacks, Mostlyharmless, Lambiam, Harryboyles, Eriatarka, EnumaElish, Michael Kinyon, Loadmaster, Mscalculus, SandyGeorgia, Rschwieb, Markan~enwiki, Danielh~enwiki, Newone, AGK, Spindled, Paul Matthews, CRGreathouse, CmdrObot, CBM, Rawling, Myasuda, WillowW, Mike Christie, Dr.enh, Kozuch, Xantharius, Thijs!bot, Epbr123, Braveorca, Markus Pössel, Konradek, Headbomb, Paxinum, Cj67, RobHar, EdJohnston, Escarbot, Sekky, Allanhalme, JAnDbot, Ricardo sandoval, Rush Psi, East718, Magioladitis, WolfmanSF, Swpb, Ling.Nut, Jakob.scholbach, Brusegadi, SwiftBot, DAGwyn, Giggy, David Eppstein, Fbaggins, Lvwarren, Olsonist, Robin S, Pbroks13, Pomte, David Callan, IPonomarev, DrKay, RJBotting, Cspan64, Cpiral, Dispenser, Indeed123, Trumpet marietta 45750, Nwbeeson, Bobrek~enwiki, Ginpasu, Treisijs, OktayD, LokiClock, TheOtherJesse, Philip Trueman, GimmeBot, JasonASmith, Anonymous Dissident, VictorMak, Skylarkmichelle, Geometry guy, Eubulides, BigDunc, Synthebot, Pjoef, AlleborgoBot, Teresol, Drschawrz, SieBot, Calliopejen1, YonaBot, Gerakibot, Soler97, Antzervos, Kareekacha, Thehotelambush, JackSchmidt, Skippydo, Jorgen W, Anchor Link Bot, S2000magician, Randomblue, CBM2, Peiresc~enwiki, A legend, Felizdenovo, Amahoney, Nergaal, Classicalecon, ClueBot, Alksentrs, Nsk92, Piledhigheranddeeper, Eeekster, Brews ohare, Cenarium, Jotterbot, Hans Adler, Wikidsp, Thingg, Dank, Qwfp, Johnuniq, TimothyRias, Basploeger, Marc van Leeuwen, Alecobbe, Kakila, GabeAB, Porphyro, CàlculIntegral, Addbot, DOI bot, Delaszk, LinkFA-Bot, Ozob, Ettrig, Luckas-bot, Yobot, WikiDan61, TaBOT-zerem, Pcap, AnomieBOT, WinoWeritas, Jarmiz, Frankenpuppy, Xqbot, Farvin111, X Pacman X, X Fallout X, Mee26, AYSH AYSH AY AY AY AY, Isheden, Point-set topologist, RibotBOT, Charvest, Harry007754, FrescoBot, Citation bot 1, HRoestBot, Wikitanvir, Jujutacular, RjwilmsiBot, Jowa fan, EmausBot, M759, Slawekb, ZéroBot, Quondum, D.Lazard, Git2010, Wayne Slam, Mentibot, ChuispastonBot, ClueBot NG, IfYouDoIfYouDon't, Tideflat, Frietjes, Mesoderm, Widr, Bibcode Bot, Brad7777, Nadapez~enwiki, ChrisGualtieri, Dexbot, Mark L MacDonald, Jochen Burghardt, Mark viking, CsDix, Itc editor2, Blackbombchu, Schwatzwutz, Khuramawais, JAaron95, Anrnusna, Sansam131192, Monkbot, Levi12349, MissouriOzark1947, IPalpedia, Y2N1-09631 and Anonymous: 211

- **Field (mathematics)** *Source:* https://en.wikipedia.org/wiki/Field_(mathematics)?oldid=678421230 *Contributors:* AxelBoldt, Bryan Derksen, Zundark, The Anome, Andre Engels, Josh Grosse, XJaM, Toby Bartels, Miguel~enwiki, Lir, Patrick, Michael Hardy, Wshun, DIG~enwiki, TakuyaMurata, Karada, Looxix~enwiki, Rossami, Andres, Loren Rosen, Revolver, RodC, Dysprosia, Jitse Niesen, Prumpf, Tero~enwiki, Phys, Philopp, R3m0t, Jmabel, Mattblack82, MathMartin, P0lyglut, Wikibot, Tobias Bergemann, Unfree, Marc Venot, Giftlite, Highlandwolf, Gene Ward Smith, Lethe, Zigger, Fropuff, Millerc, Waltpohl, Python eggs, Gubbubu, CSTAR, Pmanderson, Barnaby dawson, PhotoBox, Mormegil, Jørgen Friis Bak, Discospinster, Guanabot, Sperling, Paul August, Zaslav, Elwikipedista~enwiki, El C, Rgdboer, EmilJ, Touriste, Army1987, Giraffedata, Obradovic Goran, OoberMick, Msh210, Mlm42, Olegalexandrov, RJFJR, Oleg Alexandrov, Woohookitty, Linas, Arneth, Bkkbrad, Hypercube~enwiki, MarkTempeit, Damicatz, MFH, Isnow, Palica, Graham87, FreplySpang, Chenxlee, Josh Parris, Rjwilmsi, Hiberniantears, Salix alba, R.e.b., FlaBot, Codazzi~enwiki, Jrtayloriv, R160K, Chobot, Abu Amaal, Algebraist, Wavelength, Dmesg, Eraserhead1, Hairy Dude, KSmrq, Grubber, Archelon, Rintrah, Rat144, Rick Norwood, Trovatore, DYLAN LENNON~enwiki, Crasshopper, RaSten, DavidHouse~enwiki, Mgnbar, Children of the dragon, SmackBot, Mmernex, Melchoir, Gilliam, Nbarth, Charlotte Hobbs, Lesnail, Cybercobra, Acepectif, Slawekk, Bidabadi~enwiki, Lambiam, Jim.belk, Schildt.a, Mets501, DabMachine, Rschwieb, WAREL, Newone, Vaughan Pratt, CRGreathouse, Kupirijo, Tiphareth, DEWEY, Eulerianpath, Pedro Fonini, Goldencako, BobNiichel, Xantharius, KLIP~enwiki, JLISP, Headbomb, RobHar, Nick Number, Turgidson, Kprateek88, Martinkunev, Magioladitis, Bongwarrior, VoABot II, JamesBWatson, Jakob.scholbach, SwiftBot, Catgut, Lukeaw, MORI, Cpiral, Maproom, Gombang, Policron, Barylior, Umarekawari, LokiClock, Red Act, Anonymous Dissident, Hesam7, Joeldl, Dave703, Zermalo, Shellgirl, Cwkmail, Soler97, JackSchmidt, Jorgen W, Anchor Link Bot, Willy, your mate, Oekaki, UKe-CH, ClueBot, Mild Bill Hiccup, Tcklein, Niceguyedc, He7d3r, Bender2k14, Squirreljml, Palnot, ZooFari, Addbot, Gabriele ricci, Download, Unzerlegbarkeit, Cesiumfrog, Yobot, Ht686rg90, TaBOT-zerem, Zagothal, AnomieBOT, UBJ 43X, DSisyphBot, Depassp, Danielschreiber, MegaMouthBolt123, Point-set topologist, Charvest, KirarinSnow, FrescoBot, Mjmarkowitz, RandomDSdevel, Ebony Jackson, D stankov, Girish.ponkiya2007, Kunle102, DASHBot, Sedrikov, Tom.kemp90, Tommy2010, Wikipelli, Shishir332, Lfrazier11, Quondum, D.Lazard, JimMeiss, ClueBot NG, Ankur1vi, Wcherowi, Frietjes, MerlIwBot, Helpful Pixie Bot, !mcbloobyenstein!!, Or elharar, Fabio.nsantos, Rjs.swarnkar, Topgraph28, Deltahedron, Sanipriya, GigaGerard, CsDix, YiFeiBot, JAaron95, Teddyktchan, GeoffreyT2000, Charlotte Aryanne, Vluczkow and Anonymous: 133

- **Graph (mathematics)** *Source:* https://en.wikipedia.org/wiki/Graph_(mathematics)?oldid=685230929 *Contributors:* The Anome, Manning Bartlett, XJaM, Tomo, Stevertigo, Patrick, Michael Hardy, W~enwiki, Zocky, Wshun, Booyabazooka, Karada, Ahoerstemeier, Den fjättrade ankan~enwiki, Jiang, Dcoetzee, Dysprosia, Doradus, Zero0000, McKay, BenRG, Robbot, LuckyWizard, Mountain, Altenmann, Mayooranathan, Gandalf61, MathMartin, Timrollpickering, Bkell, Tobias Bergemann, Tosha, Giftlite, Dbenbenn, Harp, Tom harrison, Chinasaur, Jason Quinn, Matt Crypto, Neilc, Erhudy, Knutux, Yath, Joeblakesley, Tomruen, Peter Kwok, Aknorals, Chmod007, Abdull, Corti, PhotoBox, Discospinster, Rich Farmbrough, Andros 1337, Paul August, Zaslav, Gauge, Tompw, Crisófilax, Yitzhak, Kine, Bobo192, Jpiw~enwiki, Mdd, Jumbuck, Zachlipton, Sswn, Liao, Rgclegg, Paleorthid, Super-Magician, Mahanga, Joriki, Mindmatrix, Wesley Moy, Oliphaunt, Brentdax, Jwanders, Tbc2, Cbdorsett, Ch'marr, Davidfstr, Xiong, Marudubshinki, Tslocum, Magister Mathematicae, Ilya, SixWingedSeraph, Sjö, Rjwilmsi, Salix alba, Bhadani, FlaBot, Nowhither, Mathbot, Gurch, MikeBorkowski~enwiki, Chronist~enwiki, Silversmith, Chobot, Peterl, Siddhant, Borgx, Karlscherer3, Hairy Dude, Gene.arboit, Michael Slone, Gaius Cornelius, Shanel, Gwaihir, Dtrebbien, Dureo, Doetoe, Wknight94, Arthur Rubin, Netrapt, RobertBorgersen, Cjfsyntropy, RonnieBrown, Burnin1134, SmackBot, Nihonjoe, Stux, McGeddon, BiT, Algont, Ohnoitsjamie, Chris the speller, Bluebot, TimBentley, Theone256, Cornflake pirate, Zven, Anabus, Can't sleep, clown will eat me, Tamfang, Cybercobra, Jon Awbrey, Kuru, Nat2, Tomhubbard, Dicklyon, Cbuckley, Quaeler, BranStark, Wandrer2, George100, Ylloh, Vaughan Pratt, Repied, CRGreathouse, Citrus538, Jokes Free4Me, Requestion, Myasuda, Danrah, Robertsteadman, Eric Lengyel, Headbomb, Urdutext, AntiVandalBot, Hannes Eder, JAnDbot, MER-C, Dreamster, Struthious Bandersnatch, JNW, Catgut, David Eppstein, JoergenB, MartinBot, Rettetast, R'n'B, J.delanoy, Hans Dunkelberg, Yecril, Pafcu, Ijdejter, Deor, ABF, Maghnus, TXiKiBoT, Sdrucker, Someguy1221, PaulTanenbaum, Lambyte, Ilia Kr., Jpeeling, Falcon8765, RaseaC, Insanity Incarnate, Zenek.k, Radagast3, Debamf, Debeolaurus, SieBot, Minder2k, Dawn Bard, Cwkmail, Jon har, SophomoricPedant, Oxymoron83, Henry Delforn (old), Ddxc, Svick, Phegyi81, Anchor Link Bot, Jarauh, ClueBot, Vacio, Nsk92, JuPitEer, Huynl, JP.Martin-Flatin, Xavexgoem, UKoch, Mitmaro, Editor70, Watchduck, Hans Adler, Suchap, Wikidsp, Muro Bot, 3ICE, Aitias, Versus22, Djk3, Kruusamägi, SoxBot III, XLinkBot, Marc van Leeuwen, Libcub, WikiDao, Tangi-tamma, Addbot, Gutin, Athenray, Willking1979, Royerloic, West.andrew.g, Tyw7, Zorrobot, LuK3, Luckas-bot, Yobot, TaBOT-zerem, THEN WHO WAS PHONE?, E mraedarab, Tempodivalse, Пика Пика, Ulric1313, RandomAct, Materialscientist, Twri, Dockfish, Anand jeyahar, Miym, Prunesqualer, Andyman100, VictorPorton, JonDePlume, Shadowjams, A.amitkumar, Kracekumar, Edgars2007, Citation bot 1, Maggyero, DrilBot, Amintora, Pinethicket, Calmer Waters, RobinK, Barras, Tgv8925, DARTH SIDIOUS 2, Powerthirst123, DRAGON BOOSTER, Mymyhoward16, Kerrick Staley, Ajraddatz, Wgunther, Benthnm, Akutagawa10, White Trillium, Josve05a, D.Lazard, L Kensington, Maschen, Inka 888, Chewings72, ClueBot NG, Wcherowi, MelbourneStar, Kingmash, O.Koslowski, Joel B. Lewis, Andrewsky00, Timflutre, Helpful Pixie Bot, HMSSolent, BG19bot, Grolmusz, John Cummings, Stevetihi, Канеюку, Void-995, MRG90, Vanischenu, Tman159, BattyBot, Ekren, Lugia2453, Jeff Erickson, CentroBabbage, Nina Cerutti, Chip Wildon Forster, Yloreander, Manul, JaconaFrere, Monkbot, Hou710, Anon124, Aryan5496 and Anonymous: 355

- **Mathematical logic** *Source:* https://en.wikipedia.org/wiki/Mathematical_logic?oldid=677903575 *Contributors:* AxelBoldt, Michael Hardy, Modster, Dominus, TakuyaMurata, Tregoweth, Renamed user 4, Charles Matthews, Dcoetzee, Nohat, Mwoolf, Dysprosia, OkPerson, Piolinfax, Hyacinth, David Shay, Aleph4, Robbot, Romanm, Gandalf61, MathMartin, Ojigiri~enwiki, Tobias Bergemann, Giftlite, Recentchanges, Lethe, Fleminra, Jason Quinn, Matt Crypto, Edcolins, Beefalo, Kntg, Eduardoporcher, Barnaby dawson, HedgeHog, Smimram, EugeneZelenko, Guanabot, Leibniz, Wclark, Ivan Bajlo, Paul August, Spayrard, Lycurgus, Chalst, Kwamikagami, Irrᵃtiᵍnal, Art LaPella, Atomique~enwiki, Nicke Lilltroll~enwiki, Jojit fb, Obradovic Goran, Mdd, Tsirel, Msh210, Chira, CuriousOne, Sligocki, Samohyl Jan, Lebob (renamed), Jak86, Mindmatrix, Guardian of Light, Ruud Koot, Graham87, BD2412, Salix alba, R.e.b., Reinis, Nigosh, Mathbot, Tillmo, Jersey Devil, Frank-Tobia, Roboto de Ajvol, Wavelength, Hairy Dude, RussBot, Icarus3, SpuriousQ, Polyvios, KSchutte, Rick Norwood, Meloman, Trovatore, Musteval, Tony1, Bota47, Robertbyrne, Reyk, Claygate, MaNeMeBasat, Aeosynth, Otto ter Haar, Sardanaphalus, JJL, SmackBot, SaxTeacher, Bomac, Jagged 85, RockRockOn, Logic2go, Ohnoitsjamie, Chris the speller, Persian Poet Gal, MK8, MalafayaBot, Sholto Maud, Nixeagle, Maksim-bot, Allan McInnes, Stevenmitchell, Jon Awbrey, DMacks, Sammy1339, Bidabadi~enwiki, Tkos, Byelf2007, The undertow, Lambiam, Rsimmonds01, Bjankuloski06en~enwiki, Snem, Nabeth, Dan Gluck, Iridescent, Stotr~enwiki, Francl, Mrdthree, JRSpriggs, Atomobot, CRGreathouse, CBM, Thomasmeeks, Myasuda, Gregbard, Gogo Dodo, Julian Mendez, Thijs!bot, Brian G. Wilson, Blah3, Dgies, Malcolm, Alphachimpbot, VictorAnyakin, JAnDbot, Avaya1, Meeples, Ling.Nut, Robin S, Metamusing, Maurice Carbonaro, NewEnglandYankee, Policron, Uhai, DavidCBryant, Treisijs, Alan U. Kennington, VolkovBot, Hotfeba, JohnBlackburne, Am Fiosaigear~enwiki, TXiKiBoT, Ontoraul, The Tetrast, Martin451, Kowsari, Davin, Cremepuff222, Popopp, Palaeovia, Ohiostandard, Flyer22 Reborn, Hxhbot, OKBot, Valeria.depaiva, IsleLaMotte, CBM2, Butane Goddess, LarRan, ClueBot, LAX, The Thing That Should Not Be, Smithpith, WikiSBTR, Compellingelegance, Razimantv, ScNewcastle, Alexbot, Hans Adler, PergolesiCoffee, Skolemizer, Good Olfactory, Addbot, NjardarBot, MrOllie, Ozob, TeH nOmInAtOr, Legobot, Math Champion, Ddzhafar, Pcap, AnomieBOT, Götz, Citation bot, Xqbot, RJGray, GrouchoBot, The Wiki ghost, Sophus Bie, Tales23, Mark Renier, VS6507, Orhanghazi, Qiemem, Zhentmdfan, Citation bot 1, Tkuvho, I dream of horses, Jonesey95, Sa'y, TheIndianWikiEditor, Rover6891, Steve2011, Ham and bacon, No One of Consequence, Onel5969, Jmencisom, Anirudh Emani, Future ahead, Karthikndr, 28bot, Maxdlink, ClueBot NG, SusikMkr, Movses-bot, O.Koslowski, Masssly, Joel B. Lewis, MerlIwBot, Helpful Pixie Bot, Theotherscripto12, Virus2801, Brad7777, Spasoev, Sfarney, Vedsuthar, None but shining hours, Khazar2, Houbn, Dtotoo, Deltahedron, Cwobeel, Jochen Burghardt, BrooksMaxwell, Greenjello77, Xaqron, MarshalWalter, Kirstenlovesyouu, The Annoyed Logician, Dunditschia, Editor of the wiki swag, Aarsh A Chotalia, StephenDunker, Greatingsworld, Tony the but hole, KasparBot, Chery Ann Arguelles, Shahryar1976, Gari chetty, Lr0^^k and Anonymous: 191

- **Semantics** *Source:* https://en.wikipedia.org/wiki/Semantics?oldid=686038516 *Contributors:* The Anome, Youssefsan, Vaganyik, Ortolan88, Ben-Zin~enwiki, Hannes Hirzel, Heron, Ryguasu, Netesq, Stevertigo, Michael Hardy, Pit~enwiki, Gdarin, Rp, Kku, Looxix~enwiki, Glenn, Rossami, Andres, Hectorthebat, Jitse Niesen, Mjklin, Haukurth, Shizhao, Fvw, Jens Meiert, Jon Roland, Seriv, Robbot, Lambda, Pigsonthewing, Jakohn, Kiwibird, Sverdrup, Rursus, Moink, Spellbinder, Marc Venot, Gwalla, Markus Krötzsch, Jpta~enwiki, HHirzel, Everyking, Zhen Lin, Eequor, Khalid hassani, Jackol, Javier Carro, JoJan, Mukerjee, Augur, Kntg, Bornslippy, Urhixidur, Yuriz, Lucidish, Rich Farmbrough, Cacycle, Rama, Slipstream, Kzzl, Dbachmann, Paul August, Jaberwocky6669, Evice, El C, Chalst, Joanjoc~enwiki, Linkoman, Enric Naval, Nortexoid, Jonsafari, Jooyoonchung, Helix84, Anthony Appleyard, Mark Dingemanse, Sligocki, Cdc, Sabrebattletank, Ish ishwar, Tycho, EvenT, Jason L. Gohlke, Redvers, Simlorie, Galaxiaad, Ott, Jtauber, Velho, Woohookitty, Mindmatrix, Kokoriko, Kelisi, Analogisub, SDC, Mandarax, Graham87, Imersion, Rjwilmsi, Mayumashu, Koavf, Jivecat, Dmccreary, Brighterorange, Mlinar~enwiki, NeoAmsterdam, FlaBot, Sinatra, Isotope23, Ben Babcock, Vonkje, Comiscuous, Lambyuk, Chobot, Sonic Mew, Roboto de Ajvol, YurikBot, Wavelength, Hairy Dude, Retodon8, Stephenb, Anomalocaris, NawlinWiki, Maunus, MarkBrooks, JECompton, WAS 4.250, Light current, G. Lakoff, Lt-wiki-bot, Donald Albury, SMcCandlish, JuJube, Pred, AGToth, NickelShoe, Sardanaphalus, SmackBot, Zerida, Unyoyega, Shamalyguy, Gilliam, Lindosland, Chris the speller, MasterofUnvrs314, MK8, MalafayaBot, Droll, Jerome Charles Potts, A. B., Scwlong, Zsinj, Frap, Ioscius, Chlewbot, SundarBot, Khoikhoi, Cybercobra, Iblardi, Battamer, Jon Awbrey, Byelf2007, SashatoBot, 16@r, Hvn0413, Nabeth, Kvng, Hu12, Gandalf1491, J Di, DEddy, Ziusudra, George100, Stifynsemons, Wolfdog, Sir Vicious, Kensall, Gregbard, FilipeS, Cydebot, Warhorus, ST47, Quibik, Nickleus, Gimmetrow, Thijs!bot, Wikid77, Runch, Mbell, Dalahäst, Azymuthca, X201, Nick Number, Mentifisto, AntiVandalBot, Shawn wiki, Gioto, Widefox, TimVickers, Dylan Lake, Danny lost, JAnDbot, MER-C, Shermanmonroe, Jmchambers90, Dcooper, .anacondabot, Daveh1, AndriesVanRenssen, Tmusgrove, Nicodemus13, Mahitgar, Revery~enwiki, Mechanismic, Ekotkie, MartinBot, J.delanoy, Cyborg Ninja, Piercetheorganist, Dbiel, Rod57, AKA MBG, Lygophile, Erick.Antezana, Lrunge, RasputinJSvengali, Macedonian, LokiClock, Philip Trueman, Amos Han, TXiKiBoT, Purpose Observatory, Aaeamdar, Goberiko~enwiki, HillarySco, Merijn2, Synthebot, Lova Falk, Cnilep, Jimbo2222, Logan, Botev, SieBot, Nubiatech, Kgoarany, Asderff, PaulColby, Jerryobject, Yerpo, ScAvenger lv, Strife911, Bguest, MiNombreDeGuerra, Doc honcho, CharlesGillingham, Emptymountains, Martarius, ClueBot, Bbadree, Tanglewood4, Eklir, Niceguyedc, DragonBot, Awi007, PixelBot, Vanisheduser12345, Rhododendrites, MacedonianBoy, Cenarium, Aleksd, Micmachete, MystBot, Alanthehat, Addbot, Rdanneskjold, The singapore ministry of education sucks, AVand, Guoguo12, Landon1980, Friginator, K1US, Aboctok, Ayatniazi, CanadianLinuxUser, Pirtskhalava, CarsracBot, Numbo3-bot, Erutuon, Tide rolls, JAHendler, Krixou, Legobot, Luckas-bot, TaBOT-zerem, Vanished user rt41as76lk, AnakngAraw, 8ung3st, Molewood6, Rockypedia, Rjanag, Govindmaheswaran, Jim1138, Materialscientist, Citation bot, Lil-Helpa, Xqbot, Hyggelig, Lynch9000s, Aenioc, JustinCope82, Omnipaedista, Benjamin Dominic, FrescoBot, Levalley, Citation bot 1, Mundart, Smithonian, Harold Philby, Pinethicket, Joost.b, RedBot, MastiBot, Nora lives, FoxBot, عقیل فشاک, Jonkerz, Lotje, Theyetiman12345, RobotQuistnix, 2bluey, Mchcopl, Zegarad, EmausBot, Jefffi, Active Banana, Hpvpp, Alexey.kudinkin, Llamas4drama'10, Unreal7, SporkBot, Gabnh, Eric Biggs, Edunoramus, Kgsbot, Ready, Odysseus1479, Tijfo098, Manytexts, ClueBot NG, Squarrels, Aniketdalal, Movses-bot, Helpful Pixie Bot, BG19bot, BenSmak, Boblibr, Lawandeconomics1, MusikAnimal, Davidiad, Tom Pippens, Semantia, UnconsciousInferno, Darylgolden, Suraduttashandilya, Dave5702, Kevin12xd, Faizan, Bienmanchot, Ahernandez33, Didigodot, Noizy Boy, Sarahjane212013, Pavel Stankov, Csusarah, FelixRosch, Good afternoon, Nøkkenbuer, Spyker247, KasparBot, Vjpand and Anonymous: 280

- **Syntax** *Source:* https://en.wikipedia.org/wiki/Syntax?oldid=687194859 *Contributors:* Mav, RoseParks, Andre Engels, Karl E. V. Palmen, Youssefsan, Vaganyik, William Avery, Ben-Zin~enwiki, Hannes Hirzel, Hirzel, Stevertigo, K.lee, Michael Hardy, Kwertii, Looxix~enwiki, Ahoerstemeier, Darkwind, Glenn, Cadr, Hectorthebat, Chronotox, Charles Matthews, Haukurth, Hyacinth, Ed g2s, PuzzletChung, Robbot, RedWolf, Altenmann, Dittaeva, Sverdrup, Academic Challenger, Rursus, Ojigiri~enwiki, Hippietrail, Hadal, Spellbinder, Rik G., Giftlite, Marnanel, Raymond Meredith, Linguizic, Jdavidb, Neilc, Physicist, Andycjp, Sonjaaa, Beland, OverlordQ, MarkSweep, Mukerjee, Oneiros, Karl-Henner, Tooki, Robin klein, N-k, D6, Poccil, Haggen Kennedy, Skal, Discospinster, Rich Farmbrough, El C, Cherry blossom tree, Joanjoc~enwiki, Bobo192, Beige Tangerine, Viriditas, Dungodung, Kappa, Joe Jarvis, Kjkolb, Morganiq, Jonsafari, Anthony Appleyard, Mark Dingemanse, Atlant, CR7, Ciceronl, Cromwellt, Ish ishwar, RJFJR, Mikeo, Axeman89, Stemonitis, Nuno Tavares, Angr, Simetrical, Woohookitty, Mindmatrix, Bellenion, Kokoriko, JeremyA, Keta, Eilthireach, Pasta Salad, Palica, Mrcool1122, Graham87, FreplySpang, Mayumashu, NatusRoma, TheRingess, Wooddoo-eng, Krash, Nguyen Thanh Quang, Whimemsz, Malhonen, CJLL Wright, Chobot, Martin Hinks, KEJ, YurikBot, Wavelength, RobotE, RussBot, Hyad, Pigman, Polyvios, Anomalocaris, Cquan, Zarel, Bayle Shanks, Pdblues, Action potential, Maunus, Wknight94, Donald Albury, Miguelmrm~enwiki, JoanneB, CWenger, Thomas Blomberg, TuukkaH, Torgo, SmackBot, David Kernow, Jasy jatere, KnowledgeOfSelf, Zerida, David.Mestel, Unyoyega, Aurista25, C.Fred, Mgreenbe, Neutral-en, BiT, Niro5, Jwestbrook, Gilliam, Jcarroll, Rmosler2100, Bluebot, Mazeface, Thom2002, MalafayaBot, J. Spencer, Rlevse, Can't sleep, clown will eat me, Mr.Z-man, Allan McInnes, SundarBot, Stevenmitchell, Ghiraddje, RandomP, Jon Awbrey, Fuzzypeg, FlyHigh, Byelf2007, SashatoBot, Derek farn, Rijkbenik, Tim Q. Wells, Minna Sora no Shita, 16@r, J Crow, Slakr, Novangelis, Kvng, Joseph Solis in Australia, Joshuagross, Stifynsemons, Macetw, Thomasmeeks, Gregbard, FilipeS, Icek~enwiki, Cydebot, Kallerdis, Robzy213, Thijs!bot, Epbr123, Jobber, ClosedEyesSeeing, John254, Adw2000, Nick Number, Angryafghan, Pprabhakarrao, KrakatoaKatie, AntiVandalBot, Luna Santin, Prabhakar P Rao, Comhreir, Wayiran, Bogger, JAnDbot, Leuko, FromFoamsToWaves, MER-C, Rollred15, Freedomlinux, VoABot II, Pleckaitis, Rivertorch, Eldumpo, Yaxu, Glen, DerHexer, CapnPrep, Kornfan71, Anaxial, R'n'B, Dwspig2, Smokizzy, J.delanoy, Hippasus, EscapingLife, Bogey97, Numbo3, Jerry, McSly, Rwessel, SJP, Hulten, Idioma-bot, VolkovBot, Semmelweiss, Johan1298~enwiki, Jeff G., Gbouzon, TXiKiBoT, Guillaume2303, PaulTanenbaum, ^demonBot2, Yannis1962, Synthebot, Cnilep, Wiredrabbit, Jimbo2222, SieBot, Mycomp, Nihil novi, Space Dracula, Sky1er, Yerpo, Hexham, Oxymoron83, Tautologist, ClueBot, JonnybrotherJr, Pi zero, Mike Klaassen, CyrilThePig4, Wutsje, ChandlerMapBot, Zack wadghiri, Rhododendrites, Divespluto, Fattyjwoods, SchreiberBike, Acabashi, Aitias, Versus22, Will Hen, Cc116, XLinkBot, Jbeans, Starfire777, Nyoro n, Addbot, Xp54321, ConCompS, Willking1979, Rmalouf, Vishnava, Coffeeassured, Numbo3-bot, Lpjurca, Erutuon, Tide rolls, OlEnglish, Jarble, Legobot, Luckas-bot, Yobot, Ptbotgourou, Denispir, Raimundo Pastor, AnomieBOT, Quangbao, Rjanag, Springindd, IRP, JackieBot, Glenfarclas, Materialscientist, Citation bot, ArthurBot, TwigsCogito, Jchthys, Xqbot, Triplejo2, Dr Oldekop, Omnipaedista, RibotBOT, MarkuP, Acb4341, LucienBOT, Airborne84, Mundart, Sopher99, Pinethicket, Jonesey95, Full-date unlinking bot, Dude1818, Jauhienij, Æk, Lotje, Tjo3ya, Reaper Eternal, Kajervi, Brian the Editor, Tbhotch, DARTH SIDIOUS 2, 2bluey, Ripchip Bot, Jmonk95, EmausBot, RA0808, K6ka, Mainstreet27, Goudron, Neddy1234, Tijfo098, ClueBot NG, AK IM OP, Iloveandrea, Helpful Pixie Bot, BG19bot, CityOfSilver, Solomon7968, AdventurousSquirrel, CitationCleanerBot, Qetuth, Nuyhij, ChrisGualtieri, Grahas02, ComfyKem, BigBangTheoryLad, Kevin12xd, Beatrice57, BreakfastJr, JKJasmineWongLaiKwan, Cmckain14, Wikiuser13, Werddemer, ANALYN AYCOCHO, Monkbot, Boblamus, Crystallizedcarbon, YeOldeGentleman and Anonymous: 305

- **Universal algebra** *Source:* https://en.wikipedia.org/wiki/Universal_algebra?oldid=685007015 *Contributors:* AxelBoldt, Bryan Derksen, Zundark, Andre Engels, Toby~enwiki, Toby Bartels, Youandme, Michael Hardy, GTBacchus, Andres, Revolver, Charles Matthews, Dysprosia, Aleph4, Robbot, Fredrik, Sanders muc, Kowey, Fuelbottle, Giftlite, APH, Sam Hocevar, AlexChurchill, Zaslav, Tompw, Rgdboer, EmilJ, Chasmo, AshtonBenson, Msh210, ABCD, Linas, Mindmatrix, Smmurphy, Isnow, Magidin, Jrtayloriv, Wavelength, Hairy Dude, Chaos, Wiki alf, Arthur Rubin, RonnieBrown, SmackBot, Gilliam, El Fahno, Alink, Nbarth, Zvar, Spakoj~enwiki, Henning Makholm, WillowW, HStel, Sam Staton, Sadeghd, Rlupsa, Knotwork, JAnDbot, Sean Tilson, Twisted86, David Eppstein, JaGa, Pavel Jelínek, Trusilver, TheSeven, JohnBlackburne, AllS33ingI, Popopp, Synthebot, Nicks221, SieBot, Sneakfast, Tkeu, Excirial, He7d3r, Hans Adler, Kaba3, Algebran, Addbot, Delaszk, Loupeter, Legobot, Yobot, Ptbotgourou, AnomieBOT, RibotBOT, Oursipan, Thehelpfulbot, Ilovegrouptheory, D stankov, Yaddie, Rausch, Eivuokko, Nascar1996, GoingBatty, Quondum, ClueBot NG, Bezik, Frietjes, MerlIwBot, Beaumont877, Brad7777, Duxwing, Jochen Burghardt, NQ, JMP EAX, JohnAGough and Anonymous: 63

- **Algebraic geometry** *Source:* https://en.wikipedia.org/wiki/Algebraic_geometry?oldid=684529647 *Contributors:* AxelBoldt, Bryan Derksen, Zundark, Youssefsan, Omenge, Youandme, Edward, Patrick, Michael Hardy, Ixfd64, TakuyaMurata, BigFatBuddha, Poor Yorick, Mxn, Ideyal, Charles Matthews, Dysprosia, Phys, David.Monniaux, Robbot, MathMartin, DHN, Wikibot, Fuelbottle, Mlk, Tosha, Giftlite, Geeoharee, Schopenhauer, Dratman, Waltpohl, Ezhiki, Jason Quinn, Just Another Dan, Bobblewik, Reilly, Antandrus, APH, Sam Hocevar, Klemen Kocjancic, Grstain, D6, Rich Farmbrough, Guanabot, Paul August, Bender235, Obradovic Goran, Geschichte, Patsw, Msh210, Alansohn, Pion, Melaen, Deacon of Pndapetzim, Dirac1933, Linas, JATerg, Mandarax, BD2412, Salix alba, Andrei Polyanin, R.e.b., Wars, Sodin, Masnevets, Chobot, Bgwhite, Kummi, YurikBot, Wavelength, Michael Slone, Bhny, Lenthe, Grubber, Bachrach44, TVilkesalo~enwiki, Crasshopper, Ms2ger, Bo Jacoby, Sardanaphalus, RupertMillard, SmackBot, InverseHypercube, Unyoyega, Jagged 85, Alsandro, Wikikris, Hmains, Chris the speller, Domthedude001, Silly rabbit, Zoran.skoda, Nbarth, RyanEberhart, Acepectif, Gleuschk, Christoffel K~enwiki, Robofish, Jim.belk, Slakr, CRGreathouse, CBM, Myasuda, Mattbuck, Ntsimp, WillowW, After Midnight, Thijs!bot, Headbomb, Newton2, Nick Number, AntiVandalBot, Exoriat, Forgetfulfunctor, The Transhumanist, Meeples, Magioladitis, Sishaman, Jakob.scholbach, Singularitarian, Cocoaguy, TomyDuby, C quest000, Jacksonwalters, Gogobera, JohnBlackburne, LokiClock, Philip Trueman, Anonymous Dissident, Hesam7, Arcfrk, Ishboyfay, Quietbritishjim, Stca74, Flyer22 Reborn, Oxymoron83, Macy, AMackenzie, OKBot, Mr. Stradivarius, ClueBot, Historychecker, Marsupilamov, NuclearWarfare, SchreiberBike, Johnuniq, Goodvac, Download, LaaknorBot, Delaszk, Ozob, Tassedethe, Legobot, Luckas-bot, Yobot, THEN WHO WAS PHONE?, Nallimbot, AnomieBOT, Citation bot, ArthurBot, Xqbot, Capricorn42, Almabot, Miym, Charvest, FrescoBot, Lagelspeil, D'ohBot, Sławomir Biały, Citation bot 1, Tkuvho, Tcnuk, TobeBot, Trappist the monk, Fayedizard, Aditya Bawane, Bj norge, TjBot, EmausBot, Marcus0107, Dinkelburg 21, GoingBatty, K6ka, Slawekb, Werieth, Chewbacca51, JSquish, Hilbertthm90, Neechalkaran, D.Lazard, L Kensington, Harapp, ClamDip, ClueBot NG, Satellizer, Dylan Moreland, LJosil, Uni.Liu, Islifenm, Helpful Pixie Bot, Aidanrockstar, Teika kazura, Brad7777, Minsbot, ChrisGualtieri, Jonier.a.a, Brirush, Marcela louis, Mark viking, Purnendu Karmakar, Delphenich, Bloorain, SakeUPenn, Dgervini, Monkbot, Adult Traveler One$, Loraof, Lucywyman, KasparBot, SilverSurfingSerpent, Batman's butler and Anonymous: 155

- **Finite model theory** *Source:* https://en.wikipedia.org/wiki/Finite_model_theory?oldid=682558077 *Contributors:* Michael Hardy, Charles Matthews, Dcoetzee, Hyacinth, Giftlite, Waltpohl, Creidieki, AshtonBenson, Linas, Mathbot, YurikBot, Ott2, Sterling, SmackBot, CBM, Gregbard, Ronfagin, David Eppstein, Daniel5Ko, Tatrgel, JohnBlackburne, Saibod, Jamelan, Synthebot, ImageRemovalBot, Hans Adler, Addbot, Yobot, Citation bot, FrescoBot, Arthur MILCHIOR, Skyerise, Jonkerz, John of Reading, Tijfo098, Modelpractice, Brad7777, Deltahedron, Cubism44 and Anonymous: 10

- **Higher-order logic** *Source:* https://en.wikipedia.org/wiki/Higher-order_logic?oldid=686624292 *Contributors:* Evercat, Charles Matthews, David.Monniaux, Altenmann, Stephan Schulz, Gandalf61, MathMartin, Jason Quinn, Christopherlin, Saugart, Lucidish, Pmetzger, AshtonBenson, Ruud Koot, Rjwilmsi, Mathbot, Tillmo, Eskimbot, Mhss, FordPrefect42, Jon Awbrey, Byelf2007, Physis, Mets501, CBM, Sdorrance, Gregbard, Thijs!bot, David Eppstein, Crisperdue, Camrn86, Sapphic, Dale A Miller, Paradoctor, IsleLaMotte, Addbot, Atethnekos, Pcap, AnomieBOT, FrescoBot, Mfwitten, Chronulator, Gf uip, Klbrain, Quondum, Spirarel, Master ecclesias, Jochen Burghardt, Mark viking, J.F1p., Monkbot and Anonymous: 21

- **Infinitary logic** *Source:* https://en.wikipedia.org/wiki/Infinitary_logic?oldid=684554380 *Contributors:* Michael Hardy, Charles Matthews, Tobias Bergemann, Jason Quinn, Barnaby dawson, Smimram, Ben Standeven, AshtonBenson, RussBlau, BD2412, R.e.b., Mathbot, Ian-Manka, Trovatore, Arthur Rubin, Cybercobra, Zero sharp, CRGreathouse, Tamarkot, Sniffnoy, CBM, Gregbard, Headbomb, Sapphic, Addbot, AnomieBOT, Citation bot, Godelian, Citation bot 1, WikitanvirBot, Kephir and Anonymous: 13

- **Gödel's completeness theorem** *Source:* https://en.wikipedia.org/wiki/G%C3%B6del'{}s_completeness_theorem?oldid=656456908 *Contributors:* AxelBoldt, Mav, Michael Hardy, Modster, Mdebets, Tim Retout, Charles Matthews, Timwi, Dysprosia, Hyacinth, Aleph4, Gandalf61, Giftlite, Lethe, Lupin, Siroxo, Spoirier~enwiki, Mike Rosoft, Rich Farmbrough, Guanabot, Maksym Ye., Ben Standeven, El C, Chalst, Teorth, 3mta3, Gene Nygaard, Drbreznjev, Eric Qel-Droma, Oleg Alexandrov, Ott, Pdn~enwiki, Dzordzm, Flamingspinach, Dionyziz, Tim!, Mathbot, NavarroJ, Reetep, Algebraist, YurikBot, Tony1, Scope creep, SmackBot, RDBury, BeteNoir, Betacommand, Mhss, Dr. de Seis, Taggart Transcontinental, SashatoBot, Dan Gluck, Joseph Solis in Australia, Hilverd, Zero sharp, JRSpriggs, CBM, Myasuda, Gregbard, Spewin, Thijs!bot, Headbomb, Jbaranao, Morphriz, Meeples, TXiKiBoT, EuTuga, Da Joe, IsleLaMotte, Martarius, Hans Adler, El bot de la dieta, Thingg, Hugo Herbelin, Marc van Leeuwen, MrOllie, Lightbot, Legobot, OrgasGirl, Omnipaedista, BrideOfKripkenstein, Citation bot 1, WikitanvirBot, Jaydiem, Brad7777, ChrisGualtieri, Jochen Burghardt, Epicgenius, Salspaugh, 2.71828182845904523austen, Whiterray and Anonymous: 33

- **Computable model theory** *Source:* https://en.wikipedia.org/wiki/Computable_model_theory?oldid=635892079 *Contributors:* Greenrd, Tompw, BD2412, Cydebot, Grahamec, Headbomb, David Eppstein, Mark lee stillwell, JohnBlackburne, Matthew Yeager, CBM2, Helpful Pixie Bot, Brad7777 and Brirush

- **Binary operation** *Source:* https://en.wikipedia.org/wiki/Binary_operation?oldid=651651472 *Contributors:* AxelBoldt, Tarquin, Andre Engels, Danny, Toby~enwiki, Toby Bartels, Patrick, Michael Hardy, Wshun, TakuyaMurata, Delirium, Yaronf, Poor Yorick, Rotem Dan, Andres, Emperorbma, Charles Matthews, Dysprosia, Greenrd, Robbot, Mattblack82, Romanm, Giftlite, Jason Quinn, Avaragado, C4~enwiki, Ornil, Mormegil, Guanabot, Paul August, Smalljim, Adrian.benko, Alexrudd, Linas, Isnow, Waldir, SixWingedSeraph, Josh Parris, Salix alba, Noya Watan, PlatypeanArchcow, Chobot, Hairy Dude, Hede2000, Piet Delport, NawlinWiki, Maerk, Klutzy, BraneJ, Bo Jacoby, SmackBot, RD-Bury, Mmernex, Melchoir, NoJoy, Octahedron80, Nbarth, DHN-bot~enwiki, Chendy, Richard L. Peterson, Mets501, CmdrObot, Georg Peter, Xtv, Xantharius, Mhaitham.shammaa, Salgueiro~enwiki, JAnDbot, Magioladitis, Arno Matthias, David Eppstein, Tommy Herbert, R'n'B, Ps ttf, Trumpet marietta 45750, AntiSpamBot, Jrugordon, VolkovBot, LokiClock, Am Fiosaigear~enwiki, Skylarkmichelle, PaulTanenbaum, AJRobbins, Bernstein2291, SieBot, Ivan Štambuk, Bentogoa, Aravindk editing, Jdaloner, ClueBot, Razimantv, JP.Martin-Flatin, Chininazu12, DumZiBoT, Kintaro, Addbot, EjsBot, CarsracBot, BepBot, PV=nRT, Zorrobot, Legobot, Luckas-bot, Yobot, TaBOT-zerem, Götz, Spiros Bousbouras, Xqbot, Sixequalszero, Howard McCay, Erik9bot, Sae1962, 00Ragora00, Orenburg1, Ripchip Bot, Bodhisvaha, ZéroBot, Chharvey, Quondum, Maschen, Superion maximus, ClueBot NG, Mohanapriya94, Wcherowi, MerllwBot, Brad7777, Vijeenroshpw, Proxyma, Wik2kassa, Ro0800, Deadlyblight and Anonymous: 50

- **Unary operation** *Source:* https://en.wikipedia.org/wiki/Unary_operation?oldid=684757810 *Contributors:* AxelBoldt, Danny, Arvindn, Patrick, Wshun, Voidvector, Andres, Seth Arlington, Charles Matthews, Robbot, Fredrik, Giftlite, Paul Pogonyshev, Macrakis, Jacob grace, Pmanderson, Tzarius, Andreas Kaufmann, Mormegil, Noisy, Rich Farmbrough, Forderud, Oleg Alexandrov, Japanese Searobin, Gsnxn, Ruud Koot, MFH, Isnow, Salix alba, Tardis, JPD, YurikBot, Bota47, Kjak, SmackBot, RDBury, Gelingvistoj, Bluebot, Octahedron80, Dfletter, Richard L. Peterson, Eassin, Pointlessness, Lark ascending, PamD, Thijs!bot, Headbomb, Mhaitham.shammaa, Ianare, Salgueiro~enwiki, JAnDbot, Burga, JoergenB, Gwern, R'n'B, BlGene, TXiKiBoT, Anonymous Dissident, Gerakibot, Alksentrs, JP.Martin-Flatin, Jedihawk, Addbot, Ghettoblaster, Btx40, Download, Zorrobot, Legobot, Bunnyhop11, Ztianjin, Erik9bot, Alph Bot, EmausBot, Quondum, Kodefuguru, BG19bot, Creation FrOH, Pratyya Ghosh, Quenhitran, Arjunmmurali007 and Anonymous: 43

- **Arity** *Source:* https://en.wikipedia.org/wiki/Arity?oldid=682753049 *Contributors:* Damian Yerrick, AxelBoldt, Zundark, Tarquin, Edward, Patrick, Michael Hardy, Kaczor~enwiki, Wapcaplet, Andres, Charles Matthews, Dcoetzee, Aleph4, Robbot, Fredrik, Sanders muc, Merovingian, Bkell, Wikibot, Giftlite, Mshonle~enwiki, Sohailstyle, Antandrus, Beland, Vina, Jacob grace, Naff89, Zaslav, Elwikipedista~enwiki, Gauge, Theinfo, CanisRufus, Kwamikagami, Obradovic Goran, Mdd, Marc van Woerkom, Shawn K. Quinn, Ceyockey, Forderud, Oleg Alexandrov, Feezo, Madmardigan53, MFH, Mangojuice, Bluemoose, BD2412, Qwertyus, NatusRoma, MarSch, Salix alba, Stuwee, Metacarpus, Mathbot, Roboto de Ajvol, YurikBot, Hairy Dude, Kiscica, RussBot, Twin Bird, Modify, Otto ter Haar, SmackBot, Melchoir, Jjalexand, Nbarth, Colonies Chris, Javalenok, Tsca.bot, Frap, Cybercobra, Jon Awbrey, Luís Felipe Braga, Autopilot, Geoinline, EdC~enwiki, CR-Greathouse, Difluoroethene, Torc2, Kilva, Ablonus, Liquid-aim-bot, JAnDbot, Deflective, Soulbot, David Eppstein, JaGa, Jimka, Gwern, Jim.henderson, R'n'B, Cpiral, Trumpet marietta 45750, Daniel5Ko, Boute, AlleborgoBot, S8333631, AHMartin, Mild Bill Hiccup, Watchduck, Hans Adler, Libcub, Addbot, AnomieBOT, Rubinbot, RokerHRO, Xqbot, TechBot, Charvest, FrescoBot, The Utahraptor, Drake921, Hpvpp, ZéroBot, Tijfo098, Sindikat, Frietjes, Helpful Pixie Bot, AvocatoBot, QuarkyPi and Anonymous: 57

- **Ring (mathematics)** *Source:* https://en.wikipedia.org/wiki/Ring_(mathematics)?oldid=686239035 *Contributors:* Damian Yerrick, AxelBoldt, Bryan Derksen, Zundark, Tarquin, Youssefsan, Toby Bartels, Miguel~enwiki, Patrick, Michael Hardy, Wshun, Dominus, TakuyaMurata, Stevan White, Stan Shebs, Suisui, Angela, AugPi, Rotem Dan, Andres, Clausen, Vargenau, Schneelocke, Ideyal, Ffransoo, Loren Rosen, Revolver, Charles Matthews, Dysprosia, Jitse Niesen, Kbk, Prumpf, Itai, Taxman, VeryVerily, Aleph4, Robbot, Romanm, MathMartin, Henrygb, Puckly, Bkell, Tobias Bergemann, Tosha, Giftlite, BenFrantzDale, Tom harrison, Fropuff, Berjoh, FunnyMan3595, Michael Devore, Jorend, Ssd, Jorge Stolfi, Python eggs, Chowbok, Sigfpe, Profvk, Pmanderson, Mschlindwein, Frenchwhale, Barnaby dawson, Vivacissamamente, PhotoBox, D6, Smimram, Discospinster, Hydrox, Mecanismo, Xezbeth, Paul August, Gauge, Syp, Remag12@yahoo.com, El C, Rgdboer, EmilJ, O18, Touriste, La goutte de pluie, Nk, Obradovic Goran, Msh210, AidanH, Emvee~enwiki, Oleg Alexandrov, Joriki, Linas, Aaron McDaid, Robert K S, Halcatalyst, Mandarax, Graham87, Ilya, Jclemens, Chenxlee, Radomir, Rjwilmsi, Pako, Staecker, Salix alba, Mathbot, R160K, Sunil.nandihalli, Chobot, Bgwhite, Algebraist, Jayme, Wavelength, Hairy Dude, Dmharvey, Michael Slone, Piet Delport, KSmrq, Grubber, Chrispounds, ENeville, Rick Norwood, NickBush24, David Pierce, Crasshopper, Pooryorick~enwiki, 2over0, Arthur Rubin, Netrapt, Pred, Bo Jacoby, Jsnx, SmackBot, Selfworm, InverseHypercube, CapitalSasha, Cazort, Hmains, Anastasios~enwiki, JasonMR, Snori, PrimeHunter, MalafayaBot, Silly rabbit, Octahedron80, Don Hosek, AdamSmithee, Kjetil1001, Cybercobra, Diocles, Tilin, Ninte, Soumyasch, Rschwieb, Ojan, CRGreathouse, DavidFHoughton, Nightwriter50, Ironmagma, Myasuda, Mct mht, Marqueed, Dl573, Gogo Dodo, Goldencako, Xantharius, Thijs!bot, Wikid77, Andri Egilsson, Headbomb, RobHar, OrenBochman, CZeke, Papipaul, JAnDbot, Drizzd~enwiki, Magioladitis, JamesBWatson, Swpb, Jakob.scholbach, Alvian, Twsx, WhatamIdoing, JJ Harrison, David Eppstein, JaGa, Tejon~enwiki, Pomte, Laurusnobilis, TomyDuby, Jeepday, OliverHarris, Yecril, VolkovBot, JohnBlackburne, LokiClock, AlnoktaBOT, Cbigorgne, TXiKiBoT, Malsaqer,

VanishedUserABC, Kgoarany, RJaguar3, ConcernedScientist, Lord British, Ljf255, SouthLake, Kumioko (renamed), DesolateReality, Anchor Link Bot, Wireless99, Randomblue, CBM2, NoBu11, Francvs, Classicalecon, Phyte, NicDumZ, Jan1nad, Gherson2, Mild Bill Hiccup, Dkf11, Nanobear~enwiki, Nanmus, Watchduck, Cacadril, Hans Adler, Djk3, Willhig, Palnot, WikHead, Subversive.sound, Sameer0s, Addbot, Norman Ramsey, Histre, Pdibner, Tassedethe, ב דניאל., Snaily, Legobot, Yobot, Ht686rg90, Cloudyed, Pcap, AnakngAraw, AnomieBOT, Citation bot, TitusCarus, Grim23, Ejars, FrescoBot, Hobsonlane, Mark Renier, Liiiii, Citation bot 1, Tkuvho, DrilBot, Sh Najd, 34jjkky, Rlcolasanti, Diannaa, Reach Out to the Truth, Lauri.pirttiaho, WildBot, Gf uip, Carbo1200, Be hajian, Chharvey, Sampletalk, Bulwersator, Jaseemabid, Tijfo098, Templatetypedef, ClueBot NG, Johannes Schützel, MerlIwBot, Daviddwd, BG19bot, Lifeformnoho, Dhruvbaldawa, Virago250, Solomon7968, Rjs.swarnkar, Sanpra1989, Deltahedron, Gabefair, Jochen Burghardt, Hoppeduppeanut, Cptwunderlich, Seppi333, Holyseven007, Wilbertcr, Threerealtrees, Immanuel Thoughtmaker, Jwinder47, Mario Castelán Castro, Purgy Purgatorio, Comp-heur-intel, Broswald and Anonymous: 249

- **Signature (logic)** *Source:* https://en.wikipedia.org/wiki/Signature_(logic)?oldid=618254276 *Contributors:* Fibonacci, Giftlite, Creidieki, Ntmatter, EmmetCaulfield, Oleg Alexandrov, Baccala@freesoft.org, Ott2, Tobixen, Physis, Zero sharp, CRGreathouse, CBM, Gregbard, AlleborgoBot, Shellgirl, Brenont, JerroldPease-Atlanta, PixelBot, Hans Adler, Addbot, Yobot, Prijutme4ty, Liiiii, Monkbot and Anonymous: 8

- **Quasiidentity** *Source:* https://en.wikipedia.org/wiki/Quasiidentity?oldid=618191838 *Contributors:* Michael Hardy, Charles Matthews, Oleg Alexandrov, SmackBot, Vaughan Pratt, CBM, Tikiwont, Hans Adler, Paine Ellsworth and Monkbot

- **Atomic formula** *Source:* https://en.wikipedia.org/wiki/Atomic_formula?oldid=675306991 *Contributors:* Hyacinth, Aleph4, Creidieki, Kaustuv, Paul August, Spayrard, Linas, BD2412, Tizio, Trovatore, Hughitt1, Mhss, Jon Awbrey, Mets501, E-boy, CBM, HenningThielemann, Gregbard, Cydebot, Thijs!bot, MartinBot, Naohiro19, Gurchzilla, The best gamer, Dessources, VolkovBot, Morenooso, Kyle the bot, Philogo, RatnimSnave, Curtdbz, Addbot, Charletan, Yobot, Citation bot, Xqbot, D'ohBot, Dinamik-bot, RjwilmsiBot, Classerre, Alpha Quadrant (alt), Tijfo098, Helpful Pixie Bot, Wasbeer, Jianhui67 and Anonymous: 14

- **List of logic symbols** *Source:* https://en.wikipedia.org/wiki/List_of_logic_symbols?oldid=686603690 *Contributors:* Boud, Jitse Niesen, Hyacinth, Benwing, Ancheta Wis, Petershank, Abdull, Paul August, EmilJ, SpeedyGonsales, PWilkinson, Chira, Dirac1933, Mindmatrix, DePiep, Koavf, R.e.b., YurikBot, Wavelength, KSmrq, KSchutte, Grafen, Trovatore, Mkouklis, Dbmag9, Arthur Rubin, Jbalint, SmackBot, InverseHypercube, Melchoir, Isaac Dupree, Mhss, Bluebot, Da nuke, Pioto, Alphathon, DGerman, Cybercobra, DemosDemon, Ihatetoregister, Dqb124, SashatoBot, Lambiam, Eric76, Vanisaac, CBM, Gregbard, Cydebot, Julian Mendez, Cartesian1, Robocracy, JJ Harrison, David Eppstein, It Is Me Here, Anonymous Dissident, Bentogoa, Hello71, Kumioko (renamed), Francvs, Leranedo, Dead10ck, Philosophy.dude, Peter.C, Saeed.Veradi, Addbot, Melab-1, Luckas-bot, GorgeUbuasha, AnomieBOT, Xqbot, Coretheapple, Oscarjquintana, Lagelspeil, Mscdancer, Cailean8, Erelen, George Richard Leeming, EmausBot, ZéroBot, Derekleungtszhei, Quondum, Rfontiveros, Donner60, BrendanLarvor, RockMagnetist, Delusion23, BG19bot, Rm1271, Fylbecatulous, The Mol Man, Tcrosley, Victor lesyk, Gufosowa, MetazoanMarek and Anonymous: 65

- **T-schema** *Source:* https://en.wikipedia.org/wiki/T-schema?oldid=544243364 *Contributors:* Charles Matthews, Adam78, Frencheigh, Chalst, PWilkinson, Kzollman, Marudubshinki, Trovatore, Mhss, Physis, Mets501, CBM, Gregbard, Cydebot, Kenneth M Burke, Grammarmonger, Philogo, Addbot, Ht686rg90, Hriber, Tijfo098, Philofet, Helpful Pixie Bot and Anonymous: 6

- **Theory (mathematical logic)** *Source:* https://en.wikipedia.org/wiki/Theory_(mathematical_logic)?oldid=683264889 *Contributors:* Vkuncak, Hyacinth, Timrollpickering, Jabowery, Uffish, Woohookitty, Linas, Waldir, Tillmo, Algebraist, Trovatore, Arthur Rubin, Cícero, Lambiam, CBM, Myasuda, Gregbard, David Eppstein, Philogo, Hans Adler, Iranway, Addbot, Numbo3-bot, Yobot, Buenasdiaz, MastiBot, RockMagnetist, ClueBot NG, Wcherowi, Masssly and Anonymous: 10

- **Consistency** *Source:* https://en.wikipedia.org/wiki/Consistency?oldid=669765886 *Contributors:* Michael Hardy, Dominus, Kku, Tim Retout, Charles Matthews, Hyacinth, Chealer, Chancemill, Robinh, Tobias Bergemann, Filemon, Giftlite, Beland, Luqui, Chalst, Dreish, Oleg Alexandrov, Ylem, Isnow, Jobnikon, Amorrow, NatusRoma, Jevon, Rbonvall, Intgr, Robbyslaughter, Tillmo, Gslin, SmackBot, Radagast83, Jon Awbrey, Tilin, PhiJ, Lambiam, Wvbailey, Levineps, Iridescent, Zero sharp, DavidBalde, JRSpriggs, CRGreathouse, CBM, Myasuda, Gregbard, DumbBOT, Boemanneke, Thijs!bot, Dqd, JAnDbot, Magioladitis, MetsBot, Vinograd19, Natsirtguy, Regicollis, Kumioko (renamed), ClueBot, The Thing That Should Not Be, Qwfp, Bleeve, Feministo, Addbot, Maschelos, Tide rolls, LuK3, Luckas-bot, Yobot, Vanished user rt41as76lk, Ningauble, AnomieBOT, Akhran, Galoubet, VladimirReshetnikov, Noamz, Whassan, Antares5245, FrescoBot, RedBot, MastiBot, Gopher p, Morton Shumway, EmausBot, John of Reading, Shshahryari, Bomazi, Tijfo098, Wcherowi, JimsMaher, Masssly, Helpful Pixie Bot, Lowercase sigmabot, Justincheng12345-bot, Tapped-out, BurkeFT and Anonymous: 40

- **Löwenheim–Skolem theorem** *Source:* https://en.wikipedia.org/wiki/L%C3%B6wenheim%E2%80%93Skolem_theorem?oldid=662490561 *Contributors:* Michael Hardy, David Martland, Schneelocke, Charles Matthews, Dysprosia, Jitse Niesen, Markhurd, Hyacinth, Aleph4, Robbot, Josh Cherry, Psychonaut, Giftlite, KelvSYC, Lupin, Mboverload, Jabowery, Pjacobi, EmilJ, Nortexoid, Gene Nygaard, FlaBot, Mathbot, Korg, Roboto de Ajvol, Archelon, David Pierce, RDBury, BeteNoir, Mhss, Cícero, Viebel, Turms, Esoth~enwiki, Tesseran, Lambiam, Vanished user v8n3489h3tkjnsdkq30u3f, Loadmaster, Tyrrell McAllister, Dan Gluck, Zero sharp, Jason22~enwiki, JRSpriggs, CBM, Chrisahn, Ksoileau, Gregbard, Mattbuck, Thijs!bot, JAnDbot, Wasell, Ttwo, Alan U. Kennington, VolkovBot, Jamelan, Alexbot, Hans Adler, Addbot, דניאל ב., Luckas-bot, Yobot, Ptbotgourou, Citation bot, ArthurBot, Undsoweiter, FrescoBot, D'ohBot, Sławomir Biały, Rhalah, Citation bot 1, EmausBot, Set theorist, Tijfo098, BG19bot, CitationCleanerBot, Minsbot, Deltahedron, Jochen Burghardt and Anonymous: 19

- **Cardinal number** *Source:* https://en.wikipedia.org/wiki/Cardinal_number?oldid=686002442 *Contributors:* AxelBoldt, Tobias Hoevekamp, Derek Ross, Calypso, Mav, Bryan Derksen, Zundark, The Anome, Iwnbap, Andre Engels, Danny, XJaM, Christian List, Gritchka, Toby Bartels, Olivier, Patrick, JohnOwens, Michael Hardy, Wshun, Nixdorf, Rp, TakuyaMurata, Mpagano, Poor Yorick, Jiang, Vargenau, Schneelocke, Charles Matthews, Dysprosia, Jitse Niesen, Maximus Rex, Hyacinth, Gutsul, Mosesklein, Bloodshedder, Dmytro, Hmackiernan, Robbot, Fredrik, Sverdrup, Henrygb, Choni, Paul Murray, Fuelbottle, Wile E. Heresiarch, Tobias Bergemann, Giftlite, Ævar Arnfjörð Bjarmason, Lethe, Fropuff, Snowdog, Ajgorhoe, Mboverload, Mobius, Fuzzy Logic, Beland, Pmanderson, Karl-Henner, Sam Hocevar, Neutrality, Splatty, Paul August, Dmr2, Grick, Randall Holmes, PWilkinson, Minghong, Jumbuck, Sligocki, Caesura, Spambit, DV8 2XL, Oleg Alexandrov, Needmoredinosaur, Linas, Miaow Miaow, Aaron McDaid, Dionyziz, Graham87, ArchonOFages, Qwertyus, Golden Eternity, Salix alba, FlaBot, Chobot, Algebraist, YurikBot, Alpt, Jacques Antoine, Gaius Cornelius, Trovatore, PAStheLoD, Expensivehat, Figaro, Arthur Rubin, Clair de Lune, Fram, Teply, Bo Jacoby, SmackBot, YellowMonkey, InverseHypercube, Although, Jagged 85, Tim Pierce, Nakon, Dreadstar, Mmehdi.g, Lambiam, John H, Morgan, Loadmaster, JHunterJ, Gabn1, Jason.grossman, Joseph Solis in Australia, Easwaran, JRSpriggs, The Letter J,

- **Reduct** *Source:* https://en.wikipedia.org/wiki/Reduct?oldid=618205741 *Contributors:* John Baez, Spring Rubber, SmackBot, Vaughan Pratt, Tikiwont, Hans Adler, Backslash Forwardslash, Gf uip, EmausBot, Bgeron, Monkbot and Anonymous: 3

- **Interpretation (model theory)** *Source:* https://en.wikipedia.org/wiki/Interpretation_(model_theory)?oldid=604304885 *Contributors:* TexasAndroid, Martijn Hoekstra, Lambiam, Courcelles, Gregbard, Kumioko (renamed), ClueBot, Hans Adler, Alexey Muranov, Citation bot, Citation bot 1, DrilBot and Anonymous: 6

- **Decidability (logic)** *Source:* https://en.wikipedia.org/wiki/Decidability_(logic)?oldid=677174213 *Contributors:* AugPi, Hyacinth, Romanm, Tobias Bergemann, Nomeata, Rich Farmbrough, Ben Standeven, EmilJ, Nortexoid, Obradovic Goran, Oleg Alexandrov, Linas, Nahabedere, Margosbot~enwiki, Dmharvey, Arthur Rubin, Imz, Pokipsy76, Tibuy2008, Henning Makholm, Lambiam, Schildt.a, WAREL, CapitalR, JR-Spriggs, CBM, Gregbard, Julian Mendez, Thijs!bot, Headbomb, JustAGal, Nick Number, Erxnmedia, JAnDbot, Glivi, Jamelan, Sapphic, Stanning, Kumioko (renamed), Hans Adler, Addbot, Yobot, Xqbot, GrouchoBot, Bytbox, Aliotra, Arthur MILCHIOR, BG19bot, Kaltenmeyer, Kiewbra, Jochen Burghardt, Naereen, LieutenantLatvia, NicolasPott, Velvel2 and Anonymous: 19

- **Boolean algebra (structure)** *Source:* https://en.wikipedia.org/wiki/Boolean_algebra_(structure)?oldid=675939438 *Contributors:* AxelBoldt, Mav, Bryan Derksen, Zundark, Tarquin, Taw, Jeronimo, Ed Poor, Perry Bebbington, XJaM, Toby Bartels, Heron, Camembert, Michael Hardy, Pit~enwiki, Shellreef, Justin Johnson, GTBacchus, Ellywa, Александър, DesertSteve, Samuel~enwiki, Charles Matthews, Timwi, Dcoetzee, Dysprosia, Jitse Niesen, OkPerson, Maximus Rex, Imc, Fibonacci, Mosesklein, Sandman~enwiki, Johnleemk, JorgeGG, Robbot, Josh Cherry, Fredrik, Romanm, Voodoo~enwiki, Robinh, Ruakh, Tobias Bergemann, Ancheta Wis, Giftlite, Markus Krötzsch, Lethe, MSGJ, Elias, Eequor, Pvemburg, Macrakis, Gauss, Ukexpat, Eduardoporcher, Barnaby dawson, Talkstosocks, Poccil, Guanabot, Cacycle, Slipstream, Ivan Bajlo, Mani1, Paul August, Bunny Angel13, Plugwash, Elwikipedista~enwiki, Chalst, Nortexoid, Jojit fb, Wrs1864, Masashi49, Msh210, Andrewpmk, ABCD, Water Bottle, Cburnett, Alai, Klparrot, Woohookitty, Linas, Igny, Uncle G, Kzollman, Graham87, Magister Mathematicae, Ilya, Qwertyus, SixWingedSeraph, Rjwilmsi, Isaac Rabinovitch, MarSch, KamasamaK, Staecker, GOD, Salix alba, Yamamoto Ichiro, FlaBot, Mathbot, Alhutch, Celestianpower, Scythe33, Chobot, Visor, Nagytibi, YurikBot, RobotE, Hairy Dude, Baccala@freesoft.org, KSmrq, Joebeone, Archelon, Wiki alf, Trovatore, Bota47, Ott2, Kompik, StuRat, Cullinane, Jованвб, Arthur Rubin, JoanneB, Ilmari Karonen, Bsod2, SmackBot, FunnyYetTasty, Incnis Mrsi, Unyoyega, SaxTeacher, Btwied, Srnec, ERcheck, Izehar, Ciacchi, Cybercobra, Clarepawling, Jon Awbrey, Lambiam, Cronholm144, Meco, Condem, Avantman42, Zero sharp, Vaughan Pratt, Makeemlighter, CBM, Gregbard, Sopoforic, Pce3@ij.net, Sommacal alfonso, Julian Mendez, Tawkerbot4, Thijs!bot, Sagaciousuk, Colin Rowat, Tellyaddict, KrakatoaKatie, AntiVandalBot, JAnDbot, MER-C, Magioladitis, Albmont, Omicron18, David Eppstein, Honx~enwiki, Dai mingjie, Samtheboy, Policron, Fylwind, Pleasantville, Enoksrd, Anonymous Dissident, Plclark, The Tetrast, Fwehrung, Escher26, Wing gundam, CBM2, WimdeValk, He7d3r, Hans Adler, Andreasabel, Hugo Herbelin, Aguitarheroperson, Download, LinkFA-Bot, דניאל ב., Jarble, Yobot, Ht686rg90, 2D, AnomieBOT, RobertEves92, Citation bot, ArthurBot, LilHelpa, Gonzalcg, RibotBOT, SassoBot, Charvest, Constructive editor, FrescoBot, Irmy, Citation bot 1, DixonDBot, EmausBot, KHamsun, Thecheesykid, D.Lazard, Tijfo098, ChuispastonBot, Anita5192, ClueBot NG, Delusion23, Jiri 1984, Widr, Helpful Pixie Bot, Solomon7968, ChrisGualtieri, Tagremover, Freeze S, Dexbot, Kephir, Jochen Burghardt, GeoffreyT2000, JMP EAX and Anonymous: 153

- **Algebraically closed field** *Source:* https://en.wikipedia.org/wiki/Algebraically_closed_field?oldid=632933850 *Contributors:* Mav, Bryan Derksen, Zundark, Tarquin, Jan Hidders, Michael Hardy, Wshun, Poor Yorick, Revolver, Dysprosia, Markhurd, JensMueller, Robbot, Jmabel, Math-Martin, Tosha, Giftlite, Rich Farmbrough, Paul August, EmilJ, Wood Thrush, LutzL, Msh210, Oleg Alexandrov, -Ril-, Graham87, FlaBot, Chobot, YurikBot, Klutzy, Banus, Emeraldemon, Bluebot, JCSantos, MalafayaBot, Addshore, Kingdon, Jim.belk, JdH, Newone, CmdrObot, AndrewHowse, Wikid77, Gus8591, RobHar, JAnDbot, Magioladitis, LOTOK, STBotD, DorganBot, SieBot, DesolateReality, Bender2k14, Hans Adler, Razorflame, Popffabrik, PV=nRT, Legobot, Yobot, Nallimbot, ArthurBot, Point-set topologist, Nikiriy, Oldrrb, TobeBot, WikitanvirBot, Honestrosewater, ChrisGualtieri, K9re11 and Anonymous: 29

- **Bijection** *Source:* https://en.wikipedia.org/wiki/Bijection?oldid=683131003 *Contributors:* Damian Yerrick, AxelBoldt, Tarquin, Jan Hidders, XJaM, Toby Bartels, Michael Hardy, Wshun, TakuyaMurata, GTBacchus, Karada, Александър, Glenn, Poor Yorick, Rob Hooft, Pizza Puzzle, Hashar, Hawthorn, Charles Matthews, Dcoetzee, Dysprosia, Hyacinth, David Shay, Ed g2s, Bevo, Robbot, Fredrik, Benwing, Bkell, Salty-horse, Tobias Bergemann, Giftlite, Jorge Stolfi, Alberto da Calvairate~enwiki, MarkSweep, Tsemii, Vivacissamamente, Guanabot, Guanabot2, Quistnix, Paul August, Ignignot, MisterSheik, Nickj, Kevin Lamoreau, Obradovic Goran, Pearle, HasharBot~enwiki, Dallashan~enwiki, ABCD, Schapel, Palica, MarSch, Salix alba, FlaBot, VKokielov, RexNL, Chobot, YurikBot, Michael Slone, Member, SmackBot, RDBury, Mmernex, Octahedron80, Mhym, Bwoodacre, Dreadstar, Davipo, Loadmaster, Mets501, Dreftymac, Hilverd, Johnfuhrmann, Bill Malloy, Domitori, JRSpriggs, CmdrObot, Gregbard, Yaris678, Sam Staton, Panzer raccoon!, Kilva, AbcXyz, Escarbot, Salgueiro~enwiki, JAnDbot, David Eppstein, Martynas Patasius, Paulnwatts, Cpiral, GaborLajos, Policron, Diegovb, UnicornTapestry, Yomcat, Wykypydya, Bongoman666, SieBot, Paradoctor, Flyer22 Reborn, Paolo.dL, Smaug123, MiNombreDeGuerra, JackSchmidt, I Spel Good~enwiki, Peiresc~enwiki, Classicalecon, Adrianwn, Biagioli, Watchduck, Hans Adler, Humanengr, Neuralwarp, Baudway, FactChecker1199, Kal-El-Bot, Subversive.sound, Tanhabot, Glane23, PV=nRT, Meisam, Legobot, Luckas-bot, Yobot, Ash4Math, Shvahabi, Omnipaedista, RibotBOT, Thehelpfulbot, FrescoBot, MarcelB612, CodeBlock, MastiBot, FoxBot, Duoduoduo, Xnn, EmausBot, Hikaslap, TuHan-Bot, Cobaltcigs, Wikfr, Karthikndr, Anita5192, Wcherowi, Widr, Strike Eagle, PhnomPencil, Knwlgc, Dhoke sanket, Victor Yus, Dexbot, Cerabot~enwiki, JPaestpreornJeolhlna, Yardimsever, CasaNostra, KoriganStone, Whamsicore, JMP EAX, Kiwifist, Sweepy and Anonymous: 91

- **Constructible topology** *Source:* https://en.wikipedia.org/wiki/Constructible_topology?oldid=649415047 *Contributors:* TakuyaMurata, ShelfSkewed, RobHar, Junior Wrangler, RjwilmsiBot, Bomazi, Helpful Pixie Bot, BG19bot, Qetuth, K9re11 and Anonymous: 1

- **Zariski topology** *Source:* https://en.wikipedia.org/wiki/Zariski_topology?oldid=687296131 *Contributors:* AxelBoldt, Zundark, TakuyaMurata, Schneelocke, Charles Matthews, Phys, Saforrest, Tosha, Giftlite, Waltpohl, Jason Quinn, Klemen Kocjancic, 4pq1injbok, Kundor, Pearle, Ryan Reich, MarSch, R.e.b., Krishnavedala, YurikBot, Eraserhead1, Crasshopper, Woscafrench, KnightRider~enwiki, RDBury, Melchoir, Bluebot, Nbarth, Colonies Chris, Phatom87, Thijs!bot, RobHar, Jakob.scholbach, JJ Harrison, Stwitzel, LokiClock, Mr. Granger, PixelBot, SilvonenBot, Bail Fetter, Addbot, Ozob, Luckas-bot, Yobot, D'ohBot, Tkuvho, RedBot, Trappist the monk, Rickhev1, EmausBot, Marcus0107, D.Lazard, Anita5192, Deltahedron, Mark L MacDonald, Enyokoyama, Stephan Alexander Spahn, Brirush and Anonymous: 28

53.6.2 Images

- **File:6n-graf.svg** *Source:* https://upload.wikimedia.org/wikipedia/commons/5/5b/6n-graf.svg *License:* Public domain *Contributors:* Image: 6n-graf.png simlar input data *Original artist:* User:AzaToth

- **File:A_syntactic_parse_of_\char"0022\relax{}Alfred_spoke\char"0022\relax{}_under_the_dependency_formalism..png** *Source:* https://upload.wikimedia.org/wikipedia/commons/b/bc/A_syntactic_parse_of_%22Alfred_spoke%22_under_the_dependency_formalism..png *License:* CC BY-SA 4.0 *Contributors:* Own work *Original artist:* Christian Nassif-Haynes

- **File:Aleph0.svg** *Source:* https://upload.wikimedia.org/wikipedia/commons/e/e0/Aleph0.svg *License:* Public domain *Contributors:* Own work *Original artist:* PNG made by Maksim, SVG made by Amada44

- **File:Ambox_important.svg** *Source:* https://upload.wikimedia.org/wikipedia/commons/b/b4/Ambox_important.svg *License:* Public domain *Contributors:* Own work, based off of Image:Ambox scales.svg *Original artist:* Dsmurat (talk · contribs)

- **File:Ammonia-3D-balls-A.png** *Source:* https://upload.wikimedia.org/wikipedia/commons/0/05/Ammonia-3D-balls-A.png *License:* Public domain *Contributors:* Own work *Original artist:* Ben Mills

- **File:Bijection.svg** *Source:* https://upload.wikimedia.org/wikipedia/commons/a/a5/Bijection.svg *License:* Public domain *Contributors:* enwiki *Original artist:* en:User:Schapel

- **File:Bijective_composition.svg** *Source:* https://upload.wikimedia.org/wikipedia/commons/a/a2/Bijective_composition.svg *License:* Public domain *Contributors:* ? *Original artist:* ?

- **File:C60a.png** *Source:* https://upload.wikimedia.org/wikipedia/commons/4/41/C60a.png *License:* CC-BY-SA-3.0 *Contributors:* Transferred from en.wikipedia to Commons. *Original artist:* The original uploader was Mstroeck at English Wikipedia Later versions were uploaded by Bryn C at en.wikipedia.

- **File:CardContin.svg** *Source:* https://upload.wikimedia.org/wikipedia/commons/7/75/CardContin.svg *License:* Public domain *Contributors:* en:Image:CardContin.png *Original artist:* en:User:Trovatore, recreated by User:Stannered

- **File:Chapitel_IX._of_Die_Theorie_der_algebraischen_Zahlkörper.png** *Source:* https://upload.wikimedia.org/wikipedia/commons/5/52/Chapitel_IX._of_Die_Theorie_der_algebraischen_Zahlk%C3%B6rper.png *License:* Public domain *Contributors:* https://jscholarship.library.jhu.edu/handle/1774.2/34070 *Original artist:* David Hilbert

- **File:Circle_as_Lie_group2.svg** *Source:* https://upload.wikimedia.org/wikipedia/commons/d/de/Circle_as_Lie_group2.svg *License:* Public domain *Contributors:* self-made with en:Inkscape *Original artist:* Oleg Alexandrov

- **File:Clock_group.svg** *Source:* https://upload.wikimedia.org/wikipedia/commons/a/a4/Clock_group.svg *License:* CC-BY-SA-3.0 *Contributors:* Transferred from en.wikipedia to Commons. *Original artist:* The original uploader was Spindled at English Wikipedia

- **File:Commons-logo.svg** *Source:* https://upload.wikimedia.org/wikipedia/en/4/4a/Commons-logo.svg *License:* ? *Contributors:* ? *Original artist:* ?

- **File:Complete_graph_K5.svg** *Source:* https://upload.wikimedia.org/wikipedia/commons/c/cf/Complete_graph_K5.svg *License:* Public domain *Contributors:* Own work *Original artist:* David Benbennick wrote this file.

- **File:Cubane-3D-balls.png** *Source:* https://upload.wikimedia.org/wikipedia/commons/1/18/Cubane-3D-balls.png *License:* Public domain *Contributors:* Own work *Original artist:* Ben Mills

- **File:Cyclic_group.svg** *Source:* https://upload.wikimedia.org/wikipedia/commons/5/5f/Cyclic_group.svg *License:* CC BY-SA 3.0 *Contributors:*

- Cyclic_group.png *Original artist:*

- derivative work: Pbroks13 (talk)

- **File:DFAexample.svg** *Source:* https://upload.wikimedia.org/wikipedia/commons/9/9d/DFAexample.svg *License:* Public domain *Contributors:* Own work *Original artist:* Cepheus

- **File:Dedekind.jpeg** *Source:* https://upload.wikimedia.org/wikipedia/commons/c/ca/Dedekind.jpeg *License:* Public domain *Contributors:* http://dbeveridge.web.wesleyan.edu/wescourses/2001f/chem160/01/Photo_Gallery_Science/Dedekind/FrameSet.htm *Original artist:* not found

- **File:Directed.svg** *Source:* https://upload.wikimedia.org/wikipedia/commons/a/a2/Directed.svg *License:* Public domain *Contributors:* ? *Original artist:* ?

- **File:Edit-clear.svg** *Source:* https://upload.wikimedia.org/wikipedia/en/f/f2/Edit-clear.svg *License:* Public domain *Contributors:* The *Tango! Desktop Project.* *Original artist:*

  The people from the Tango! project. And according to the meta-data in the file, specifically: "Andreas Nilsson, and Jakub Steiner (although minimally)."

- **File:Exponentiation_as_monoid_homomorphism_svg.svg** *Source:* https://upload.wikimedia.org/wikipedia/commons/a/a3/Exponentiation_as_monoid_homomorphism_svg.svg *License:* CC BY-SA 3.0 *Contributors:* Own work *Original artist:* Jochen Burghardt

- **File:Folder_Hexagonal_Icon.svg** *Source:* https://upload.wikimedia.org/wikipedia/en/4/48/Folder_Hexagonal_Icon.svg *License:* Cc-by-sa-3.0 *Contributors:* ? *Original artist:* ?

- **File:Fundamental_group.svg** *Source:* https://upload.wikimedia.org/wikipedia/commons/b/ba/Fundamental_group.svg *License:* CC BY-SA 3.0 *Contributors:* en:Image:Fundamental group.png *Original artist:* en:User:Jakob.scholbach (original); Pbroks13 (talk) (redraw)

- **File:Greek_letter_uppercase_Phi.svg** *Source:* https://upload.wikimedia.org/wikipedia/commons/9/96/Greek_letter_uppercase_Phi.svg *License:* GPLv3 *Contributors:* A character from the font Linux Libertine. *Original artist:*

- SVG by Tryphon

- **File:Group_D8_180.svg** *Source:* https://upload.wikimedia.org/wikipedia/commons/6/64/Group_D8_180.svg *License:* Public domain *Contributors:* Own work *Original artist:* TimothyRias

- **File:Group_D8_270.svg** *Source:* https://upload.wikimedia.org/wikipedia/commons/3/33/Group_D8_270.svg *License:* Public domain *Contributors:* Own work *Original artist:* TimothyRias

- **File:Predicate_logic;_2_variables;_example_matrix_e1a2.svg** *Source:* https://upload.wikimedia.org/wikipedia/commons/0/02/Predicate_logic%3B_2_variables%3B_example_matrix_e1a2.svg *License:* Public domain *Contributors:* ? *Original artist:* ?
- **File:Predicate_logic;_2_variables;_example_matrix_e2a1.svg** *Source:* https://upload.wikimedia.org/wikipedia/commons/3/3a/Predicate_logic%3B_2_variables%3B_example_matrix_e2a1.svg *License:* Public domain *Contributors:* ? *Original artist:* ?
- **File:Predicate_logic;_2_variables;_implications.svg** *Source:* https://upload.wikimedia.org/wikipedia/commons/5/5b/Predicate_logic%3B_2_variables%3B_implications.svg *License:* Public domain *Contributors:* Own work *Original artist:* Watchduck (a.k.a. Tilman Piesk)
- **File:Prop-tableau-4.svg** *Source:* https://upload.wikimedia.org/wikipedia/commons/2/21/Prop-tableau-4.svg *License:* CC-BY-SA-3.0 *Contributors:* Transferred from en.wikipedia; transferred to Commons by User:Piquart using CommonsHelper. *Original artist:* Original uploader was Tizio at en.wikipedia. Later version(s) were uploaded by RobHar at en.wikipedia.
- **File:Question_book-new.svg** *Source:* https://upload.wikimedia.org/wikipedia/en/9/99/Question_book-new.svg *License:* Cc-by-sa-3.0 *Contributors:*
  Created from scratch in Adobe Illustrator. Based on Image:Question book.png created by User:Equazcion *Original artist:*
  Tkgd2007
- **File:Quintic_polynomial.svg** *Source:* https://upload.wikimedia.org/wikipedia/commons/8/8a/Quintic_polynomial.svg *License:* CC0 *Contributors:* Own work *Original artist:* Krishnavedala
- **File:Rubik'{}s_cube.svg** *Source:* https://upload.wikimedia.org/wikipedia/commons/a/a6/Rubik%27s_cube.svg *License:* CC-BY-SA-3.0 *Contributors:* Based on Image:Rubiks cube.jpg *Original artist:* This image was created by me, Booyabazooka
- **File:Rubik'{}s_cube_v3.svg** *Source:* https://upload.wikimedia.org/wikipedia/commons/b/b6/Rubik%27s_cube_v3.svg *License:* CC-BY-SA-3.0 *Contributors:* Image:Rubik'{}s cube v2.svg *Original artist:* User:Booyabazooka, User:Meph666 modified by User:Niabot
- **File:Sixteenth_stellation_of_icosahedron.png** *Source:* https://upload.wikimedia.org/wikipedia/commons/e/e7/Sixteenth_stellation_of_png *License:* CC BY-SA 3.0 *Contributors:* This image was generated by Vladimir Bulatov's Polyhedra Stellations Applet: http://bulatov.org/polyhedra/stellation_applet *Original artist:* Jim2k
- **File:Slanted_circle.png** *Source:* https://upload.wikimedia.org/wikipedia/commons/9/90/Slanted_circle.png *License:* CC-BY-SA-3.0 *Contributors:* Transferred from en.wikipedia; transferred to Commons by User:Common Good using CommonsHelper. *Original artist:* Original uploader was Jakob.scholbach at en.wikipedia
- **File:Speakerlink-new.svg** *Source:* https://upload.wikimedia.org/wikipedia/commons/3/3b/Speakerlink-new.svg *License:* CC0 *Contributors:* Own work *Original artist:* Kelvinsong
- **File:Spec_Z.png** *Source:* https://upload.wikimedia.org/wikipedia/commons/9/94/Spec_Z.png *License:* Public domain *Contributors:* tranfered from en.wikipedia *Original artist:* user:Jakob.scholbach
- **File:T_30.svg** *Source:* https://upload.wikimedia.org/wikipedia/commons/1/10/T_30.svg *License:* CC0 *Contributors:* Own work *Original artist:* Mini-floh
- **File:Text_document_with_red_question_mark.svg** *Source:* https://upload.wikimedia.org/wikipedia/commons/a/a4/Text_document_with_red_question_mark.svg *License:* Public domain *Contributors:* Created by bdesham with Inkscape; based upon Text-x-generic.svg from the Tango project. *Original artist:* Benjamin D. Esham (bdesham)
- **File:Togliatti_surface.png** *Source:* https://upload.wikimedia.org/wikipedia/commons/e/e0/Togliatti_surface.png *License:* CC BY-SA 3.0 *Contributors:* Own work *Original artist:* Claudio Rocchini
- **File:Undirected.svg** *Source:* https://upload.wikimedia.org/wikipedia/commons/b/bf/Undirected.svg *License:* Public domain *Contributors:* ? *Original artist:* ?
- **File:Uniform_tiling_73-t2_colored.png** *Source:* https://upload.wikimedia.org/wikipedia/commons/1/14/Uniform_tiling_73-t2_colored.png *License:* CC BY-SA 3.0 *Contributors:* Created by myself *Original artist:* Jakob.scholbach (talk)
- **File:Venn1001.svg** *Source:* https://upload.wikimedia.org/wikipedia/commons/4/47/Venn1001.svg *License:* Public domain *Contributors:* ? *Original artist:* ?
- **File:Venn_A_intersect_B.svg** *Source:* https://upload.wikimedia.org/wikipedia/commons/6/6d/Venn_A_intersect_B.svg *License:* Public domain *Contributors:* Own work *Original artist:* Cepheus
- **File:Wallpaper_group-cm-6.jpg** *Source:* https://upload.wikimedia.org/wikipedia/commons/8/8d/Wallpaper_group-cm-6.jpg *License:* Public domain *Contributors:* *The Grammar of Ornament* (1856), by Owen Jones. Persian No 1 (plate 44), image #1. *Original artist:* Owen Jones
- **File:Wiki_letter_w_cropped.svg** *Source:* https://upload.wikimedia.org/wikipedia/commons/1/1c/Wiki_letter_w_cropped.svg *License:* CC-BY-SA-3.0 *Contributors:*
- Wiki_letter_w.svg *Original artist:* Wiki_letter_w.svg: Jarkko Piiroinen
- **File:Wikibooks-logo-en-noslogan.svg** *Source:* https://upload.wikimedia.org/wikipedia/commons/d/df/Wikibooks-logo-en-noslogan.svg *License:* CC BY-SA 3.0 *Contributors:* Own work *Original artist:* User:Bastique, User:Ramac et al.
- **File:Wikiquote-logo.svg** *Source:* https://upload.wikimedia.org/wikipedia/commons/f/fa/Wikiquote-logo.svg *License:* Public domain *Contributors:* ? *Original artist:* ?
- **File:Wiktionary-logo-en.svg** *Source:* https://upload.wikimedia.org/wikipedia/commons/f/f8/Wiktionary-logo-en.svg *License:* Public domain *Contributors:* Vector version of Image:Wiktionary-logo-en.png. *Original artist:* Vectorized by Fvasconcellos (talk · contribs), based on original logo tossed together by Brion Vibber

53.6.3   Content license

- Creative Commons Attribution-Share Alike 3.0